INTERNATIONAL HERITAGE INSTRUMENTS AND CLIMATE CHANGE

RAE SHERIDAN
JOHN SHERIDAN

INTERNATIONAL HERITAGE INSTRUMENTS AND CLIMATE CHANGE

RAE SHERIDAN
JOHN SHERIDAN

Common Ground

First published in Champaign, Illinois in 2013
by Common Ground Publishing LLC
as a part of Inclusive museum series

Library of Congress Cataloging-in-Publication Data

International heritage instruments and climate change / Rae Sheridan, John Sheridan.

pages cm. -- (Inclusive museum series)

Includes bibliographical references.

ISBN 978-1-61229-071-3 (pbk. : alk. paper) -- ISBN 978-1-61229-072-0 (pdf : alk. paper)

1. Cultural property--Protection--Political aspects. 2. Climatic changes--Political aspects. I. Sheridan, John (John W.), 1941- II. Title.

CC135.S453 2012
174'.99301--dc23

2012028444

Cover Photograph: Cham dance performer, Ladakh Festival. John Sheridan
Series Editor: Amareswar Galla

Dedicated to the people of Ladakh.
May their *joie de vivre* ever remain infectious.

Table of Contents

Acknowledgements

We welcome the chance to express our appreciation.

To my (Rae's) Supervisor, Professor Amareswar Galla, I wish to express my gratitude, first for such a challenging topic and secondly for the hours of inspired and generous teaching both in Queensland and in Vietnam which preceded it. I have come to appreciate a world beyond museums through Professor Galla's broad knowledge and experience of the human 'cultural' condition.

To our good friend and student comrade-in-arms Joycelin Leahy, we say thank you for your kind and continually cheerful support.

Nuns, Lamas, Mullahs, NGO workers, cows, cars, local shoppers, merchants, street sellers, and students enliven the Main Bazar of Leh in winter. The 17th Century Jamia Masjid (central mosque) stands at the head of the Bazar with the Namgyal Chorten (Stupa), the Guru Lhakhang (Buddhist Shrine), Lonpo House and the Leh Palace (1616-1642) above. John Sheridan

Leh after the catastrophic flood and mud slide of 6th August 2010...the result of a climate change related extreme weather event which killed over 300 and left suburbs and life giving fertile terraces smothered. LEDeG

All unattributed photographs within this book, including the cover photograph, have been taken by the authors

Acronyms

AOSIS	Alliance of Small Island States
BAU	Business as usual
CAML	Central Asian Museum Leh
CBDR	Common but differentiated responsibilities
CBO	Community based organisation
COMEST	World Commission on the Ethics of Scientific Knowledge and Technology
CRMD	Chief Roi Mata's Domain
DHDR	Declaration of Human Duties and Responsibilities
DRM	Disaster Risk Management
EDA	Education for all
ESD	Education for sustainable development
FIP	Forest Investment Programme
GFC	Global financial crisis
GHF	Global Humanitarian Forum
GHG	Greenhouse gases
GISS	Goddard Institute for Space Science
GNP	Gross National Product
HDI	Human Development Index
HR	Human Rights
IADG	Internationally Agreed Development Goals
ICCROM	International Centre for the Study of the Preservation and Restoration of Cultural Property
ICESCR	International Covenant on Economic, Social and Cultural Rights
ICH	Intangible cultural heritage
ICHC	Convention for the Safeguarding of the Intangible Cultural Heritage, usually called the Intangible Cultural Heritage Convention
ICHCAP	Intangible Cultural Heritage Centre for Asia and the Pacific
ICOMOS	International Council on Monuments and Sites
IHDP	International Human Dimensions Programme on Global Environmental Change
IJIH	International Journal of Intangible Heritage
IPCC	Intergovernmental Panel on Climate Change
IUCN	International Union for Conservation of Nature
LAHDC	Ladakh Autonomous Hill Development Council
LDC	Least Developed Country
LEDeG	Ladakh Ecological Development Group
LINKS	Local and Indigenous Knowledge Systems
LOT	Leh Old Town
LOTI	Leh Old Town Initiative
MOST	Management of Social Transformations
NASA	US National Aeronautics and Space Administration
NP	National park
NGO	Non-governmental organisation
NIRLAC	Namgyal Institute for Research on Ladakhi Art and Culture

NOAA	US National Oceanic and Atmospheric Administration
OHCHR	Office of the High Commissioner for Human Rights
PACT	Partnerships for Conservation Initiative
PCDP	Pacific Century Premium Developments
REDD	Reducing Emission from Deforestation and Forest Degradations
STI	Science, technology and innovation
SD	Sustainable development
SECMOL	Students' Educational and Cultural Movement of Ladakh
SIDS	Small Island Developing States
SLR	Sea level rise
SP	State Party
THF	Tibet Heritage Fund
UCCAF	UNESCO Climate Change Adaptation Forum
UDHR	Universal Declaration of Human Rights
UN	United Nations
UNDESA	United Nations Department of Economic and Social Affairs
UNDP	United Nations Development Programme
UNEP	United Nations Environment Programme
UNESCO	United Nations Educational, Scientific and Cultural Organisation
UNFCCC	United Nations Framework Convention on Climate Change
UNHCHR	United Nations High Commissioner for Human Rights
VEC	Village Education Committee
WH	World Heritage
WHC	Convention concerning the protection of the World Cultural and Natural Heritage, usually referred to as the World Heritage Convention
WHIN	World Heritage Information Network
WHO	World Health Organisation
WMO	The World Meteorological Organisation

Foreword

International Heritage Instruments and Climate Change addresses two of our most renowned UNESCO Conventions: the 1972 World Heritage Convention and the 2003 Convention for the Safeguarding of Intangible Cultural Heritage. What potential do these international legal instruments have for keeping our cultural heritage safe in a climate confronted world? The authors use a multifaceted approach, setting the context, posing the questions, exploring the capabilities and then proposing recommendations that breathe fresh 'fight' into the Conventions and into those that wrestle with their implementation.

The surreal lunar landscape of Ladakh in northern India is the setting for a practical look at how the Conventions could address climate change impacted tangible and intangible heritage. The authors' photographs have given a presence to a people that are at the frontline of climate change vulnerability, yet hold a reserve of strength, even under calamitous impact, that is in part due to their entrenched culture.

The timing of the book is 'precarious'. Rio+20, the United Nations Conference on Sustainable Development has just weathered an unprecedented degree of challenges and disappointments with multilateral agreement making. This book does not offer a diluted steady as you go for alternative means of tackling climate change to prevent it from damaging our heritage and fragmenting and impoverishing our heritage custodians. It squares up the actions needed of the Conventions to increase the possibility of stabilising the climate and keeping a heritage worthy of 'world' status.

As I write news is breaking of the plight, a 50% decline in 50 years, of the world's largest World Heritage Area - despite it being the best protected, best known and best loved Marine Park on our planet. 2,600 marine scientists are declaring that human induced climate change is irreversibly destroying The Great Barrier Reef. The nub of the problem is that the science can open our minds to the emergency but cannot protect our heritage, natural or cultural from this planetary threat.

This study has had its genesis in research conducted under my supervision. It consumes the lives of both authors who have been concerned about 'the herd of (climate change) elephants in the room' for well over twenty five years. The book is in the spirit of the 'museum as a process' in all dimensions of global change that we must confront to bestow a better future for posterity.

Common Ground and its President Professor Dr. Bill Cope are proud to present the book to the readers. We are also grateful to Jamie Burns and Brian Kornell, former and current Managing Editors of Common Ground Publishing for making the publication possible.

Professor Amareswar Galla, PhD
Editor, Inclusive Museum Series, Common Ground Publishing, Champaign
Executive Director, International Institute for the Inclusive Museum, Copenhagen & Hyderabad
21 July 2012

Part I

Climate Change, Cultural Heritage and the UNESCO Conventions

Chapter 1

Culture, Last Cab off the Climate Rank

What we need now is good information and careful thinking, because in the years to come this issue (climate change) will dwarf all the others combined. It will become the only issue. We need to re-examine it in a truly sceptical spirit – to see how big it is and how fast it's moving – so that we can prioritise our efforts and resources in ways that matter".

Written 3 years later: "Clearly, the task of combating the climate crisis is far larger than conventional wisdom assumed. .....In order to rise to this challenge, humanity will need to implement clean energy technologies around ten times faster than projected by the most ambitious of the IPCC scenarios. Tim Flannery (Flannery 2008 p38)

The Climate is Changing

For the first time civilisation is on a trajectory to informed self-destruction. The trajectory has nothing to do with overt conflict or zealous fingers on bomb triggers, but everything to do with the developed world's life style. The industrialised consumer has adopted a standard of living which threatens the global climate. Consumption is increasingly contributing to greenhouse gas levels; this in turn is resulting in accelerating climate change.

The science underlying this is unequivocal. The dangers have been explicitly stated in scientific terms by the UN peak body on climate change, **The Intergovernmental Panel on Climate Change (IPCC)**; directly in economic terms by Nicholas Stern in the United Kingdom and Ross Garnaut in Australia (Stern 2008; Garnaut 2008) and shockingly in human terms in the Report by the **Global Humanitarian Forum (GHF)**: *Climate Change - The Anatomy of a Silent Crisis* (GHF 2009a).

Disappointingly, the GHF, the leading world body outside the UN courageous enough to expose this human cost, has been shut down, a victim of the global financial crisis. The GHF was initiated by the Swiss Government in 2007 and presided over by Kofi Annan. It provided an independent platform for debate and collaboration on global humanitarian issues. The Forum's first programme focus was the human impact of climate change and the boosting of assistance to the poor and vulnerable who suffer the brunt of its impact {Global Humanitarian Issues, 2010 p706) rather than on emissions or purely environmental aspects.

Climate change is the ultimate moral challenge to our species because of the tragic impacts from which millions in developing nations are already suffering. It is also profoundly different from previous challenges. It is the great correction we now have to have. It will demand that we redefine our relationship with our planetary home to live within its capacity to support us. And this will require that we think and act beyond the present economic growth paradigm. Up till two decades ago climate change was the unintentional doing of rich nations. Now climate science has removed the 'un'.

Most in the well-to-do nations find the science too confronting, offensive even, to squarely face, so do not force their leaders to make both the necessary emissions reductions to lessen the trajectory and to shoulder the responsibility to assist poor and climate change afflicted and susceptible nations to 'develop' within sustainable parameters. Mindful scientists, economists, academics, global humanitarians and religious leaders are repeatedly told by the political leaders they entreat to act that the scientific imperatives lie beyond the politically possible (Leahy 2009b; Hansen 2010a).

Climate change is a contagion of fossil fuelled, resource hungry, business promoting, share-holder oriented and consumer driven behaviour; what Tim Flannery has termed 'future eating' (Flannery 2002) that is eroding ecosystems, jeopardising health, corrupting political leadership and consigning the poor to greater poverty. In so doing, climate change is threatening our tangible and **intangible cultural heritage** (TCH and **ICH)**, the very wealth we share and a source of our resilience. Our cultural heritage, from home community to international level - *the common heritage of humanity* (UNESCO 2003a) - comprises the diversity that is our greatest possession to inherit and transmit. Cultural heritage imbues identity and gives meaning to life. While the international voice of the scientifically enlightened consistently calls for climate change mitigation, the same voice also calls for the preparation for and adaptation to the inevitable worsening impacts of climate change (ACJP, CANA, and FOE Australia 2008). How is the cultural heritage of humankind going to weather the brunt?

Under-recognised Threat to Culture

It is well known that natural heritage icons such as the Great Barrier Reef are threatened by global warming. It is less well known that heritage such as historic Prague is threatened by severe flash flooding and the three great mosques of Timbuktu are being overwhelmed by desert sand, both as a result of climate change. As yet it remains largely unrecognised by industrialised society that millions in Africa, Asia and the Pacific are climate displaced people. These diverse victims are bearers of intangible cultural heritage. Their traditions are under threat of fragmentation and loss. This human dimension of climate change has for years been conveniently ignored or incorrectly ascribed. Enclaves of the displaced are generally far from world centres of power. Their plight fails to attract a media all too ready to panda to the desire for ignorance from a compassion fatigued public. For political reasons, this massive dislocation has been described in more palatable poverty and racial conflict-driven terms rather than in more accurate climate change-induced terms. It is only in the last few years that the human face of the climate impacted is finally beginning to gain international recognition. However the threat of climate change to intangible cultural heritage - the customs, arts, rituals, festivals, craftsmanship, that is the diversity and strength of our humanity, has not been recognised, let alone its role in the transition to a sustainable future. "The research required to inform and support a major societal transformation lies primarily in the domains of the humanities and social sciences, which have been much less prominent in the climate change discourse than natural sciences and economics. Nevertheless, their insights into human cultures, behaviours and organisation are crucial to meeting the climate change challenge" (Richardson et al. 2009 p32). Furthermore it is likely that the transformation will not only inform and support but be scaled up by the recognition of the cultural dimensions of the humanities. When people see their culture threatened, temples washed away by cyclones or cathedrals destroyed by flooding, the whole climate change scenario will be heightened. Culture, the last cab off the climate rank may be the one that galvanises collective behaviour change.

UNESCO - the United Nations Educational, Scientific and Cultural Organisation works with nations to promote culture, education and science. UNESCO Conventions set the most ethical and highest international standards in permanent legal, financial and administrative frameworks for international cooperation. Two of the UNESCO Cultural Conventions oblige their member states to globally and nationally protect and safeguard both living and historic tangible and intangible cultural heritage. To these ambitious instruments falls the best chance and means to internationally protect cultural heritage from climate change forces. The two Conventions have worldwide stature despite their legal enforcement powers being limited to 'requests', 'encouragement', 'urging' and moral leverage and calls for cooperation and monitoring. As the Conventions are the most influential safeguarding and protective of the available instruments, it is timely that their capacity for protective and safeguarding potential against the impacts of climate change be scrutinised.

This book examines two of the UNESCO Cultural Conventions. These are the

- **Convention Concerning the Protection of the World Cultural and Natural Heritage, 1972, popularly known as the World Heritage Convention (WHC),** and the
- **Convention for the Safeguarding of the Intangible Cultural Heritage, 2003, often shortened to the Intangible Cultural Heritage Convention (ICHC)**

The concern is that there are gaps between the emerging threat of damaging climate change impacts on the world's cultural heritage and the capacity of the two Conventions to continue to lead in its protection. There are a number of facets to this situation.

The 1972 WHC was written before the phenomenon of climate change was widely recognised. Today we know much about the physical and biological dimensions such as temperature rise, increase in frequency of extreme weather events and the loss of species. Less is known about the economic, health and security dimensions, less still about the human dimension and very little at all about the cultural dimension. Reflecting this pattern, it is not unsurprising to find that despite its age, the WHC has been the more responsive of the two Conventions to climate change. This is because its natural sites have attracted considerable scientific attention which has led to changes in the Convention's operations and initiated adaptive interventions. Convention reports though, admit to limited knowledge on how to protect cultural world heritage in the short, mid and long term and at site, national and international level (UNESCO 2006a).

The 2003 ICHC is an exciting community centred convention embracing broader cultural components. It acts as a counter to the conservative and elite monumentalist impressions some have of the WHC. However despite it being recent and inclusive, it appears to lack climate change safeguarding processes. While the cultural implications of climate change may have in part eluded the Convention gatekeepers, they have been all too evident to those in the marginal environments of Africa, the **Small Island Developing States (SIDS)** and the Arctic. UNESCO, aware that these peoples have the least powerful voices internationally, established in 2010 an online forum: *Climate Frontlines. A global forum for indigenous peoples, small islands and vulnerable communities* to report climate impacts and to share knowledge (UNESCO 2010j). This initiative, where local and indigenous voices can contribute to decision-making, has brought an international audience to many personal accounts of economic hardship due to climate changes and arguably has become a 'first voice (Galla 2008) for stories of climate impacted cultural practices. The programme has generated support for the undertaking of more than 30 community-level field research projects (Tiatia 2008; UNESCO 2010j).

Critically, both Conventions face the issue that climate change has a central moral component. The most vulnerable nations to climate change impacts have contributed the least to global pollution from greenhouse gases. They are also in the weakest position to take advantage of the Conventions' protection because they lack the resources to formalise their requests for expertise and funds and hence comply fully with Convention requirements. Their cultures are being disproportionally damaged and lost as climate change impacts compound their social and economic disadvantages. Thus the problem of the Conventions' inadequacy in protecting world cultural heritage against the influences of the changing climate has historical and moral underpinnings. The challenge of rectifying the problem lies in The Conventions' capacities to remain true to their human rights obligations of bringing equity to their operations to address climate change justice. As well, the Conventions' protective operations must continue to seek to reflect global cultural diversity, to strive to balance national representativeness and geographic distribution, and to counter marginality.

The aim of examining the two Conventions is to recommend the means by which they might be modified to address the impacts of climate change on the cultural heritage of the world.

This book is not inclusive of the full scope of the Conventions. A major exclusion is discussion on the safeguarding of endangered languages. These form a critical and sizeable part of UNESCO's investment, comprising a raft of specialist programmes, partnerships and a register under the ICHC and as such deserve a dedicated study which is beyond the scope of this book. Oral traditions, though, are integral to intangible cultural heritage and are included here.

The WHC deals with categories of heritage sites called natural and cultural sites. Historian David Lowenthal has pointed out that the earth, water, plants and animals are integral to indigenous peoples' cultural diversity and that most, as primary stakeholders, have always understood that cultural heritage includes the natural as well as the cultural. Both are integral to understanding the impacts of climate change on culture. Lowenthal concludes that "the binary of nature and culture is yet to be adequately interrogated in the heritage discourse" (Lowenthal 2005 p.81). UNESCO cultural heritage law experts Patrick O'Keefe and Lyndel Prott agree "UNESCO wrestles with the issue of protecting the fundamental holism of traditional communities, which do not distinguish between nature and culture nor separate the spiritual, physical, traditional or ceremonial aspects of their practices." This study while recognising the limitations of the WHC categories for many communities will use the WHC terminology and concentrate on cultural World Heritage sites.

Today we live in a time of great crisis, confronted by the gravest challenge that humanity has ever faced: the ecological consequences of our own collective karma. The scientific consensus is overwhelming: human activity is triggering environmental breakdown on a planetary scale. The Time to Act is Now.
A Buddhist Declaration on Climate Change, 2010

View of Leh Valley at barley harvest time. The irrigation channels are fed by glacial water from the Indus River and its tributaries. They reflect the evening light forming a mosaic of the carefully tendered fields. On 6th August 2010 a torrential downpour caused mud and debris slides that suffocated much such farmland

Chapter 2
The Cultural Conventions–Standard Bearers for Protection

As the sea level rises, the tidal waters are slowly sweeping these artifacts away and that process threatens the ancestral stories and attachment to country of the Aboriginal peoples of Kowanyama, (Queensland, Australia). These biophysical changes are encroaching on spiritual and conceptual boundaries that are so integral to the Aboriginal identity. Indeed, Inherkowinginambana's reflection, "Where will we go?" extends far beyond concern for the ecological changes and forced environmental migration that would result from the encroaching floodwaters, as it becomes a question of "Who will we be?" Ameyali Ramos Castillo, 2009. (Castillo 2009)

The World Heritage Convention

The Convention Concerning the Protection of the World Cultural and Natural Heritage (World Heritage Convention, WHC) aims to protect in perpetuity, earth's natural and cultural heritage through identifying, protecting and preserving unique sites such as a reef, mountain, monument, building, city, landscape or place (See Appendix 3 for full text). **World Heritage Sites** are often referred to as properties as each remains the responsibility of the state on whose territory the site is located but their protection is recognised to be "the duty of the international community as a whole to co-operate" (Article 6). Each must have **Outstanding Universal**

Value (OUV), authenticity and integrity. The Convention is exceptional in its universal application. It links together the concepts of nature conservation and the preservation of cultural properties. This heritage – "our legacy from the past that we live with today, and that we pass on to future generations – belongs to all the peoples of the world and contributes to their common heritage" (UNESCO 2008d).

World Heritage Convention Lists

The WHC was adopted in 1972 and came into force in 1975. It has been spectacularly successful attracting ratification by 188 countries. On becoming a **State Party** to the Convention, a nation is duty bound to identify, protect and nominate properties that meet the selection criteria for inscription on the **World Heritage List**. These properties may be nominated as natural sites (e.g. Sagamatha National Park), cultural sites (e.g. Sydney Opera House) or mixedsites which meet both natural and cultural selection criteria (e.g. Uluru-kata Tjuta National Park). Since 1992 significant interconnections between people and the natural environment such as towns, canals and routes have been recognized as being eligible for inclusion on the list as cultural landscapes. Progressively, the WHC can be seen to be moving to a position of more fully recognising the way in which people interact with the natural environment and the fundamental need to preserve the balance between the two. The WHC has built a worldwide reputation, strengthening public appreciation for World Heritage properties through awareness raising and educational and information programmes.

> ***The planet is not ours; it is the treasure we hold in trust for future generations.***
> *Kofi Annan,*
> *President of the Global Humanitarian Forum, 2009*

A second list – the **List of World Heritage in Danger** lies within the World Heritage List. It records properties that are threatened by serious dangers such as development projects, the outbreak or threat of armed conflict or natural disasters.

Insert 2.1 Criteria for WH selection of properties

1. to represent a masterpiece of human creative genius;
2. to exhibit an important interchange of human values, over a span of time or within a cultural area of the world, on developments in architecture or technology, monumental arts, town-planning or landscape design;
3. to bear a unique or at least exceptional testimony to a cultural tradition or to a civilization which is living or which has disappeared;
4. to be an outstanding example of a type of building, architectural or technological ensemble or landscape which illustrates (a) significant stage(s) in human history;

5. to be an outstanding example of a traditional human settlement, land-use, or sea-use which is representative of a culture (or cultures), or human interaction with the environment especially when it has become vulnerable under the impact of irreversible change;
6. to be directly or tangibly associated with events or living traditions, with ideas, or with beliefs, with artistic and literary works of outstanding universal significance. (The Committee considers that this criterion should preferably be used in conjunction with other criteria);
7. to contain superlative natural phenomena or areas of exceptional natural beauty and aesthetic importance;
8. to be outstanding examples representing major stages of earth's history, including the record of life, significant on-going geological processes in the development of landforms, or significant geomorphic or physiographic features;
9. to be outstanding examples representing significant on-going ecological and biological processes in the evolution and development of terrestrial, fresh water, coastal and marine ecosystems and communities of plants and animals;
10. 10. to contain the most important and significant natural habitats for in-situ conservation of biological diversity, including those containing threatened species of outstanding universal value from the point of view of science or conservation.

A third list - the **Tentative List** stands outside the World Heritage List. It is an inventory of those sites which each State Party intends to consider for nomination during the following years. A cultural site must be on this list before it can be nominated for inscription. States Parties are encouraged to submit Tentative Lists and to do so with the participation of a wide variety of stakeholders, including site managers, local and regional governments, local communities, NGOs and other interested parties and partners.

By signing the WHC, each State Party pledges to "adopt a general policy that aims to give their cultural and natural heritage a function in the life of the community and to integrate the protection of that heritage into comprehensive planning programmes." States Parties agree to take "appropriate legal, scientific, technical, administrative and financial measures necessary for the identification, protection, conservation, presentation and rehabilitation of this heritage" (Article 5) and refrain from "any deliberate measures which might damage, directly or indirectly, the cultural and natural heritage" of other States Parties to the Convention as well as to help other States Parties in the identification and protection of their own properties (Article 6). Sites inscribed on the World Heritage List have to have a dedicated comprehensive management plan that sets out adequate preservation measures and monitoring mechanisms incorporating research projects and local participation. It is little understood that listing is of no legal consequence under the WHC except that it makes the site eligible for international assistance. The 'prize' of listing is seen by many countries more as being the 'esteem' of having a national site on the World Heritage List (O'Keefe and Prott 2011 p206) than the opportunity for financial help. But the real benefit to humanity is the "web of obligations and rights created by the WHC" (O'Keefe and Prott 2011 p80) that lies behind and is a prerequisite of the listing. This web was artfully written into the WHC to make States Parties legally bound to 'holistically' ensure protection of their cultural heritage from the time the Convention comes into force – usually at ratification (O'Keefe 1994).

The **World Heritage Fund** provides financial and technical assistance and professional training, both for emergencies including human made and natural disasters, and for longer term projects. Such support particularly benefits developing nations that are States Parties. The **World Heritage Committee** decides which of the annually nominated properties for the World Heritage List will be inscribed. It considers requests from States Parties for international assistance from the World Heritage Fund and advises States Parties on how they can ensure they meet their obligations under the WHC to protect World Heritage and their national heritage (UNESCO 2008d). The WHC obliges States Parties to report regularly to the World Heritage Committee on the **State of conservation** of their inscribed properties. These reports are crucial to the work of the Committee as they enable it to assess the conditions of the sites, decide on specific programme needs and resolve recurrent problems. The concept of World Heritage is so well understood that sites on the list galvanise international co-operation and attract financial assistance for heritage conservation projects from a variety of sources (UNESCO 2008d). When these are well planned and organized to respect sustainable tourism principles, they can bring important economic and cultural benefits to the site and to the local economy.

The **WHC selection criteria** listed at insert 2.1 give an overview of the range and types of properties that are and will be impacted by climate change. To be included on the List, sites must meet at least one of the selection criteria.

The WHC has evolved from a 'Western' Eurocentric perspective of concern for property over process, tangible over intangible and non living over living. In the past the WHC has been criticised for emphasising unchanging over changing, fixed over movable and non-indigenous over indigenous concerns. The adoption of the mixed heritage category reflects a maturing WHC, acknowledging a more symbolic intangible heritage. The growing move within UNESCO to integrate cultural protective instruments to address the 'bigger picture' came to fruition in the 2004 Yamanto Declaration (UNESCO 2004e). This has relevance when considering that climate change will demand such an overarching, progressive and interdisciplinary approach.

UNESCO's World Heritage mission is to support States Parties' public awareness-building activities for World Heritage conservation, to encourage international cooperation in the conservation of our world's heritage and to promote participation of the local population in the preservation of their cultural and natural heritage. The benefit of ratifying the WHC is that States Parties become members of an international community of appreciation and concern for universally significant properties that are outstanding examples of cultural diversity and natural wealth. States Parties share a commitment to preserving our legacy for future generations. Concomitantly the ramifications of the Convention – of international networking, exchange and commitment – has special application in a world challenged by a global threat.

The Intangible Cultural Heritage Convention

The **Convention for the Safeguarding of the Intangible Cultural Heritage (ICHC)** is the first binding standard setting international instrument dedicated to safeguarding intangible cultural heritage. It was adopted in 2003 and came into force in 2006. It "safeguards" by "ensuring respect" and assigning importance "for the intangible cultural heritage of communities, groups and individuals concerned". It raises "awareness", ensuring mutual appreciation "at the local, national and international levels" and by providing "for international cooperation and assistance" (Article 1). (See Appendix 4 for full text).The Convention grew out of persistence for nothing short of convention status for safeguarding intangible cultural heritage (O'Keefe and Prott 2011). It reflected not only the growing awareness of the "interdependence between the world's tangible and intangible heritage and the overall importance of safeguarding cultural diversity, but also the need to adopt an integrated approach to issues of environmental preservation and sustainable development" (Colette 2007 p4).

Intangible cultural heritage is defined as "the practices, representations, expressions, knowledge, skills, social values, traditions, customs, aesthetic and spiritual beliefs" – as well as the associated "instruments, objects, artefacts and cultural spaces.... that communities, groups and, in some cases, individuals recognize as part of their cultural heritage" (Article 2). The importance of the physical artifacts can be understood as part of the socioeconomic, political, ethnic, religious and philosophical values of a particular community.

The ICHC states that the intangible cultural heritage "is manifested, among others, in the following **domains**: oral traditions and expressions; performing arts; social practices, rituals and festive events; knowledge and practices concerning nature and the universe; and traditional craftsmanship" (Article 2). *Falconry, a living human heritage* nominated by 11 States Parties is one of the most recent inscriptions. The tradition celebrates much that characterises intangible cultural heritage. "While falconers come from different backgrounds, they share common values, traditions and practices such as the methods of training and caring for birds, the equipment used and the bonding between falconer and the bird, which are similar throughout the world. Falconry forms the basis of a wider cultural heritage, including traditional dress, food, songs, music, poetry and dance, all of which are sustained by the communities and clubs that practise it" (UNESCO 2010k).

ICHC has a universality as it provides communities and groups with a sense of identity and continuity that enhances overall social cohesion. As such it is the force energising humanity's cultural diversity and human creativity. It can serve in diverse ways such as by offering new insights into conflict resolution. For instance in 2006, UNESCO, drawing on ICHC guidelines, launched story-telling sessions in refugee camps situated in Western Tanzania to facilitate the transmission of traditional cultural knowledge from elders to the youth. This consequently, encouraged a "sustainable repatriation in a post-conflict situation where major socio-cultural challenges relating to past conflicts and to refugee conditions of living, needed to be addressed to ensure sustainable reintegration of the refugees" (UNESCO 2007h p.13).

Another way the ICHC can help is by sharing intangible cultural knowledge among stakeholders leading to more sustainable natural resource management. When integrated into curricula it can make school attendance more attractive, relevant and inclusive. By such means intangible cultural heritage has been translated by the Convention into worldwide programmes of advocacy, capacity building and training.

The ICHC celebrates intangible cultural heritage as living culture, being transmitted from generation to generation, usually orally. It strongly contends that intangible cultural heritage "is constantly recreated by communities and groups in response to their environment, their interaction with nature and their history" (Article 2). Intangible cultural heritage doesn't need to have OUV, a defined integrity or an unchanging authenticity. Significantly the Convention only aligns itself with intangible cultural heritage "as is compatible with existing international human rights instruments, as well as with the requirements of mutual respect among communities, groups and individuals, and of sustainable development" (Article 2). Through this binding association with human rights, the ICHC holds promise in having the capacity to respond to climate justice - the recognition that the developed polluting nations have a moral obligation to help the less developed relatively less-polluting climate impacted nations. As well, the ICHC, by recognising the changing nature of culture seems more aligned, than the WHC, with the underreported fact that climate change has precipitated humanity into a world that is increasingly climatically dynamic. There can never be a winding back of the climate clock, nor, given the present world condition, a climate status quo. Furthermore the ICHC points out the centrality of intangible cultural heritage to community identity, continuity and therefore resilience, a quality integral to meeting the demands of cultures under climate threat.

Article 2.3 of the Convention states that "safeguarding means measures aimed at ensuring the viability of the intangible cultural heritage including the identification, documentation, research, preservation, protection, promotion, enhancement, transmission, particularly through formal and non-formal education, as well as the revitalization of various aspects of such heritage." By comparison the WHC's operative term is protection which "shall be understood to mean the establishment of a system of international co-operation and assistance designed to support States Parties to the Convention in their efforts to conserve and identify that heritage" (Article 7). These similarities and differences indicate that the potential of the two Conventions to address climate change will likewise have commonalities and differences.

On ratification States Parties must commit to safeguarding their own intangible cultural heritage viability by making provision for the passing on of knowledge, meanings and skills from generation to generation. The social and economic value of this transmission is relevant for minority and majority groups, indigenous communities, traditional and contemporary peoples as it contributes to helping individuals to feel part of one or different communities and to feel part of society at large. States Parties must ensure their intangible cultural heritage continues to be inclusive, representative, and community-based. It must remain relevant to a culture and be regularly practised and learned within communities and between generations.

States Parties are obliged to develop national inventories by nominating esteemed practices, skills, knowledge, representations and expressions for listing after the widest possible participation of those who created, maintained and transmitted this heritage as these are the custodians who should be involved in its management. Unlike the World Heritage Lists, intangible cultural heritage inventories should "include all kinds of expressions, common or rare, how many or how few people take part in them, or how much of an effect or influence they have in that community, how under threat or under pressure they are, how widespread the manifestations and expressions are, and how weak or strong each of them is. Inventories should be regularly updated" (UNESCO 2003b p.10). Inventories are essential to the safeguarding as they can raise awareness about intangible cultural heritage and its importance for individual and collective identities.

The process of inventorying intangible cultural heritage and making those inventories accessible to the public can also encourage creativity and self-respect in the communities and individuals where the expressions and practices originate. Inventories can also provide a basis for formulating concrete plans to safeguard the intangible cultural heritage against the impact of climate change.

Developing nations have found their intangible cultural heritage is fragile and more difficult to preserve than physical objects (ICHCAP 2009 p140). Intangible cultural heritage is traditional and living at the same time. Its repository is the human mind, the human body being the main instrument for its enactment, or – literally – embodiment. The knowledge and skills are often shared within a community, and manifestations of intangible cultural heritage often are performed collectively (Khaznadar 2009).

The **Intergovernmental Committee for the Safeguarding of Intangible Cultural Heritage (Intangible Cultural Heritage Committee)** implements the Convention and the **Fund for the Safeguarding of Intan-**

gible Cultural Heritage. The Fund provides assistance for States Parties, especially of developing nations, to nominate their intangible cultural heritage.

Insert 2.2 Criteria for inclusuion in the Urgent Safeguarding List

1. The element constitutes intangible cultural heritage as defined in Article 2 of the Convention.
2. a. The element is in urgent need of safeguarding because its viability is at risk despite the efforts of the community, group or, if applicable, individuals and State(s) Party (i) concerned; (or)
   b. The element is in extremely urgent need of safeguarding because it is facing grave threats as a result of which it cannot be expected to survive without immediate safeguarding.
3. Safeguarding measures are elaborated that may enable the community, group or, if applicable, individuals concerned to continue the practice and transmission of the element.
4. The element has been nominated following the widest possible participation of the community, group or, if applicable, individuals concerned and with their free, prior and informed consent.
5. The element is included in an inventory of the intangible cultural heritage present in the territory (ies) of the submitting State(s) Party (ies), as defined in Articles 11 and 1
6. In cases of extreme urgency, the State(s) Party (ies) concerned has (have) been duly consulted regarding inscription of the element in conformity with Article 17.3.

The Intangible Cultural Heritage Convention Lists

The ICHC has 3 Lists. The **List of Intangible Heritage in Need of Urgent Safeguarding** inscribes proposed **elements** (expressions, practices etc) whose viability is endangered despite the efforts of the community concerned. Many elements suffer from the effects of globalization, lack of interest among the younger generations, uniformisation policies, and lack of means, appreciation and understanding which – taken together – may lead to the erosion of functions and values of such elements. This is the most important list since it aims at taking appropriate safeguarding measures to ensure the continuous recreation and transmission of threatened intangible cultural heritage. By inscribing an element on this list, the State Party undertakes to implement specific safeguards and may be eligible to receive financial assistance. This list has a major role in ensuring visibility of the intangible cultural heritage, in increasing awareness of its significance and also in encouraging dialogue that respects cultural diversity. The endangered element must satisfy <u>all</u> of the criteria listed at insert 2.2 to indicate how threats, including those posed by climate change are assessed, if assessed at all, by the ICHC.

Insert 2.3 Criteria for inclusion in the Representative List

1. The element constitutes intangible cultural heritage as defined in Article 2 of the Convention.
2. Inscription of the element will contribute to ensuring visibility and awareness of the significance of the intangible cultural heritage and to encouraging dialogue, thus reflecting cultural diversity worldwide and testifying to human creativity.
3. Safeguarding measures are elaborated that may protect and promote the element.
4. The element has been nominated following the widest possible participation of the community, group or, if applicable, individuals concerned and with their free, prior and informed consent.
5. The element is included in an inventory of the intangible cultural heritage present in the territory (ies) of the submitting State(s) Party (ies), as defined in Articles 11 and 12

The less important list, but the one that has attracted by far the majority of nominations is the **Representative List of the Intangible Cultural Heritage of Humanity.** It inscribes those elements submitted by States Parties which satisfy all of the criteria listed at insert 2.3. The popularity of this list is related to the enhancement it gives to the visibility and status of a State Party and its inscribed intangible cultural heritage plus the awareness raising that follows. The Intangible Cultural Heritage Committee has found it necessary to curb the enthusiasm of some States Parties by emphasising that the nominated intangible cultural heritage must benefit from measures to promote its continued practice and transmission, and must have been nominated by States Parties with the active and widest possible participation of the communities concerned, and with their free, prior and informed consent. 'Representative' can be interpreted in a number of ways. It can mean representative of the creativity of humanity or/and representative for the cultural heritage of the State Party. It can also mean a representative for the cultural heritage communities who are the bearers of the traditions.

Insert 2.4 Criteria for inclusion on the Best Practices List

a. The programme, project or activity involves safeguarding, as defined in Article 2.3 of the Convention.
b. The programme, project or activity promotes the coordination of efforts for safeguarding intangible cultural heritage on regional, subregional and/or international levels.
c. The programme, project or activity reflects the principles and objectives of the Convention.
d. If already completed, the programme, project or activity has demonstrated effectiveness in contributing to the viability of the intangible cultural heritage concerned. If still underway or planned, it can reasonably be expected to contribute substantially to the viability of the intangible cultural heritage concerned.

e. The programme, project or activity has been or will be implemented with the participation of the community, group or, if applicable, individuals concerned and with their free, prior and informed consent.

f. The programme, project or activity may serve as a subregional, regional or international model, as the case may be, for safeguarding activities.

g. The submitting State(s) Party (ies), implementing body (ies), and community, group or, if applicable, individuals concerned are willing to cooperate in the dissemination of best practices, if their programme, project or activity is selected.

h. The programme, project or activity features experiences that are susceptible to an assessment of their results.

i. The programme, project or activity is primarily applicable to the particular needs of developing countries.

The third list is the relatively small **Best Practices List** which registers programmes, projects and activities that best reflect the principles and objectives of the Convention. These programmes may serve as safeguarding examples and be disseminated as good practices. The low number of nominations probably reflects the unfamiliarity of the opportunity to contribute in this innovative way. The Committee selects from submissions from States Parties whose programmes best satisfy all of the criteria listed at insert 2.4. The criteria have been included here to indicate the extent of community engagement encouraged by the Convention.

The Leh Festival. Performers dancing on the polo field stage in front of the floodlit Leh Palace. Ladakhis have a rich cultural life which they share with visitors at this annual September festival

Both Conventions are sets of basically general principles. Their implementation is guided by, in the case of the WHC, **Operational Guidelines** and in the case of ICHC, **Operational Directives.** These are revised as interpretations of the Conventions are brought to bear on emerging issues such the inclusion of cultural landscapes as World Heritage properties and climate change.

Chapter 3

The Science of Climate Change

Causes and the Physical and Biological Impacts

The scientific debate and the political debate are like two ships passing in the night with no communication.
Ian Dunlop, Deputy Convenor, Australian Association for the Study of Peak Oil

Yaks and dzos on the move in Zanskar. Yaks and dzos are high altitude multi-purpose animals that provide milk for cheese and butter, power for tilling fields and threshing grain, a hardy means of transport through snow and a source of hair and leather

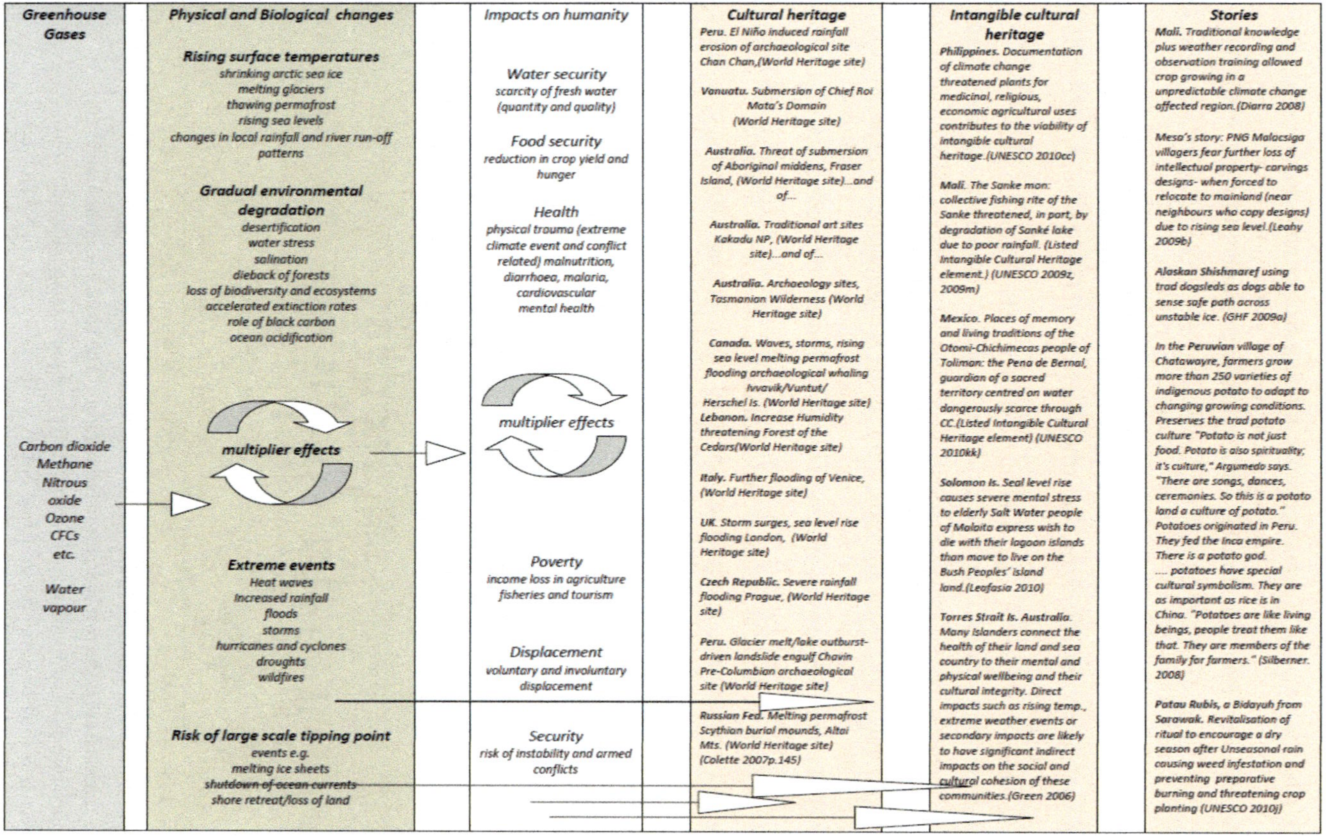

Greenhouse Gases
Carbon dioxide
Methane
Nitrous oxide
Ozone
CFCs
etc.
Water vapour
Physical and Biological changes
Rising surface temperatures
shrinking arctic sea ice
melting glaciers
thawing permafrost
rising sea levels
changes in local rainfall and river run-off patterns
Gradual environmental degradation
desertification
water stress
salination
dieback of forests
loss of biodiversity and ecosystems
accelerated extinction rates
role of black carbon
ocean acidification
multiplier effects
Extreme events
Heat waves
Increased rainfall
floods
storms
hurricanes and cyclones
droughts
wildfires
Risk of large scale tipping point
events e.g.
melting ice sheets
shutdown of ocean currents
shore retreat/loss of land
Impacts on humanity
Water security
scarcity of fresh water (quantity and quality)
Food security
reduction in crop yield and hunger
Health
physical trauma (extreme climate event and conflict related) malnutrition, diarrhoea, malaria, cardiovascular mental health
multiplier effects
Poverty
income loss in agriculture fisheries and tourism
Displacement
voluntary and involuntary displacement
Security
risk of instability and armed conflicts
Cultural heritage
Peru. El Niño induced rainfall erosion of archaeological site Chan Chan,(World Heritage site)
Vanuatu. Submersion of Chief Roi Mata's Domain (World Heritage site)
Australia. Threat of submersion of Aboriginal middens, Fraser Island, (World Heritage site)...and of...
Australia. Traditional art sites Kakadu NP, (World Heritage site)...and of...
Australia. Archaeology sites, Tasmanian Wilderness (World Heritage site)
Canada. Waves, storms, rising sea level melting permafrost flooding archaeological whaling Ivvavik/Vuntut/ Herschel Is. (World Heritage site)
Lebanon. Increase Humidity threatening Forest of the Cedars(World Heritage site)
Italy. Further flooding of Venice, (World Heritage site)
UK. Storm surges, sea level rise flooding London, (World Heritage site)
Czech Republic. Severe rainfall flooding Prague, (World Heritage site)
Peru. Glacier melt/lake outburst-driven landslide engulf Chavin Pre-Columbian archaeological site (World Heritage site)
Russian Fed. Melting permafrost Scythian burial mounds, Altai Mts. (World Heritage site) (Colette 2007p.145)
Intangible cultural heritage
Philippines. Documentation of climate change threatened plants for medicinal, religious, economic agricultural uses contributes to the viability of intangible cultural heritage.(UNESCO 2010cc)
Mali. The Sanke mon: collective fishing rite of the Sanke threatened, in part, by degradation of Sanké lake due to poor rainfall. (Listed Intangible Cultural Heritage element.) (UNESCO 2009z, 2009m)
Mexico. Places of memory and living traditions of the Otomi-Chichimecas people of Toliman: the Pena de Bernal, guardian of a sacred territory centred on water dangerously scarce through CC.(Listed Intangible Cultural Heritage element) (UNESCO 2010kk)
Solomon Is. Seal level rise causes severe mental stress to elderly Salt Water people of Malaita express wish to die with their lagoon islands than move to live on the Bush Peoples' island land.(Leafasia 2010)
Torres Strait Is. Australia. Many Islanders connect the health of their land and sea country to their mental and physical wellbeing and their cultural integrity. Direct impacts such as rising temp., extreme weather events or secondary impacts are likely to have significant indirect impacts on the social and cultural cohesion of these communities.(Green 2006)
Stories
Mali. Traditional knowledge plus weather recording and observation training allowed crop growing in a unpredictable climate change affected region.(Diarra 2008)
Meso's story: PNG Malacsiga villagers fear further loss of intellectual property- carvings designs- when forced to relocate to mainland (near neighbours who copy designs) due to rising sea level.(Leahy 2009b)
Alaskan Shishmaref using trad dogsleds as dogs able to sense safe path across unstable ice. (GHF 2009a)
In the Peruvian village of Chatawayre, farmers grow more than 250 varieties of indigenous potato to adapt to changing growing conditions. Preserves the trad potato culture "Potato is not just food. Potato is also spirituality; it's culture," Argumedo says. "There are songs, dances, ceremonies. So this is a potato land a culture of potato." Potatoes originated in Peru. They fed the Inca empire. There is a potato god. .... potatoes have special cultural symbolism. They are as important as rice is in China. "Potatoes are like living beings, people treat them like that. They are members of the family for farmers." (Silberner. 2008)
Patau Rubis, a Bidayuh from Sarawak. Revitalisation of ritual to encourage a dry season after Unseasonal rain causing weed infestation and preventing preparative burning and threatening crop planting (UNESCO 2010j)

Insert 3.1 Diagrammatic summary of the relationships between climate change and cultural heritage. Chapter 3 is concerned with the changes outlined in the first two and the last three columns

Climate Change Past and Present

Climate change is not new. Climate change ranging from minor to catastrophic (with mass extinctions) has occurred on many occasions over the earth's long history. Previous climatic change has been initiated by major natural events and by combinations of lesser natural events. Natural events include collisions with asteroids, continental drift, volcanism, long term as well as cyclical changes in the sun's output and cyclical changes in the earth's axis, tilt and orbit. What is new is that present climate change is being caused by us through our effects on the atmosphere; therefore it is anthropogenic climate change.

After some 10,000 years of relative climatic stability (the Holocene) during which time atmospheric carbon dioxide level, global temperature and sea level remained stable enough for man to thrive; to establish agriculture, permanent settlements, civilisations with occupational specialisation, craftsmanship and trade; came the invention of the steam engine and the industrial revolution. This was followed by massive technical, industrial and agricultural advances and growth that supported a population explosion, almost exclusively powered, unfortunately, by fossil fuels. When burnt, fossil fuels such as coal, oil and gas, release carbon dioxide and aerosols into the atmosphere. Land clearing, agriculture, industrial and other activities have added further carbon dioxide and climate affecting pollutants.

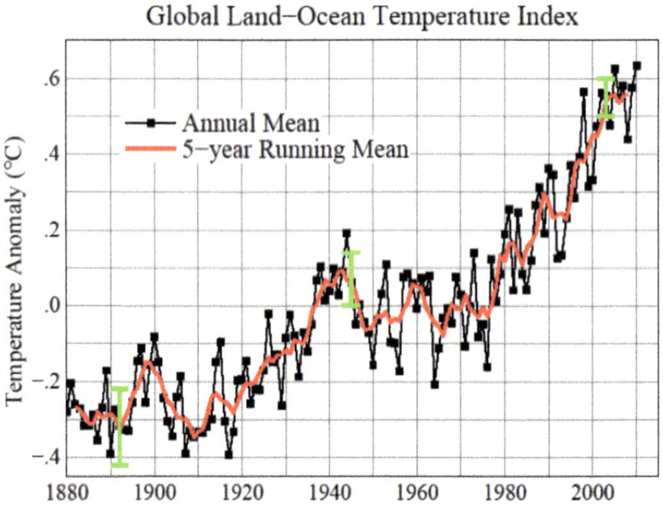

Insert 3.2 Line plot of global mean surface air temperature change, 1880-2010. The 'baseline' temperature is the 1951-1980 mean temperature. The black line is the annual mean temperature difference and the red line is the five year mean temperature difference. The green bars show uncertainty estimates. (NASA 2011b ; Leahy 2009b) NASA Goddard Institute for Space Studies (Hansen, R. Ruedy et al. 2010 p10)

Global mean surface air temperature, the averaged temperature of the world's entire surface, is calculated from data inputs from many thousands of measuring sites on land, on ocean, in the air and from satellites. The data is corrected for biases and extensively analysed by three separate research groups, the US National Aeronautics and Space Administration (NASA) Goddard Institute for Space Studies (GISS) headed by James Hansen, the US National Oceanic and Atmospheric Administration (NOAA) group, and the UK University of East Anglia (UEA) group. Despite each group using different methods of analysis their results remain in close agreement. Recent analysis shows that 2005 and 2010 were the warmest years ever recorded with 1998, 2002, 2003, 2006, 2007 and 2009 statistically tying for third warmest (NASA 2011a) "Global temperature in the past decade was about 0.8 degrees C warmer than at the beginning of the 20th century (1880-1920 mean). Two thirds of this warming occurred since 1975" (Hansen, Ruedy et al. 2010). The world is warming unevenly. Land is warming faster than ocean because ocean takes far more energy to heat than does land (Hansen, Ruedy et al. 2010). The Northern Hemisphere with its greater land surface is heating faster than the Southern Hemisphere with its greater area of ocean (NASA 2011b). In addition temperature increases at the poles are greater than the temperature increases at the equator (Hansen 2009 p38). In the last decade, regional warming, compared to global mean warming, was about 50% greater in the US, 2 to 3 times greater in Eurasia, and 3 to 4 times greater in the Arctic and on the Antarctic Peninsula. Warming of the ocean surface has been largest over the Arctic Ocean, second largest over the Indian and Western Pacific Oceans, and third largest over most of the Atlantic Ocean. Temperature changes have been small, even with cooling in some places over the North Pacific Ocean, the Southern Ocean, and the regions of upwelling off the coast of South America (Hansen, Ruedy et al. 2010).

The Greenhouse Effect

Carbon dioxide, trace amounts of several other gases, water vapour and clouds are responsible for the heat trapping, or greenhouse effect, of the atmosphere.

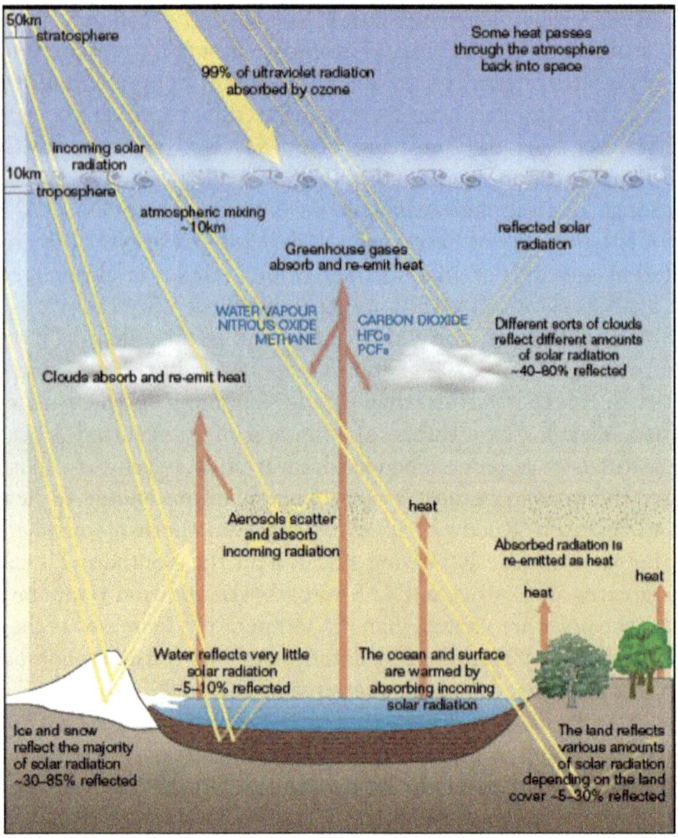

Insert 3.3 A stylised model of the natural greenhouse effect and other influences on the energy balance of the climate system. Sourced with permission Garrnaut Climate Change Review, 2008 (Garnaut 2008 Fig 2.2, p28)

Carbon dioxide (CO_2) is the most important of the **non-condensing greenhouse gases** (GHG). It is a gas that is both essential and threatening to life. In addition to its crucial role in photosynthesis, it acts both as a global thermostat and as an insulator (Schmidt et al. 2010). It is crucial to maintaining the narrow temperature range compatible with life. To successfully continue this role carbon dioxide must remain within a safe atmospheric concentration range. Despite over two decades of repeated warnings, the concentration of carbon dioxide is now well beyond its known upper safe limit and continuing to rise steeply (Hansen 2009 pxv). The pre-industrial carbon dioxide level was 280 parts per million by volume (ppm), the safe level is less than 350ppm (Hansen, Sato, and al 2008), and the current level is over 392ppm (NOAA 2011). The level now is higher than at any time in the last 800,000 years and probably in the last 20 million years (Simms, Johnson, and Chowla 2010).

Carbon dioxide *directly* contributes to about 20% of the earth's greenhouse (heat trapping) effect (Schmidt et al. 2010). Several trace gases act similarly in having both insulating and thermostat properties. Although these occur in far lesser quantities than carbon dioxide, they are much more potent on a volume for volume basis. These trace non-condensing greenhouse gases that include **methane** (CH_4), **ozone** (O_3), **nitrous oxide** (N_2O) and **man-made compounds such as chlorofluorocarbons** (CFCs) *directly* contribute a further 4-5% to the greenhouse effect (Schmidt et al. 2010). Fortunately their levels are not rising at anywhere near the rate of carbon dioxide. The heat trapping capacity of the trace greenhouse gases is measured in carbon dioxide equivalents (CO_2-e or CO_2-eq).

The remaining 75% or so of the greenhouse effect is from water vapour and clouds (Schmidt et al. 2010). Water vapour and cloud however, can only exist in the atmosphere because of the baseline warming provided by the non-condensing greenhouse gases discussed above. Without the former gases water on earth would be condensed and frozen leaving the atmosphere virtually free of water vapour and cloud (Schmidt et al. 2010).

Despite *directly* contributing only 20% to the total greenhouse effect, carbon dioxide is by far the most important of the non-condensing greenhouse gases. This is because of carbon dioxide's *very long life in the atmosphere*, and its *rapidly increasing level*. It is carbon dioxide (assisted by the trace greenhouse gases) that by partly warming the air is driving the increase in water vapour that accounts for the bulk of the warming (Schmidt et al. 2010).

Water vapour is a weak but very abundant greenhouse gas that amplifies the warming effects of the 'non-condensing' greenhouse gases. Atmospheric water vapour, although varying locally with climate and weather, is rising steeply overall (Santer et al. 2007). Increases of 1% or more per decade in water vapour have been recorded since the 1980s (Henson 2008 p190). In addition to its major contribution to the greenhouse effect, water vapour stores a very large amount of energy (latent heat); but more about that later.

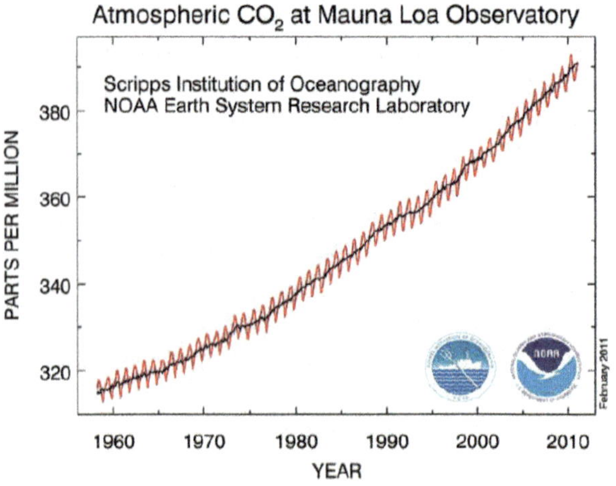

Insert 3.4 Line plots of atmospheric carbon dioxide concentrations between 1958 and 2011. The red line is a monthly plot, the saw toothed appearance being due to the seasonal influence of deciduous trees in the Northern Hemisphere; autumn leaves releasing carbon dioxide, spring growth taking up carbon dioxide. The steadily steepening black line is an annual plot, the rise of which is due to anthropogenic carbon dioxide emissions (NOAA 2011). (With reference to Dr. Pieter Tans, NOAA/ESRL (www.esrl.noaa.gov/gmd/ccgg/trends/) and Dr. Ralph Keeling, Scripps Institution of Oceanography (scrippsco2.ucsd.edu/))

Clouds have a pronounced greenhouse effect. This is obvious from a comparison of cloudy nights with clear nights. Cloudy nights are noticeably warmer than clear nights. Despite this the overall effect of clouds is to cool the earth. This net cooling of about 5 degrees C is largely caused by the shading effect of clouds during daylight hours (ISCCP 2009).

Surface reflectivity (Albedo): Albedo (whiteness) is defined as the reflective quality of a surface. Light surfaces have high albedos, dark surfaces, low albedos. Some 30-35% of arriving sunlight is reflected back into space by clouds, aerosols within the atmosphere and by the earth's surface. Snow and ice are particularly effective reflectors, lighter coloured deserts and lighter shaded vegetation also being important.

Albedo varies widely according to the surface, for example: fresh snow has an albedo of 80-90%, cloud 36-77%, ocean ice 50-70%, new concrete 55%, desert sand 40%, green grass 25%, bare soil 17%, deciduous trees 15-18%, summer conifer forest 8-15%, worn asphalt 12%, ocean 6%, and fresh asphalt 4% (Wikipedia 2011).

Clearly changes in the area of any of the major surface features contributing to the earth's albedo may influence the earth's energy balance.

Aerosols are suspensions of fine solids or liquid droplets in air. Natural aerosols include volcanic dust, windblown dust from arid soils and plant pollen. Most aerosols however are produced from burning fossil fuels. Smoke and sulphur dioxide and subsequent smog come from coal fired power stations, and photochemical smog from vehicle and industrial emissions. Aerosols reflect, scatter and absorb light. They also seed cloud formation and add to the longevity of clouds. Their contribution to albedo is sometimes referred to as causing global dimming. Their residence in the atmosphere is short lived. Despite reductions in aerosol emissions from developed countries, lower pollution standards and increased industrialisation in developing countries have seen global pollution levels grow overall. A concerted further draw down on polluting man made aerosols could lead to greatly accelerated warming through a reduction in albedo (Ramanathan and Feng 2008).

The Global Carbon Cycle; Carbon Sources and Sinks

Carbon (C) accounts for 27.3% of carbon dioxide's weight. Carbon is measured in gigatons (GtC) (or petagrams), one GtC (or petagram) being a billion tons of carbon. Since 1950 anthropogenic carbon emissions have increased from less than 2 GtC (Hansen 2009 p119) to over 11 GtC per year (Canadell 2007).

The main anthropogenic **carbon sources** are emissions from fossil fuel combustion, industrial processes, land use change, and agriculture. The majority of the emissions come from the burning of fossil fuels (coal, oil and gas), for electricity, heating, transportation and industry, with a small but significant proportion coming from fugitive emissions (coal mining, oil and gas extraction and their later processing). Some industrial processes especially cement manufacture, produce significant amounts of carbon dioxide from input raw materials like limestone. Together these account for some 75% of the increase in atmospheric carbon dioxide since pre-industrial times (IPCC, 2007, p512). The remainder comes mainly from deforestation, with a small contribution from agriculture (IPCC 2007 p511).

Less than half of the carbon dioxide going into the air stays in the air. The remainder is taken up by **carbon sinks**, both on land (mainly forests and soil) and by ocean. Worryingly these sinks are becoming less efficient with time. This has caused the proportion of carbon dioxide remaining in the air to have risen by some 4%, (from 41% to 45%) over the last 50 years (Canadell 2007). These findings are of concern because they show that the natural carbon sinks are weakening and that greater reductions in emissions will be needed to achieve chosen targets.

The main carbon sink involves the return flow of carbon to the solid earth. This occurs via the weathering of rocks and takes place over millions of years. It is not of relevance in the short term.

Maintenance of Climatic Stability

Under relatively stable conditions, such as prevailed during most of the Holocene, the amount of energy reaching the earth from the sun was closely balanced by the amount of energy reflected and re-radiated back into space. Averaged over the earth's entire surface this was about 240 watts per square metre both coming and going. Cyclical orbital, solar variations and other events during this period caused only minor climate disturbances. Because the amount of energy gained was much the same as that lost, the earth's overall temperature was able to remain within a habitable range.

Loss of Climate Stability; Global Energy Imbalance and Thermal Inertia

Now with rapidly rising carbon dioxide levels the balance between the amount of energy reaching the earth from the sun and the amount of energy reflected and re-radiated back into space has been upset. The additional carbon dioxide (moderated by the cooling effects of aerosol pollution) is trapping a little more of the outgoing energy resulting in a **global energy imbalance** (Hansen 2009 p102-103). Currently this imbalance or **climate forcing** is of the order of + 0.75 watts per square metre of the earth's surface averaged over the 11 year solar cycle (Hansen 2011b p287) With less energy being lost to space than is arriving from the sun, the earth is warming. Warming will continue till the earth has reached a point where it is once again losing as much energy to space as it is gaining from the sun. This is a very slow process that takes decades and centuries to fully play out. This slowness is due to the earth's **thermal inertia**. The earth's thermal inertia is due mainly to the ocean's enormous capacity to absorb heat and the stabilising effects of ice. By comparison little energy is needed to warm the atmosphere and the upper ten metres of the ground (Hansen 2010b).

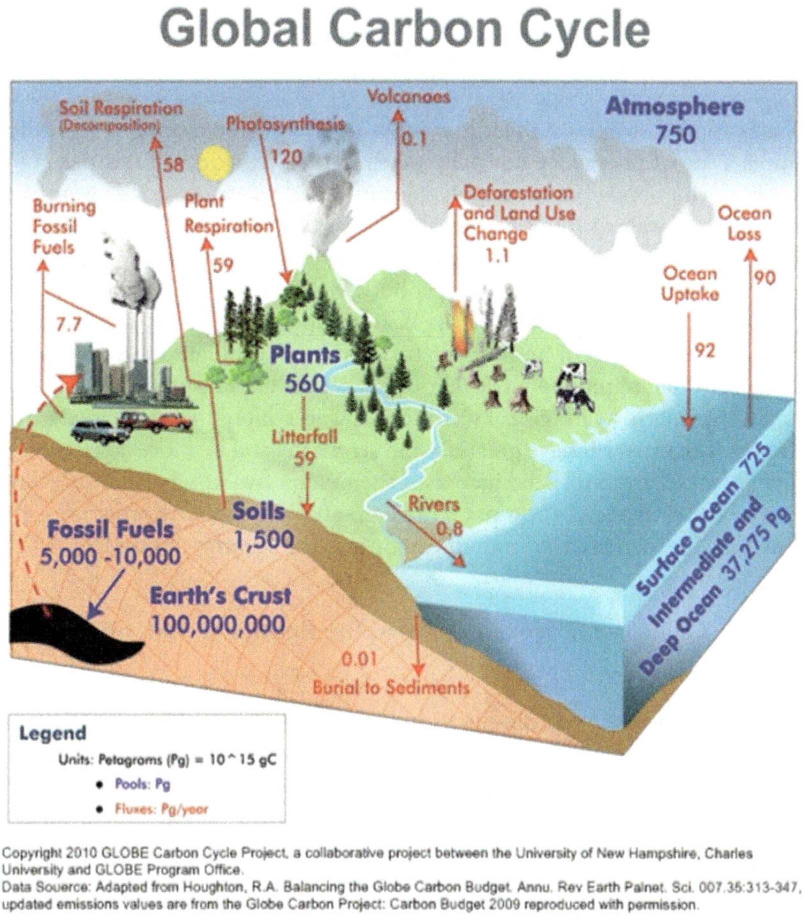

Insert 3.5 Global carbon cycle (units in gigatons (petagrams), each equal to a billion metric tons of carbon)

Thermal Inertia is a great threat, because by slowing the momentum of global warming, it allows greenhouse gas to accumulate in the absence of public concern. The effect of this may be to delay action to beyond a time when mitigation is feasible and adaptation can be anything but temporary.

Ocean surface temperature takes several decades to warm halfway, and many more decades, even centuries, to fully respond to an energy imbalance. The ocean surface has now warmed a little over halfway in its committed response to the added greenhouse gases of the 20th century (Hansen 2009 p72). In other words today's added warmth is the result of events prior to the 1970s or 1980s. The payback for the massive increase in carbon dioxide that has occurred more recently is in the pipeline. That payback will be the *climatic consequences* of the further 0.6-0.7 degrees C rise that is committed to occur in mean global temperature.

Water Vapour; the Force Behind Storms

Water vapour stores a very large amount of energy. To evaporate water requires a lot of energy. It normally takes more than 500 calories per gram to break the bonds between molecules in liquid water and make water vapour. When water vapour condenses (for instance as cloud) the stored energy (latent heat) is released as heat that is potentially available to fuel the violence of a storm. Increased atmospheric water vapour can therefore have a supercharging effect on storms. Water vapour level does not however dictate the strength of an individual storm, meteorological circumstances do this. What water vapour level does is set an upper potential limit to a storm's strength. With global warming, as water vapour levels increase, the strongest storms will have greater wind strengths and be more damaging. A 10% increase in wind speed increases the destructive potential of the wind by one third (Hansen 2009 p253). The greater atmospheric moisture content will also increase the amount of rain and the severity of floods.

Climate Change and Extreme Events; Predictions and Attribution of Cause

"Climate is defined not simply as average temperature and precipitation but also by the type, frequency and intensity of weather events. Human-induced climate change has the potential to alter the prevalence and severity of extremes such as heat waves, cold waves, storms, floods and droughts. Though predicting changes in these types of events under a changing climate is difficult, understanding vulnerabilities to such changes is a crucial part of estimating vulnerabilities and future climate change impacts on human health, society and the environment" (US Environmental Protection Authority 2009).

It is a widespread perception that the frequency and intensity of extreme events in recent years is increasing. Insurance companies can confirm this trend. They have no illusions about the occurrence of climate change. "Globally, loss-related floods have more than tripled since 1980, and windstorm natural catastrophes more than doubled, with particularly heavy losses from Atlantic hurricanes. This rise can only be explained by global warming" (Munich RE 2010).

The experience of the United Nations High Commissioner for Refugees (UNHCR) is similar: "Over the last two decades the number of recorded natural disasters has doubled from some 200 to over 400 per year. Nine out of ten natural disasters today are climate related" (UNHCR 2009).

The **Intergovernmental Panel on Climate Change (IPCC)** has studied the change in climate: "Since 1950, the number of heat waves has increased and widespread increases have occurred in the numbers of warm nights. The extent of regions affected by droughts has also increased as precipitation over land has marginally decreased while evaporation has increased due to warmer conditions. Generally, numbers of heavy daily precipitation events that lead to flooding have increased, but not everywhere. Tropical storm and hurricane frequencies vary considerably from year to year, but evidence suggests substantial increases in intensity and duration since the 1970s. In the extratropics, variations in tracks and intensity of storms reflect variations in major features of the atmospheric circulation, such as the North Atlantic Oscillation" (IPCC 2007) IPCC FAQ 3.3.

As yet it is impossible to conclusively attribute a specific storm or flood to global warming. However with a combination of climate models, weather observations and probability theory it may be possible to determine how climate change affects the odds. For instance it has been established that global warming at least doubled the likelihood of the 2003 European heatwave (Stott, Stone, and Allen 2004). It can also be said with some confidence that the increased rainfall intensity in the Northern Hemisphere in the latter half of the twentieth century cannot be explained by estimates of internal climate variability (Min et al. 2011). In other words what was considered a once in a century event in a stationary climate may occur more frequently in future. Another study (Pall et al. 2011) links climate change to a specific event: damaging floods in England and Wales in 2000. Here it was found that climate change may have almost doubled the risk of the extreme weather that caused the floods. Similar attribution studies are underway for flood and drought risk in Europe, meltwater availability in the Western US and drought in Southern Africa. These studies are typical of the research needed to develop effective climate-adaption policies (Schiermeier 2011). An earlier study (Webster et al. 2005) found evidence that global warming has increased the probabilities of category 4 and 5 hurricanes.

Warmer temperatures appear to be increasing the duration and intensity of the wildfire season in the western USA. Since 1986, longer warmer summers have resulted in a fourfold increase of major wildfires and a sixfold increase in the area of forest burned, compared to the period from 1970 to 1986 (Westerling et al. 2006). A similar increase in wildfire activity has occurred in Canada from 1920 to 1999 (Running 2006).

The devastating south east Australian Ash Wednesday and Black Saturday bushfires of 1983 and 2007 were in each case preceded by a prolonged drought caused by a strong positive dipole pattern across the Indian Ocean (Cai, Cowan, and Raupach 2009). This pattern, far more common in the last 30 years than in the previous 70, is consistent with IPCC (2007) climate change predictions.

Positive Feedbacks, Tipping Points and Abrupt Change; the Story to Date and Future Concerns

With continued warming, climatic events will change more rapidly and abruptly than up to now. A warming of 3 degree C, for instance will be far more than twice as damaging as a warming of 1.5 degree C. The augmented danger comes from **positive feedbacks**, processes that amplify change. See Insert 3.6. Positive feedbacks create disproportionately large changes. An increase in global mean temperature of 3 degrees C would be expected to result from a doubling in atmospheric carbon dioxide concentration.

Some positive feedback processes once underway have the potential to gain so much momentum as to become unstoppable. Once this has occurred they can be said to have passed a **tipping point**, a point of no return. An irritating but harmless example of a tipping point occurs when a microphone is placed too close to a speaker system, allowing the microphone to pick up background noise from the speaker, then amplify and re-amplify this ad nauseam, until the noise becomes unbearable (Hansen 2009 pix, 285).

Described below are some disturbing changes observed already. Most constitute positive feedback processes. Some have the potential to pass tipping points.

The melting of glaciers, snow fields, ice sheets and sea ice is increasing because of global warming. As warming continues surface snow darkens due to age-related changes in texture. The accumulation of deposited black soot on both snow and on ice adds further to the darkening. As a result snow and ice absorb more heat and melt more quickly. More dramatically, underlying vegetation, soil and rock, or ocean, become exposed. Where this occurs a great deal more heat is absorbed by the far darker surface. This further accelerates the loss of snow and ice. *These changes, which cause melting to accelerate, constitute a powerful positive feedback.*

Mountain glaciers are receding throughout the world. Loss of glaciers has a practical impact beyond reduced albedo. "In the driest months more than half of the water in major rivers such as the Indus and Brahmaputra, is provided by glacier melt water. Without glaciers, floods from spring snowmelt will be greater and rivers will tend to run dry in the drier months" (Hansen 2010b).

Arctic sea ice area in late summer is now 30% less than in the late 1970s when satellite observations began. This is largely because the earth is warming unevenly, warming being about twice as great towards the poles than global average. "Unless we can restore the planets energy imbalance we can expect to lose all late-summer sea ice within the next few decades" (Hansen 2010b). The loss of reflective sea ice exposes more dark heat-absorbing-ocean *thus accelerating warming and further melting*. "It is difficult to imagine how the Greenland ice shelf could survive if the arctic sea ice is lost entirely" (Hansen 2009 p164-165).

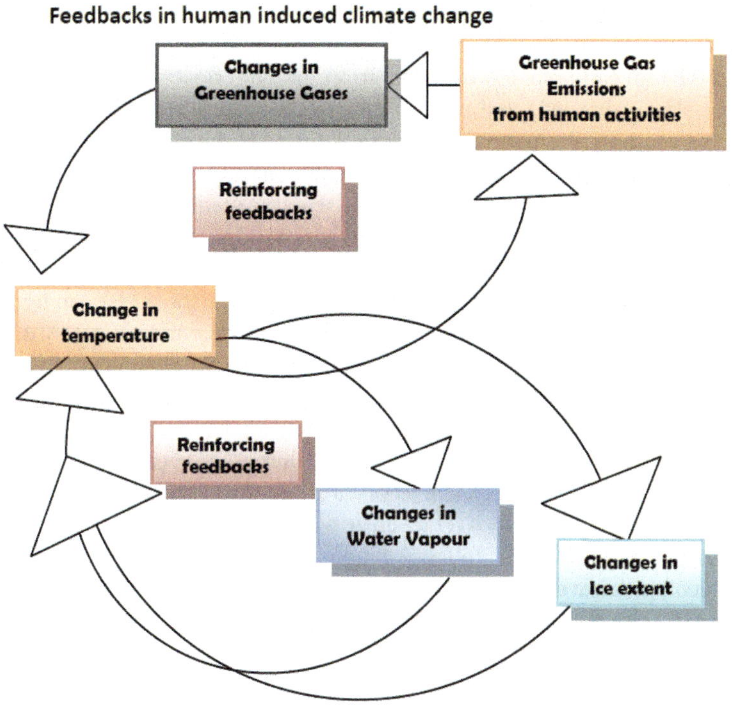

Insert 3.6 Feedbacks in human induced climate change. (Based on diagram from 'The Science of Climate Change'. Questions and answers. August 2010. Australian Academy of Science)

Sea level rise is an emerging major threat. Data available from the US National Oceanic and Atmospheric Administration (NOAA) National Climatic Data Center (NCDC) indicate that between 1900 and 2000 the average global sea level rose some 17cm in response to 0.8 degrees C of global warming. Since satellite measurements began in 1992, sea level has risen at around 3.4mm/annum (Rahmstorf 2010) giving a total rise, since 1900, of around 20cm. The explanation for most of the post-1992 sea level rise is the increasing rate of melting of the great land based ice sheets of Greenland and Antarctica (Richardson et al. 2009). The remaining sea level rise is mainly due to the thermal expansion of water.

Satellite gravity measurements show Greenland to be losing mass at a rate equivalent to about 250 cubic kilometres of ice per year with Antarctica losing about half as much a year. The rate of loss has doubled in the past ten years. If the rate of ice sheet loss doubles again in the coming decade it will suggest that we are passing a tipping point, and that ice loss and sea level rise may continue to *increase more and more rapidly*. If that is the case, by the end of the century we will already be faced by a sea level rise of several metres (Hansen 2011b p287).

The heat transferring capacity of water makes it far more effective at melting and disintegrating ice than air and/or sunlight (Hansen 2009 p82). With the ocean-assisted disintegration of the floating buttressing ice shelves, the grounded ice sheets are able to flow far more rapidly. "This buttressing opposes the natural plastic flow of ice towards the ocean which is driven by the weight of the ice sheet as snowfall piles up in the interior. If a warming ocean melts ice shelves, the ice streams coming from the ice sheet, which discharge giant icebergs into the ocean, begin to move more rapidly, discharging more ice. The West Antarctic ice sheet is especially vulnerable to removal of its ice shelves, because much of the ice sheet rests on bedrock several hundred metres below sea level. Loss of the entire West Antarctic ice sheet would raise sea level 6 to 7 metres and eventually open a path to the ocean for part of the much larger East Antarctic ice sheet" (Hansen 2009 p83). In addition, ice sheet collapse and melting will prove cataclysmic for yet another reason. The mixing of large volumes of ocean that have been cooled by melt water with warm lower latitude water will result in devastating storms and storm surges (Hansen 2009 p250).

"Once the ice sheets' collapse begins, global coastal devastations and their economic reverberations may make it impractical for humanity to take actions to rapidly reverse climate forcings. Thus if we trigger the collapse of the West Antarctic ice sheet, sea level rise may continue to even much higher levels via contributions from the Greenland and East Antarctic ice sheets" (Hansen 2009 p83). By contrast the IPCC in both its 2001 third assessment report, and its 2007 fourth assessment report, were strikingly conservative in their estimates of sea level rise. In their 2007 report the IPCC predicted sea level rises ranging from 18cm to 59cm for the 105 years from 1980-99 to 2090-99, the magnitude of the rise depending on the chosen emission scenario.

In both reports the IPCC assumed a near-zero net contribution to sea level rise from the Greenland and Antarctic ice shelves. They assumed that any melting would be fully compensated for by an increase in Antarctic snowfall. They did not include disappearing ice shelves, ice stream dynamics and iceberg melting in their global climate models. This was despite access to the satellite gravity determinations described above. Also surprising was the IPCC's *reduced* estimate for the anticipated sea level rise attributable to thermal expansion in their 2007 report compared with their 2001 estimate (Hansen 2009 p88).

Shortly prior to the release of the IPCC's 2007 report Rahmstorf (2007) published a study in the authoritative journal *Science*, that was to show that the IPCC's sea level rise estimates were seriously flawed. His estimates, which were based purely on the heat-dependent expansion of water, predicted sea level increases nearly three times those later found in the 2007 IPCC report. When the IPCC report was released the report's serious underestimates of predicted sea level rise led to consternation and indignation on the part of a number of scientists. This resulted in IPCC policy makers becoming more aware of the threat of sea level rise.

The reliance of Hansen's group on objective measurement and historical palaeoclimate data was undoubtedly the major reason for the differences between their predictions and those of the IPCC which relied heavily upon climate models. "Climate models do a decent job of demonstrating certain feedbacks, such as water vapour and sea ice, even though they failed to predict the recent rapid Arctic sea ice loss" (Hansen 2009 p44). "Models, at best, produce answers consistent with the assumptions put into them" (Hansen 2009 p81).

Whilst there can be no doubt about the enormous information content, value and integrity of the scientific information collated and presented by the IPCC, the IPCC is by its nature conservative in its estimates. The conservative estimates of the IPCC may be partly the consequence of the politicised nature of the IPCC; participating scientists being selected by their governments, the dependence of many scientists on government employment and/or funding, and for those working within US federal agencies, incidences of interference with their work and/or freedom of publication and communication. There has also been an inability to incorporate recent findings due to the slowness and caution of the refereeing process, a lack of clarity over the means by which the IPCC selected lead authors, and an unfortunate censorship of the Summary for Policy Makers by attending politicians e.g. Monbiot (2007).

Hansen (2009 p.39), indicates that the current global temperature is less than 1 degree C below the peak of the last interglacial period, the Eemian (125–130,000 years ago). At that time ocean levels were 6 – 9 metres higher than today (Hearty et al. 2007) and carbon dioxide under 290ppm (Petit et al. 1999). "The additional water (at that time) must have come from Antarctica, Greenland, or some combination of the two" (Hansen 2009 p143). This is alarming news especially when we take into account the further warming in the pipeline. Taking into account the greater than 0.6 degree C temperature rise to which the Earth is already committed, it seems likely that we are perilously close to triggering a similar sea level rise.

"Global chaos will be difficult to avoid if we allow the ice sheets to become unstable" (Hansen 2009 p85).

For the people of Tuvalu, Kiribati and the Maldives the prospect of survival on their nations' atolls is all but over. They will be forced to move within decades. Also at particular risk are the people of the densely populated low-lying Ganges Delta of India and Bangladesh and the people of Egypt's Nile Delta. As previously explained, rising sea level will be accompanied by far more violent weather, severer storms, greater storm surges and larger floods than are seen today. It will be the combination of events that will be so threatening. Continuing sea level rises and accompanying extreme weather could devastate vast areas throughout the world leading to food shortages, displacement, conflict and death.

Many of the world's cities are situated on or near coast lines. More than a billion people live within 25 metres of sea level. If ice sheets begin to disintegrate there won't be a new stable sea level on any foreseeable time scale. Change will continue for centuries. Local catastrophes will occur in association with powerful regional storms. It is most important to note that sea level changes to heights several metres greater than today's level have occurred in previous interglacial periods that were at most, 1-2 degrees C warmer than now (Hansen 2009 p85).

Regrettably the conservative estimates of the IPCC have enabled politicians to formulate upper limit carbon dioxide and temperature targets that favour short term fossil fuel, economic, and re-election interests. The survival of the world's ecosystems on which we all depend has received scant consideration. This is a recipe for disaster.

Permafrost is thawing in the northern latitudes. Permafrost is ground that remains frozen year round. Permafrost starts anywhere from a few centimetres to several tens of metres below the ground surface. It can extend downwards hundreds of metres until continued freezing is prevented by warmth from the earth's core. Above the permafrost is an **active layer** that typically freezes and thaws every year. The deepening of this active layer through thawing of the adjacent permafrost is now causing concern in parts of the Arctic. Within the permafrost is water, some as fine ice crystals but some as much larger ice beds or wedges. If a thawing extends into areas containing larger pockets of ice, their melting can create an underground lake or **thermokarst**. If this water drains or is displaced, subsidence may cause trees and buildings to undergo a slow motion fall and surface lakes to either appear or disappear. This is a particular problem across the fast warming lower Arctic (Henson 2008 p83-84).

About one third of the world's soil-bound carbon is in taiga (boreal forest) and tundra areas. When permafrost melts, it releases carbon dioxide and methane. Methane's potency as a greenhouse gas, together with the vast reserves of methane hydrates in permafrost, makes further significant warming in the polar region very concerning. In the 1970s the tundra was a carbon sink, but today, it is a carbon source (Oechel et al. 1993). *This is another worrying example of a positive feedback.*

Arid subtropical climate zones are expanding poleward faster (50-60km/decade), but otherwise as predicted of global warming. Already an average expansion of around 400km has occurred, affecting the southern USA, the Mediterranean region, Australia and southern Africa. These changes are accompanied by a drying of lakes and an increased frequency, area and intensity of wildfires (Hansen 2009 p165); (Westerling et al. 2006). *Wildfires contribute to rising carbon dioxide levels and thus further temperature rise.*

Animal and plant species are already stressed and becoming extinct as a direct result of human activities. Such activities include deforestation, the replacement of biologically diverse grasslands and forests with monoculture crops, and the introduction of foreign invasive plant and animal species some of which wipe out native ones (Hansen 2009 p144). The **rapid poleward shift of climate zones** will be the primary cause of continuing extinctions. Many species, both plant and animal, cannot migrate rapidly. As some species falter, ecosystems can collapse because of species interdependencies (Hansen 2010c). Ecosystems are based on interdependencies (between for example, flower and pollinator, hunter and hunted, grazer and plant life) so that the less mobile species have an impact on the survival of others (Hansen 2009 p145). Lindenmayer (2007) estimates the current extinction rate as 1000 times above the natural rate. Ten to 30% of mammal, bird and amphibian species are currently threatened with extinction and 15-37% of known plant and animal species committed to extinction by 2050 if the world warms by between 1.5 and 2.5 degrees C (Thomas 2004). Past activity has already committed the earth to a 1.4 degree C temperature increase above the 1880-1920 mean, with the rise likely to be at least twice this by century's end. It is thus apparent that the **loss of biodiversity will be overwhelming**. The sixth mass extinction may have already commenced (Hansen 2009 p147).

The ocean is now both warmer and 30% more acidic (a fall of 0.11 of a pH unit from pH 8.179 to pH 8.069) than it was in the 18th century. These changes, the result of rising atmospheric carbon dioxide levels and carbon dioxide's solubility, are harmful to marine species. Coral reefs, where a quarter of all marine species live, are particularly vulnerable. Coral death not only affects the marine species coral's harbour but also human communities that depend on the reef for physical protection and for food. Marine species with carbonate shells are also threatened, indeed sufficiently acidic water dissolves carbonates e.g., (Hansen 2009 p165-166).

Forest loss is a huge concern. Healthy tropical rainforest is a vast carbon sink storing nearly half the plant carbon on earth. In addition to replacing atmospheric carbon dioxide with oxygen during photosynthesis, rainforests have a local cooling effect which is thought to attract rain (Henson 2008 p11,158-159).

Some 20% or more of today's emissions come from deforestation. Till the present most deforestation has been caused purposefully by man. This situation may be starting to alter in an even more alarming manner. For example consider the situation developing in the Amazon.

In 2005 the Amazon experienced a once in a century drought that turned it from a carbon sink into a carbon source. Analysing the impact of the drought, a team of 68 researchers across 13 countries and 40 institutions found evidence that rainfall-starved tropical forests lose massive amounts of carbon due to reduced plant growth and dying trees. The 2005 drought, triggered by warming in the tropical North Atlantic, resulted in a net flux of 5 gigatons of carbon dioxide (1.4 GtC) into the atmosphere, more than the combined annual emissions of Japan and Europe. In earlier years the Amazon was a net sink for 2 gigatons of carbon dioxide (0.5 GtC) a year. The findings suggest that in the face of a warming climate, relying on tropical forests as a massive carbon sink is a perilous proposition, raising questions about the effectiveness of schemes to offset industrial emissions by protecting rainforests without also curbing fossil fuel use. Should droughts worsen on a global scale, *forests could become a net source of emissions, exacerbating climate change* (Phillips et al. 2009).

A second drought in 2010 was even worse. "If drought events continue, the era of intact Amazon forest buffering the increase in atmospheric carbon dioxide may be passed" (Lewis et al. 2011).

Wildfires are becoming more frequent, more extensive and more intense in many parts of the world. Tropical rainforest is frequently burnt to clear land. Such fires can get out of control. In 1997 devastating fires in Kalimantan and Sumatra poured up to 2.6 gigatons of carbon dioxide (0.7 GtC), and vast amounts of health damaging particulates, into the atmosphere. Additional to the initial carbon dioxide and smoke, nitrous oxide, a trace greenhouse gas, is released from the soil through bacterial action after fires (Henson 2008 p11, 158).

Mention has already been made to climate change related fires in the western United States, in Canada, and in Australia. Across western North America from Alaska to Mexico drought and warmer winters have favoured forest fires, increased beetle attacks and forest die back. In 2002 alone, British Columbia lost 100,000 square km to fire and disease. Still more recently there were widespread fires in Russia during the 2010 heatwave, in Bolivia also in 2010, and elsewhere.

Warmer drier summers are becoming increasingly common in the Arctic. Siberia which has half of the world's evergreen forest has seen a tenfold increase in fires in the last few decades. In 2004 alone, 220,000 square km, an area almost the size of Britain was swept by fire. In Alaska some 49,000 square km, or more than 10% of Alaska's tundra and forest was burnt in 2004 and 2005 (Henson 2008 p85).

Although not all bushfires need be climate change related, all fires contribute to atmospheric greenhouse gas levels thus *promoting further global warming.*

If we continue to burn all the available fossil fuel, the ice sheets will surely melt, the sea level will rise 75 metres within centuries, and sea bed **methane hydrates** will destabilise *causing catastrophic warming* (Hansen 2009 p236).

Over many millions of years organic material has settled into oceans, lakes and bogs, where some of it, under appropriate conditions, has eventually been transformed into methane hydrate. Methane hydrate is a frozen form of methane in which each methane molecule is enclosed in a cage or crystal of water ice. Large amounts of methane hydrate are found within and below permafrost, and more abundantly within the top several hundred metres of sediments on the sea floor. This is particularly so in the Arctic. Up to 5,000 gigatons of carbon (GtC) are believed to be trapped in cold ocean seabeds as methane hydrates (Hansen 2009 p235). Methane hydrates remain stable so long as conditions are sufficiently cold and/or enough pressure is exerted on them (as might be the case at the bottom of the sea) (Henson 2008 p212). An increase in ocean temperature could be enough to destabilize the methane releasing it to the atmosphere with catastrophic results.

This has probably happened in the past, the last major release being at the Palaeocene-Eocene Thermal Maximum (PETM) some 55 million years ago (Hansen 2009 p162-164; Henson 2008 p212-213).

The PETM was a consequence of continental drift. The Indian plate ploughed north to Asia through what is now the Indian Ocean, subducting enormous amounts of calcium carbonate and organic material that had been deposited there by major river systems. (**Subduction** is the process that takes place when one tectonic plate moves under another tectonic plate, sinking into the earth's mantle as the plates converge.) The subduction of these sediments resulted in the release of vast quantities of carbon dioxide through volcanoes, seltzer springs and gas vents. This additional atmospheric carbon dioxide warmed the ocean, as is happening today, until heating and changes in ocean circulation were sufficient to destabilize the seabed methane and send it pouring into the atmosphere. Altogether some 3,000 gigatons of carbon (GtC) in the form of methane; as much carbon as is in all of today's oil, gas and coal; entered the atmosphere. The methane and the carbon dioxide that formed as the methane decomposed, provided a positive feedback that resulted in the large PETM temperature spike of some 5-9 degrees C in global temperature (Hansen 2009 p150-164). Arctic Ocean sea surface temperatures during the PETM soared as high as 23 degrees C (Henson 2008 p212-213). Hansen (Hansen, Sato, and al 2008 p.xv) warns: This "is comparable to the warming that may occur in the next century or so if business-as-usual (BAU) greenhouse gas emissions continue."

"In a nutshell, a problem has emerged. Climate inertia and climate amplifying feedbacks, as humans rapidly increase greenhouse gases, spell danger for future generations—big danger. Yet the public is largely unaware of an impending crisis. The obliviousness of the public is not surprising—global warming, as yet, is slight compared with the day to day weather fluctuations. How in the world can a situation like this be communicated credibly?" (Hansen 2009 p89).

Mitigation, Adaptation and Palliation

The global response to climate change must be one of mitigation, adaptation and possibly palliation.

Mitigation aims to stabilize then reduce emissions, ultimately re-establishing an equable sustainable climate outcome. Measures to achieve sustainability include a rapid transition to a fossil fuel free future, one powered by renewable and possibly nuclear energy. In addition will be the need to draw down on existing greenhouse gas levels using all available means of carbon sequestration. Whilst the internationally most widely accepted CO_2 concentration target for 2100 is 450ppm, an optimistically realistic projection is about 650ppm (Anderson and Bows 2008). The challenge of returning CO_2 to a safe level of less than 350ppm would take herculean international political will, immense human determination and cooperation, unprecedented technical innovation and application and centuries to achieve.

Adaptation, although limited in effectiveness, is an unavoidable response to climate change. It is already occurring and will inevitably become more important with further climate change. Examples of adaptation include increasingly more stringent building requirements in cyclone/hurricane prone areas, flood protection measures (or relocation) in areas liable to inundation, and protective measures (or relocation) in wildfire prone areas. Migration as climate refugees might be seen as an extreme and generally unsatisfactory form of adaptation. Ultimately mankind's ability to adapt is likely to prove inadequate and societal breakdown with a global population crash may occur.

Palliation may be defined as that which serves to cloak or conceal, or that which provides superficial or temporary relief without addressing the underlying cause. The term is commonly used in relation to pain relief in the terminally ill.

Certain proposed forms of geoengineering may be regarded as offering planetary palliative care. These include measures to reduce incoming solar radiation such as the placement of massive amounts of sulphate in the stratosphere or the deployment of huge arrays of mirrors or lenses far out in space, between the sun and the earth. Such measures whilst addressing the energy imbalance problem would carry the risk of unintended consequences. For example the combination of reduced sunlight and increased CO_2 might cause unpredictable changes in regional weather patterns with serious consequences especially to vegetation and ocean life. Worst of all such measures might reduce the impetus for rapid mitigation leading to the otherwise avoidable extended use of fossil fuels and continued unnecessary ocean acidification. Such concerns were expressed by well over one hundred conservation and humanitarian organisations in an open letter to the IPCC over a joint working group expert meeting on geoengineering held in Peru in June 2011.

Solar radiation management through geoengineering can be expected to receive strong support from those with vested interests in the fossil fuel economy. Unfortunately as conditions on earth deteriorate palliation by means of solar radiation management through geoengineering is likely to become an essential. We can only hope that future policy makers will think through the matter with great caution and be less influenced by fossil fuel lobbyists than has been the case in the past.

Buddhist Monks creating a mandala, a highly decorative spiritually powerful coloured sand painting. Upon completion it is destroyed and the sand washed away in a stream to spread the blessings of the mandala and to symbolise impermanence

Chapter 4
Gaia Theory and Mother Earth Rights

A new consciousness is developing which sees the earth as a single organism and recognizes that an organism at war with itself is doomed. We are one planet. One of the great revelations of the age of space exploration is the image of the earth finite and lonely, somehow vulnerable, bearing the entire human species through the oceans of space and time. Carl Sagan, Cosmos: a personal voyage, 1990 update, Episode 13.

It is worthwhile here to consider some influential philosophical views of life on earth, how they accommodate climate change and whether there is an advantageous convergence bringing the physical, biological and the humanitarian together with culture. Is there an emerging philosophical trend that has implications for the integration (or reintegration) of culture and cultural heritage, with environmental concerns and ecology, with global rights and responsibilities? Many of the theorists are or have been eminent scientists whose most profound ideas have grown beyond the laboratory to have deep implications for culture and cultural heritage.

In 1965 British independent scientist James Lovelock, described by *New Scientist* magazine as "one of the great thinkers of our time", conceived the idea of the earth as a living self-regulating organism. Lovelock was unaware that decades previously the scientist Vladimir Vernadsky had similarly regarded the Earth as a self-contained and self-regulating system. Because Vernadsky wrote in Russian, and was isolated by politics and culture, his theory remained largely unknown for some 50 years (Smith 2007 p139).

Lovelock's hypothesis was named after the Greek goddess of earth, Gaia, on the suggestion from his novelist friend William Golding. **The Gaia hypothesis** was initially largely ignored, then in the 1970s, after further publication and publicity it met with fierce criticism from the scientific fraternity. His theory, now modified, and interpreted more broadly is gaining acceptance from academics and a wide cross section of thoughtful people.

In 2006, Lovelock defined **Gaia Theory** in his book *The Revenge of Gaia* as "A view of the Earth that sees it as a self-regulating system made up from the totality of organisms, the surface rocks, the ocean and the atmosphere tightly coupled as an evolving system. The theory sees this system as having a goal – the regulation of surface conditions so as always to be favourable as possible for contemporary life". He adds that his theory "is based on observations and theoretical models; it is fruitful and has made ten successful predictions" (Lovelock 2006 p162).

Since the 1990s Lovelock's influence through Gaia has become global. On one hand Gaia is a theory, much of which has been found valuable by scientists, being taught in universities as **Earth System Science** and enshrined in the 2001 **Amsterdam Declaration on Global Change** in the statement: "The Earth system behaves as a single, self-regulating system, comprised of physical, chemical, biological and human components. The interactions and feedbacks between the component parts are complex and exhibit multi-scale temporal and spatial variability." This scientific interpretation of Gaia has widespread acceptance.

On the other hand Gaia has become a metaphor - a powerful means of understanding how we might begin addressing current issues such as the ecological crisis and climate change in a holistic way. Lovelock's unified theory of life has become a compelling way of understanding life on our planet in an interdisciplinary context that is enriching to science, ethics, education, philosophy and politics.

Adherents to the Gaia theory vary in how much of the theory they accept. Lovelock's co-theorist Lynn Margulis, a microbiologist, emphasises the symbiotic nature of Gaia, preferring to use ecosystem rather than organism to describe Gaia. Stephan Harding, a deep ecologist who has also worked with Lovelock, emphasises the interconnectedness in the earth's great system where life – animals, plants, bacteria, fungi, algae impact the atmosphere, rocks and water stimulating feedbacks to optimise the conditions for life on earth. Harding explains that Gaia is manifest as *anima mundi - the soul of the world* - in terrestrial form-, that is, the understanding of planet earth as a living sentient being with everything on earth being full of soul, as having a psyche, beauty, intelligence, intrinsic value – as sacred. Harding calls for a return to a pre-scientific perception of the world, that of animism, a belief system shared by many traditional cultures in developing nations today and part of the world's great intangible heritage. He wants this pre-scientific perception, this respect for all life, for Gaia - to reunite with scientific understanding to heal the war on nature – that perspective driven by the western view of man as conqueror of nature.

In 2010 UK Lawyer Polly Higgins proposed to the UN that ecocide be recognised as an international Crime against Peace, triable at the International Criminal Court (Jowit 2010). Ecocide is defined as "the extensive destruction, damage to or loss of ecosystem(s) of a given territory, whether by human agency or by other causes, to such an extent that peaceful enjoyment by the inhabitants of that territory has been severely diminished" (Higgins). The crime of ecocide "arises out of human intervention." It is "a tool to enforce restorative justice" creating "responsibilities" and "sending a powerful message to the world, not just to those involved in business or during war, to take responsibility for the well being of all life". Higgins argues we need ecocide made a crime "to **stop the mass destruction** of the planet"; "to create an **international and national duty of care**" and "to ensure a shift from personal interest to **public and society interests**". Her case refers to intergenerational justice, quoting the preamble of the 1945 Charter of the UN.
The message for cultural heritage protection here is that making ecocide an international crime has the potential to stop destruction of cultural heritage through preventing climate change impacts by e.g. stopping deforestation. It could do this by recognising that cultural heritage is integral to a traditional community's ecosystem, again recalling that so many traditional communities understand themselves as belonging to their environments.

Insert 4.1 Ecocide

Australian environmentalist Tim Flannery, in his latest book *Here on earth. an argument for hope* sees the Gaia hypothesis as describing "cooperation at the highest level" (2010 p36) –"the sum of unconscious cooperation of all life that has given form to our living Earth" and comments "...most importantly, the Gaia hypothesis posits that Earth, taken as a whole possesses many of the qualities of a living thing." Flannery explores the Gaian concept further to consider the work of Bill Hamilton, the founder of socio-biology and coins the term **commonwealth of virtue** (that formed from all biodiversity) to describe a level of earth organisation that he believes can be achieved using Hamilton's research – a level that tends towards increased productivity and interdependence (Flannery 2010 p62). Flannery argues that co-evolution (natural selection that is triggered by interactions between related things) "in both a biological and a cultural sense, is critical to our hopes for sustainability" (2010 p68).

Flannery discusses the counter approach to the future, the potentially destructive Medean view, based on the Neo-Darwinism-driven ruthless selfishness – the literal survival of the fittest. He argues for a full understanding of how things are on Earth, of how ecosystems, superorganisms and Gaia itself have been built through mutual interdependency. "In this light it is absolutely clear that our future prosperity can be secured only by giving something away – cede real authority" (to what is needed to respect Gaia and live within her limits). "We stand in between, (the Gaian and Medean fates), in transition" (Flannery 2010 p272).

Cultural heritage protection and safeguarding may have an emerging role in Gaian thought. As mentioned, many traditional societies embody the Gaian understanding of our reciprocal role with the environment which includes and is inseparable from their culture. In ceding real authority and in Lovelock's words, in making a "sustainable retreat" (Lovelock 2006) from climate change driven disaster, we need to value, nurture, and take better care of our biological and cultural diversity. Gaian theory has the potential to be used as a powerful ally in arguing the importance of including culture in a holistic response to climate change. It is especially relevant, linking Gaian theory with the cultures of developing countries, many of which already perceive nature and culture as indivisible.

Dha Hanu women create colourful headdresses with a multitude of materials including fruit and flowers. This is their everyday wear

In the 1930s and 1940s Aldo Leopold, the American ecologist, forester and environmentalist pioneered the field of wildlife management and environmental ethics. He insisted that we must think of ourselves not as conquerors, but plain citizens of the land, though citizens with a unique capacity to exercise stewardship so long as our interventions are informed by a sensitive ethic. Citizenship implies the existence of civic virtues, and it is through the study of ethics that we have the best chance of remaining Gaia's partner for some time to come. A Gaian ethic might see us progress from self-centredness to enlightened humanitarianism through development of a global ecological consciousness.

Academic philosophers are also envisioning a future dependent on communities cooperating for the common good. Indeed political theorist Professor Mark Olssen argues that the only credible option for humanity is to adopt a reasoned form of *communitarianism* he describes as a 'thin communitarianism'. His 'community without unity' would be driven by the global ethic of survival, an ethic sensitive both to human solidarity and multicultural diversity. Future communities would be governed by strong democracies based on a rights culture and commitment to public life. These would address redistribution of resources and recognition of difference ensuring social equality (Olssen 2010a).

The value in considering the development of these influential ideas is to see the building of bridges across disciplines and through time, to see the centrality of cooperation, of the advantages of taking the holistic view- and of interconnectedness - physical and cultural, in forging a sustainable future, a new unity, one with another, and all with the Gaian whole (Flannery 2010 p108). In this future, human health and the health of the environmental would be inextricably entwined with each other - the double helix of life with the bonds holding it together being our universal values and diverse cultures.

There are global indicators that we are taking steps along the Gaian way. In April 2009 the General Assembly of the United Nations, adopted unanimously the draft submitted by Bolivia that 22 April be celebrated as **International Mother Earth Day.** This recognition, inspired by Gaia theory, represents a revolution in how we look at the Earth and interact with it. In Bolivia, President Evo Morales convened a **World People's Climate Conference** in 2010 that proclaimed a **Universal Declaration of the Rights of Mother Earth,** which has been taken for discussion to the UN and enacted into national legislation (World People's Conference on Climate Change and the Rights of Mother Earth 2010). The Conference also launched an international campaign for an **International Climate Justice Tribunal** which would hold countries and companies accountable for climate change.

In October 2010, the **Ad Hoc Working Group on Long-term Cooperative Action under the Convention of the United Nations Framework Convention on Climate Change (UNFCCC)** proposed that under "a shared vision for long-term cooperative action" the first paragraph of the UNFCCC have the following added : "Parties shall, in all climate change-related actions, ensure the full respect of human rights, including the inherent rights of indigenous peoples, women, children, migrants and all vulnerable sectors, and also recognize and defend the rights of Mother Earth to ensure harmony between humanity and nature" (UNFCCC 2010a p6).

These actions represent a trend towards a stage midway between formative and coming of age of a Gaian perspective. Such a perspective has the potential to make the WHC and the ICHC responsive to protecting our cultural heritage from climate change as part of a holistic approach for a sustainable future.

Part II

The Climate Change Context
How Protected is our Heritage?

Chapter 5

Physical and Biological Impacts of Climate Change on Cultural Heritage

As the average rainfall declines sharply with each passing decade in the south-west of Australia, a farmer who shares the scientific knowledge that is the common heritage of humanity will make different decisions about land use than one who thinks that a series of dry autumns is a passing phase. The regulators of power distribution in a state that has just been devastated by a bushfire during what would once have been described as once-in-a-century conditions will make different decisions if they know from science that these conditions will now arrive with awful frequency.
Ross Garnaut (Garnaut 2011 p105)

Problems Communicating Climate Change Knowledge

A grasp of climate change science is basic to understanding the global impacts of climate change, including those to cultural heritage. Through better understanding comes better preparedness and informed advocacy. Scientists and those working at the frontline on climate change problem areas are calling for the wider dissemination of knowledge (GHF 2009c; ACJP and al. 2009; UNESCO 2009j; Garnaut, 2011). As concluded by academics attending the 2009 conference, **Climate Change: Global Risks, Challenges and Decisions** (convened to update the 2007 IPCC fourth assessment), "The knowledge that human activities are influencing the climate gives contemporary society the responsibility to act. It necessitates redefinition of humanity's relationship with the Earth and – for the sake of the well-being of society – it requires management of those human activities that interfere with the climate. To support development of effective responses, however, this knowledge should be widely disseminated" (Richardson et al. 2009 p7).

Climate change science is taught to everyone on Kiribati, one of the most vulnerable **Small Island Developing States (SIDS).** Here island homes are being inundated by rising sea level and storm surges (Dekker 2011). On the present trajectory citizens will become passport holders to a virtual state, a submerged ghost nation under the Pacific Ocean by the end of the century (Henson 2008; Hamilton 2010). The response to climate change science from developed nations is not so accepting. It ranges from profound understanding and extreme concern to suspicion, through disbelief, a catatonic-like fatigue, to denial and refutation. The reasons for this are many and go to the heart of our human nature. It is appropriate to mention some obvious reasons as they will have a bearing on the difficulties of protecting world cultural heritage particularly in the developed world.

Climate change science is complex and difficult to evaluate. Beginning from the classroom, the teaching of climate change or global warming has been problematic. Leading UK scientists have criticised the practice as being impeded by "omission, simplification and misrepresentation" (Garner 2007). In the United States scientists rallied to denounce a new textbook for advanced high school classes which was so misleading as to teach students to dismiss global warming. Friends of the Earth reflected, "It is hard enough to persuade lawmakers and captains of industry to acknowledge the challenge of global warming, let alone take action. How much more difficult will our struggle for a sane response be if our own public schools are working against us?" (FOE 2008).

Climate change science may be ridiculed and censored at the highest levels by politicians and their advisers, or given illusory support 'greenwashed' by government officials and special interest groups (Hansen 2009 p112), watered down by scientists fearful of losing their jobs, of receiving death threats (Clarke 2011), of being cyber bullied (Hansen 2011a) or of being reported to their universities for 'scientific misconduct'. An unjustly infamous case of alleged scientific misconduct broke in 2009. Known as the Climatic Research Unit email controversy, now widely referred to as *Climategate*, it shook the western world. What amounted to "snipping private comments out of context" was used for a "disinformation campaign" and character assassination (Hansen 2011b p283). But the damage had been done in the eyes of the public, setting the case for emergency climate action back years. Other factors confusing the climate science message include the relentless challenges by powerful sceptics, biased think tanks (Hansen 2012a) and the lobby groups of many threatened globalised corporations and industries such as coal, oil and gas. More than one public intellectual is pointing to the acceptance of donations by political parties as being at the heart of the problem. Such practice can be malignant, allowing vested interests dangerous influence.

A mighty new factor is the power of the *blogosphere*. For the first time the public have to contend with unmoderated comment taking up companion space with peer reviewed climate science (Lyster 2010). The nature of the online world of crowd sourced information is insidiously (and helped by browser/server unseen filters) leading to a homologising of information. Internet surfers find those with similar views without being exposed to other views and even other newsworthy issues. This practice reinforces ideas without debate. Homophily meaning 'love of the same,' is the tendency of individuals to associate and bond with similar others. The presence of homophily has been discovered in a vast array of network studies.

James Hansen a leading climate scientist observed in March 2010 in relation to news media distortions over climategate, "in the last six months the gap between what is known by scientists and what is understood by the public has widened" (2010e). In 2012 he further observed "the public seems to have become less certain about the situation. Indeed, many people have begun to wonder whether the climate threat has been concocted or exaggerated (Hansen 2012a). Thus despite clear reports of climate change impacts, the results of brilliant and decisive studies (that under a health banner would be feted as breakthroughs) from village level (UNESCO 2010j) to universities, research institutions and the UN (UN 2009b) – there remains majority ignorance, denial and confusion enough to undermine the groundswell of public concern needed to goad world leaders into making climate change mitigation a first priority (Spratt 2008; Shearman and Smith 2007; Füssel 2008c; Hamilton 2010; Hansen 2010d). Hansen believes "the difficulty in communicating science to the public is related to the corrosive influence of money in politics and to increased corporate influence on the media" (2012a; 2012c).

In 1957 the Soviet Union initiated the Space Age with the launch of 'Sputnik 1', the first Earth-orbiting artificial satellite. I was among the millions who marvelled at the incandescent speck traversing the heavens. The West was shaken and practising scientists were spurred to write new school science curriculums in urgency. In Australia, as a new teacher, I was at the vanguard of this new age of science education. Weighed down with a fresh 1964 edition of the 1040 page 2.2kg textbook *Science for High School Students* I had no doubt from viewing the portraits of the distinguished line up of authors on the leading pages, that I was armed with a new concept of integrated science education that would prepare "our boys and girls for life in this scientific age"... so that they would not be "relegated in time to a second-class existence".

Sputnik 1 symbolised the West's lack of demonstrable capability in space science. Climate change science although far more important to humanity's future than space science is difficult to conceptualise and lacks the immediacy and obvious identity that space science has to most people; but not for the folk on Kiribati. This predicament is all the more schizoid because unlike Western scientists in 1957 who were on a space technology learning curve, today's experts have most of the knowledge needed to decarbonise our economies. They are aware of the urgent need to address the situation through concerted attention to renewable energy sources, efficiency, a price on carbon and much more. What they don't have are leaders who will fully inform the public of the impending crisis and stand up to vested interests.

Climate Change and Cultural Heritage

The physical and biological impacts of climate change form the basis of vulnerability and risk assessments used in making adaptive strategies for the protection of cultural heritage (Lewis 1990; Parry 2007; Füssel 2007). The rapid advances being made in climate science are necessitating ongoing re-assessment of the possible magnitude of these impacts. It is important for cultural decision-makers to understand enough of the facts and the dynamics of climate change to be able to interpret predicative models and create future scenarios that have a probability, particularly a high probability, of impacting cultural heritage. Also important is the advocacy the climate-informed cultural heritage professional can bring to the mitigation dilemma (Stern 2008; Schellnhuber et al. 2006; Schellnhuber 2008).

Climatic protection of immoveable cultural heritage is an emerging discipline. In Europe there has been a concentration on the preservation of built structures– "identifying ways of measuring the impact of climate change on the historic environment"– led by the University College London (UCL), Centre for Sustainable Heritage directed by May Cassar. In a groundbreaking scoping study, *Climate Change and the Historic Environment*, climate change impacts were identified as key climate change related factors impacting UK regional: a) heritage buildings and contents; b) parks, gardens and landscapes and; c) buried archaeology (See Insert 5.1) (Cassar 2005, 2009; Sabbioni 2006; Cassar and Hawkings 2007). The researchers explain that whilst the focus on and identification of threats was new, the preservation methods of addressing their impacts were those in common use. In 2010 they published *The Atlas of Climate Change Impact on European Cultural Heritage* to facilitate climate change programmes (Sabbioni 2010).

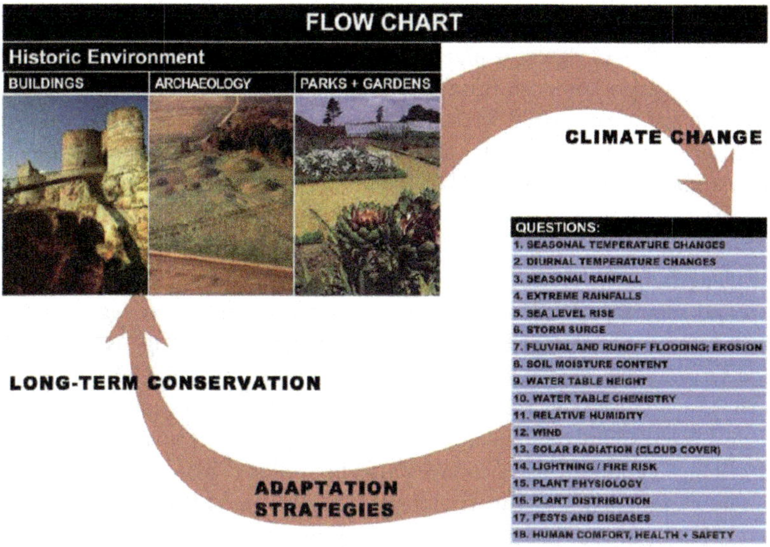

Insert 5.1 From Climate Change and the Historic Environment (2005), p70 by May Cassar. Published by the Centre for Sustainable Heritage, University College London with support from English Heritage and the UK Climate Impacts Programme. UCL Centre for Sustainable Heritage - the Bartlett School of Graduate Studies - Gower Street - London - WC1E 6BT. ISBN 0-9544830-6-5. And on the internet at www.ucl.ac.uk/sustainableheritage/climatechange/climatechangeandthehistoricenvironment.pdf Reproduced with kind permission of Prof. M. Cassar, Director, Centre for Sustainable Heritage

Their methodology involved modelling future climates; seeking information to establish climate-related factors of concern and their impact; testing the impact of climate change on these factors by feedback from a questionnaire, from site visits and from regional workshops. Results led to refinement of the factors of concern which provided feedback to government, and formed general directions for policy makers at the local level including strategies for adapting, (planning time-scales, monitoring, management and maintenance, deciding what is acceptable loss and recognising obsolescence) and policy implementation actions particularly the fostering of co-operation, research and education (Cassar 2005; Cassar and Hawkings 2007). The research required to address climate change problems that the historic environment sector is expected to encounter in the next 10–15 years has been included in Insert 5.2 to show the interdisciplinary expertise needed for cultural heritage protection.

The UCL study identifies characteristics that are becoming hallmarks of heritage protection namely: a) information gathering and sharing; b) dissemination (of knowledge and advice) and feedback; c) monitoring; d) access to and contribution to central information centres; e) management with local participation; f) valuing what can and cannot be saved; g) adaptive strategies and h) the preparation of guidelines and development of policy.

Complementary studies in Europe document meteorological 1) temperature derived parameters (range, freeze thaw, thermal shock); 2) water derived parameters (precipitation, humidity cycles, time of wetness; 3) wind derived parameters (wind, wind driven rain, sand and salt); 4) pollution derived parameters; 5) elemental carbon and 6) pH. Higher temperatures enhance fungal growth on wood. Drier summers seem likely to increase structural problems from desiccated soils and salt weathering of porous stone (Brimblecombe 2006; EU 2006). The *European multidisciplinary Joint Programming Initiative* in its 2009 final document noted "European conservation and security of cultural heritage can only be delivered against an improved understanding of the global context. Climate change is driving the need to consider the whole assemblage and its context in the 21st century so that faced with cultural and environmental challenges, our understanding and the resilience of cultural heritage will be transformed." Communities will be faced with "what to save and what to lose" (JPI 2009).

The members of the **International Council on Monuments and Sites (ICOMOS)**, aware that climate change is accelerating and amplifying the processes of deterioration to historic heritage are developing their own programmes. They have adopted much of May Cassar et al.'s protective methodology, and propose to develop measures such as a deterioration of cultural heritage (DCH) scale, a monitoring and intervention manual and incorporate the use of high tech online predicative tools for disaster risk identification. The Europe Union's *Noah's Ark* programme and ICOMOS's proposed measures are among the first to extend recognition of the link between climate change and tangible heritage to intangible cultural heritage. ICOMOS conference participants commented "As losses occur in the physical environment, intangible heritage values associated with the environment will also be lost, and hard choices will need to be made about what to try to preserve and what to let go" (ICOMOS 2009 p2; Cassar and Hawkings 2007).

May Cassar summarises the status of tangible heritage protection against the threat of climate change, "In the United Kingdom and in the European Parliament, there has been some progress in recognizing the threat of environmental change to cultural heritage. The U.S. has given the world Al Gore (and his documentary *An Inconvenient Truth*) and we must wait to see whether congressional (now in hands of Republicans) views will change or whether presidential and cabinet-level views are going to be altered by Barack Obama. Canada has given the world the **Montreal Declaration, the Global Municipal Leaders Declaration on Climate Change** signed in December 2005. Building on these developments, we must all take concerted action to persuade the IPCC that its next report should make an explicit reference to cultural heritage as an area that is vulnerable and threatened. It has not done so thus far, so the best we can do at the moment is to adopt references to the human environment, habitats, or settlements as surrogates" (Cassar. 2009 p.19). There does not appear to be any indications that IPCC AR5 is planning a chapter on cultural heritage. The outlook for cultural heritage preservation in the US is changing. In April 2010 the President set up

America's Great Outdoors Initiative to develop a 21st century conservation and recreational agenda. During the 2010 summer, 51 public listening sessions across the country took evidence from some 10,000 participants. The outcome, a report *America's Great Outdoors: A Promise to Future Generations* reflects peoples' ideas on how to reconnect with America's lands, waters, and natural and cultural treasures mindful of the threat of climate change among others (AGO 2011). The report reasons that along with other programmes, an increase in funding for the Historic Preservation Fund was needed. In February, 2011, the Obama administration's proposed 2011 budget called for austerity measures, proposing a cut of $25 million to the Historic Preservation Fund and the elimination of the *Save America's Treasures* and *Preserve America* programmes (Flechsig 2011).

Climate Change Impact

New scientific understanding of traditional materials and practices caused by extreme weather (including rain penetration, high summer temperatures and chloride loading) based on cross-field monitoring and leading to key indicators of impact in terms of scale and time and design guidance. Specific areas of need include:

- Climate change modelling and monitoring geared to the historic environment
- Predicting subsidence and heave caused by extreme weather
- Understanding damage mechanisms and remediation caused by extreme weather
- Understanding the effect of wind driven rain at a local level leading to severe damp penetration
- Understanding the effect of wind driven dust and pollutants at a local level leading to erosion
- Understanding the effect of new pest migration and infestations, e.g. termites
- Understanding water resistance of building materials and techniques
- Assessment of availability of renewable materials, stocks and development of old technologies eg lime technology
- Environmental performance of the historic fabric under extreme weather
- Interface between fragile materials and very robust construction inc. robust detailing in the historic environment

From Climate Change and the Historic Environment. (2005), p. 66 by May Cassar. Published by the Centre for Sustainable Heritage, University College London with support from English Heritage and the UK Climate Impacts Programme. UCL Centre for Sustainable Heritage - the Bartlett School of Graduate Studies - Gower Street - London - WC1E 6BT.The UCL. ISBN 0-9544830-6-5. And on the internet at www.ucl.ac.uk/sustainableheritage/climatechange/climatechangeandthehistoricenvironment.pdf Reproduced with kind permission of Prof. M. Cassar, Director, Centre for Sustainable Heritage

Insert 5.2 Gaps in information and research. Medium term actions

The Australian Experience

Australia has no comparable pioneering national or regional climate change and cultural heritage studies despite being the developed nation most vulnerable to the direct impacts of climate change (Lindenmayer 2007; ANU 2009). The status of Australian tangible and intangible cultural heritage protection (except for World Heritage sites) is parlous according to serious commentators including historians. Jane Lennon's 2006 *National and Cultural Heritage Report* is substantial. It welcomes the better acknowledgement of indigenous connection to country. However it questions the entrenched separation of natural from cultural heritage. The author puts this "unnecessary imbalance between natural, indigenous and historic heritage" at the foot of the "chauvinism of European encounter" (Lennon 2006 p41). She reports the widespread ignorance of how historic heritage has broadened beyond 'places' to incorporate intangible heritage—language, oral tradition, crafts, skills and performing arts. The most immediate need is seen as the establishing of a programme to bring together the listed heritage from across the continent under a national organisation. The Report concludes that heritage is still regarded as being 'special places' rather than as "a range of values that are found throughout the environment and encompassing stories, traditions and community associations" (Lennon 2006 p.42).

In 2007 Michael Pearson reported in his Conference paper *Climate Change and its Impacts on Australia's Cultural Heritage* the lack of any sign of an Australian climate change policy for heritage protection. He listed threats to immoveable heritage and proposed tentatively the scoping and management process recommended by the World Heritage Committee in their report *Predicting and Managing the Effects of Climate Change on World Heritage* (see chapter 15) which is based on Cassar's UK study (Pearson 2007). Intangible cultural heritage was not included within the scope of his study.

The National Cultural Heritage Forum is an advisory body to the Australian Government. In its 2008 report *Australia's Cultural Heritage: Important Needs and Proposed Responses* it confirms its 2007 finding that there was almost no research being conducted into either the assessment or identification of the potential impacts of projected climate change on cultural heritage. It regrets the continued lack of funding for tangible and intangible cultural heritage preservation, calls for a renewed National Heritage Program and proposes a *Climate Change Response Package for Cultural Heritage and Local Government Initiatives* (FORUM 2008).

The Australian academic and ICOMOS member Susan McIntyre-Tamwoy strongly presents the national case in her editorial *The impact of global climate change and cultural heritage : grasping the issues and defining the problem* adding "It seems however that there continues to be little government investment in this type of strategic research" (2008 p5). She draws attention to the scant consideration paid to the impact of climate on tangible indigenous heritage and how the potential climate impacts on the traditional custodians of Australian World Heritage properties have been overlooked. She reminds that "global climate change is likely to be among the major contributors to cultural heritage loss over the next few decades both through direct impacts such as rising sea levels and erosion of coastal sites and through secondary impacts as a result of changing tourism patterns and government settlement policies" (2008 p9).

The Australian Government's much anticipated report, *Implications of Climate Change for Australia's World Heritage Properties: a preliminary assessment 2009* (referred to as *the Assessment*) is a world first. It was carried out by the Australian National University (ANU). It drew upon the 2007 pioneering preparatory report *Impacts of Climate Change on Australia's World Heritage Properties and their Values.* The author Lance Heath found that the "built heritage is likely to be more resilient than the natural heritage to the impacts of climate change. However, it is important to recognise that our knowledge of climate change impacts on our built heritage is limited or, in some cases, non-existent" (Heath 2008 p16). His recommendations included "monitoring, reporting and mitigation of climate change effects through environmentally sound choices and decisions at a range of levels: individual, community, institutional and corporate." Heath goes so far as to include a discussion under the heading *Mitigate or lose*. His report was written for the far reaching and influential Garnaut Climate Change Review. (In 2007 Ross Garnaut, was commissioned by Australia's Commonwealth, State and Territory Governments to conduct an independent study of the impacts of climate change on the Australian economy).

The above *Assessment* closely followed the model adopted by UNESCO (to be discussed in Chapter 15) which was elaborated from the UK model. Most of Australia's World Heritage properties are 'natural'. A summary of major impacts of climate change on the six Australian World Heritage 'cultural' and 'mixed' properties presents the first national assessment of climate change impact on cultural heritage (see Insert 5.3).

The Australian *Assessment* used data from the IPCC 2007 *Assessment*, with reference to some more recent findings and the ANU's own 2006 climate measurements. It targets the physical and biological impacts of climate change, clearly identifying the limited and piecemeal knowledge of these impacts and the need for specific research. The *Assessment* does not consider mitigation measures in relation to World Heritage management. The *Assessment* finds that "There is little information available on the potential impacts of climate change on cultural values of Australia's World Heritage properties" (ANU 2009 p14). Australia now has 19 World Heritage sites, 7 with OUVs relating to culture (the 7th was listed in 2011). Four of the sites are 'mixed' with Australia having more 'mixed' sites than any other member State.

Cultural World Heritage sites	Major Climate change threats	Additional likely climate change threats	Adaptive capacity
World Heritage criteria: 3,4,5,and 6 Kakadu NP (mixed site) Tasmanian Wilderness, (mixed site) Uluru-Kata Tjuta NP (mixed site) (Human settlement and land use and Archaeological sites)	**Changed fire regimes (more frequent and intense)** **Increased erosion from extreme weather events** **Increase in temperature and changes in humidify**	More intense storms and cyclones. Sea level rise. Changes in annual and seasonal rainfall. More frequent severe droughts, affecting conditions on land. Reduced run-off into streams and creeks. Elevated CO_2 Acid rainfall	Low to Moderate
Tasmanian Wilderness, (mixed site) (Archaeological site)		Ocean acidification. Storm surges. Increase in sea surface temperatures.	Low
World Heritage criteria: 3,4,5,and 6 Willandra Lakes Region (mixed site)		Acidic rainfall. More intense storms and cyclones. Sea level rise	Moderate
World Heritage criteria 2 Royal Exhibition Building and Carlton Gardens (Built environment)	**Increase in temperature** **Drought** **Acid rainfall** **Increase in storm frequency**	Increase in pest and biological infestations.	High
World Heritage criteria 1 Kakadu NP (mixed site) (Aboriginal Rock Art) Sydney Opera House (Built Environment)	**Acid rainfall** **Fire** **Increase in storm intensity**	Increase in flash flooding. Increase in pest and biological infestations. Sea level rise	Moderate for rock art. High for Sydney Opera House

Insert 5.3 Australian World Heritage cultural properties, climate change threats and adaptive capacity. Combining tables from (ANU 2009; Heath 2008). For a description of the cultural criteria, see Chapter 2

The *Assessment* does not appear to have been written in consultation with indigenous custodians or to have drawn upon site managers' feedback. Of relevance here are Amareswar Galla's comments on World Heritage nominations which appear also to apply to climate change assessments. "The globalising tendency of World Heritage inscriptions has come under scrutiny in the past decade. The concern is that the processes of nomination and assessment and the pool of expertise, mostly derived from western countries, is resulting in a homogenising negative impact in Asia and Africa. The conservation plans....rarely engage with local communities or their living heritage" (2008).

The 2009 *Assessment* acknowledges shortcomings, recognising that in relation to indigenous properties there is a need for a wider-ranging analysis that has an interdisciplinary focus and integrates life science aspects with some exploration of cultural questions and communication with the traditional owners of the region. The overall view is: "Because there is little knowledge or understanding of potential impacts of climate change on cultural values, the extent of vulnerabilities would appear to be worthy of identification and assessment in the context of *specific site* management issues" (ANU 2009 p xiv).

Particular physical and biological impacts are listed in Insert 5.4. Weeds continue to be a major threat to some of Kakadu National Park's values which are also being undermined by the feral cane toad although its particular response to climate change is unknown. Climate impacts on rock art and archaeology are poorly understood. Changes in disease patterns in relation to native fauna with climate change are also not known. The "impact of disease on the human population is of major concern" and "one that has the potential for affecting traditional site custodians as well as their communities" (ANU 2009 p52).

Physical and Biological Impacts on Intangible Cultural Heritage

Surprisingly there has been no comprehensive study of potential climate change impacts on Uluru-Kata Tjuta's ecosystems, cultural practices and beliefs. The *Assessment* recognises that intangible cultural heritage is integral to 'mixed sites': "Particular spiritual values, for example, appear to be timeless and the loss of sacred sites can be devastating to a community. Some cultural expression would appear to be directly threatened by climate change..." "Little, if any, work has been undertaken on whether the listed cultural values are under threat from climate change and, if they are, what this could mean for the Indigenous communities that have a role in managing and maintaining the World Heritage values of the park. Research into such implications of climate change and communication between the region's traditional owners, policymakers and scientists is desirable. The landscape of Uluru Kata Tjuta is imbued with creative powers of cultural history through the Tjukurpa, an Indigenous philosophy expressed in verbal narratives, art and through the landscape itself" (ANU 2009 p106).

- rising sea level and storm surges
- changed fire regimes; heat and soot from intense fires
- more intense cyclones
- higher evaporative demand
- reduced runoff into streams and rivers
- threats from a change in disease dynamics
- threats from a change in feral animals
- increased CO_2 concentrations
- extreme weather events and flash flooding and erosion
- rainfall changes
- droughts are likely to become more frequent and continue to change vegetation
- increased sea surface temperatures,
- changes in ocean circulation and ocean acidification
- chemical and physical changes to the coastal zone, including saline intrusion
- excessive human visitation,
- exotic pests and diseases.
- species extinction (ANU 2009)

Insert 5.4 Climate change impacts on cultural properties

How fares cultural heritage in the face of climate change?

As is evident, the fate of cultural heritage under the physical impacts of climate change, while gaining recognition, is not high enough on most national policy agendas as yet to ensure some order of protection. An international response is needed.

A Changpas nomadic goat herder

Chapter 6

Climate Change Impacts on Humans, Human Impacts on Cultural Heritage

The people living around Mt Rwenzori have cultural values attached to it, thus the effects of climate on these people do not only affect their activities but cultures as well. Kule Musinguzi, a resident of Kasese believes that the snow on top of Rwenzori Mountain is key to the survival of his tribesmen. The name 'Abanyarwenzururu', he says, means 'people from the land of snow'. This makes Musinguzi and his tribesmen more worried about the melting glaciers supplying river Nyamwamba. They believe that when the snow disappears completely the people will disappear as well. The current climatic changes and behaviour of the river is believed to be controlled by the gods who live in the mountain. It is believed that when the gods become unhappy and irritated they release a lot of disastrous water at a go aimed at punishing the community for wrong doing. Residents remain uncertain of how they will live minus the river. Twebaze Paul, Uganda. (UNESCO 2010j)

The consequences of climate change impacts on humanity will, in turn, affect humanity's impact on the environment. 'Human impacts' as defined in this book relate to the impacts of climate change affected humans on cultural heritage.

Emerging 'Human Impacts' of Climate Change

The human impacts of climate change and their physical and biological drivers are summarised in Insert 6.1. This overview has been duplicated in this chapter to emphasis the human impacts. The arrows indicate how climate change destabilises environmental and social conditions jeopardizing the conservation of natural ecosystems and the sustainability of socioeconomic systems. As a result climate change has short and long term, as well as local and national implications for the protection of cultural heritage. Less research has been done on the human impacts of climate change than on the physical impacts. Much is theoretical but it serves to provoke questions about Convention potential and performance in protecting cultural heritage.

Every few months bring reports of climate related natural disasters from floods to wildfires, some unlike any in living memory. The IPCC's predictions are being verified by the observed increasing frequency and intensity of extreme weather events, some of which are unprecedented in modern human history (IPCC 2007). The expected rate of future global climate change will make adaptation particularly challenging (4-degrees-and-beyond 2009). In the past there was little understanding of climate mechanisms for example La Nina and El Nino weather patterns or means by which natural variations could be predicted. Heritage managers today have access to unmatched weather and climate change information with which to prepare (Füssel 2010). They will, though, face new methodological challenges. The groups that have traditionally managed environmental threats and resources are not well equipped to assess and address the more complex threat of climate change (Füssel 2009a). As has been noted climate change does away with the basic assumption of stationarity - that conditions will overall remain predictable, which was central to past climate related management. Decision makers will need greater understanding to cope and to exploit the emerging options which may require a combination of the traditional practices and sophisticated high technology (Fussel 2008a).

The Nature of Climate Change in Human Terms

"Scientific information, technologies and economic instruments are all part of the solution, but their interpretation and application are mediated through the cultures and worldviews of individuals and communities (see Insert 6.2). Religious and spiritual beliefs, indigenous knowledge systems, understandings of nature-society relationships, values and ethics influence how individuals and communities perceive and respond to climate change. Ultimately these human dimensions of climate change will determine whether humanity eventually achieves the great transformation that is in sight at the beginning of the 21st century or whether humanity ends the century in a "miserable existence in a +5 degree C world" (Richardson et al. 2009 p 34); (Liverman 2010 p7; Lynch 2009). Such an 'existence' has been given searing portrayal in the post apocalyptic novel *The Road* by American author, Cormac McCarthy.

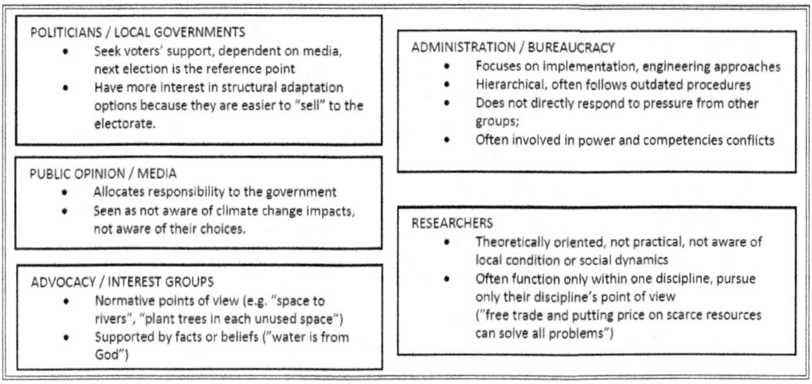

POLITICIANS / LOCAL GOVERNMENTS
• Seek voters' support, dependent on media, next election is the reference point
• Have more interest in structural adaptation options because they are easier to "sell" to the electorate.
PUBLIC OPINION / MEDIA
• Allocates responsibility to the government
• Seen as not aware of climate change impacts, not aware of their choices.
ADVOCACY / INTEREST GROUPS
• Normative points of view (e.g. "space to rivers", "plant trees in each unused space")
• Supported by facts or beliefs ("water is from God")
ADMINISTRATION / BUREAUCRACY
• Focuses on implementation, engineering approaches
• Hierarchical, often follows outdated procedures
• Does not directly respond to pressure from other groups;
• Often involved in power and competencies conflicts
RESEARCHERS
• Theoretically oriented, not practical, not aware of local condition or social dynamics
• Often function only within one discipline, pursue only their discipline's point of view ("free trade and putting price on scarce resources can solve all problems")

Insert 6.2 Groups of shared mental models. Reproduced with kind permission of Ilona Banaszak(Richardson et al. 2009 p34); (Otto-Banaszak et al. 2010)

As reported in the 2010 State of the World report *Transforming Cultures. From Consumerism to Sustainability,* "cultural heritage protection will depend on knowing about global and local drivers in order to make holistic assessments to seek optimal protection" (Assadourian 2010). Concerned scientists have been extrapolating from their findings to explain present and future human impacts (FOE 2008; ACJP, CANA, and FOE Australia 2008; Füssel 2010; Richardson et al. 2009; WBGU 2009; Hansen and Sato 2011). Water supply for households, health systems, agriculture and industry is and will be jeopardised as a result of weather extremes, altered precipitation patterns and glacial retreat. Food production is expected to decline worldwide as warming increases, with the potential to trigger regional food crises and undermine the economic productivity of affected states. The economic potential of many countries, especially in agriculture, forestry, fisheries and tourism, will be adversely affected both directly and indirectly. Global warming will accelerate the loss of biodiversity and corresponding ecosystem services which will give rise to substantial economic costs worldwide and thus create new poverty (Sukhdev 2008). In addition, in coastal areas especially, cities and vital infrastructure will be threatened by sea-level rise and weather extremes. Unrestrained climate change will result in a massive global welfare loss (WBGU 2009). Security risks will increase because in almost all regions climate change will undermine the natural life-support systems on which people depend. GHG emissions are increasing and the world population is set to grow by forty percent by 2050. Hans Joachim Schellnhuber, Director of the Potsdam Institute for Climate Impact Research warns "The abrupt change of a tipping element such as the Asian monsoon to a substantially drier state, or the eventual loss of water storage capacity in Himalayan glaciers, would lead to environmental stress of profound proportions by reducing water availability in the Indo-Gangetic plain" (Richardson et al. 2009 p16).

Lama praying to his own musical accompaniment

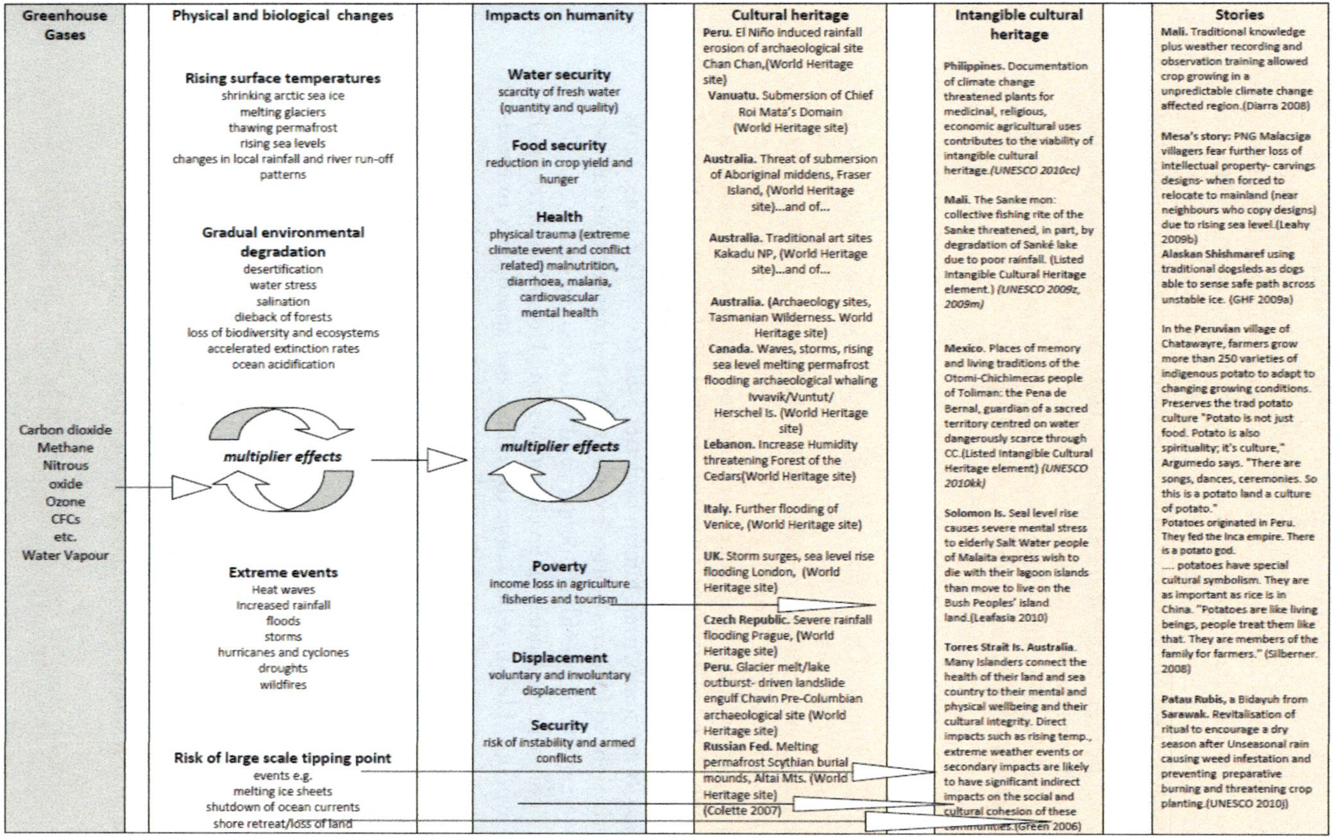

Insert 6.1 Diagrammatic summaries of the relationships between climate change and cultural heritage. Chapter 6 is concerned with the changes outlined in the first and third column and the last three shaded columns. The effects of climate change are amplified and accentuated by latitude and altitude

These changes may result in the emergence of new and unpredictable tensions and conflicts in global politics, which jeopardize international stability and security (Schellnhuber 2008; Campbell 2008). Climate change is expected to lead to increased conflict over land and maritime borders as rising seas, severe flooding and drought change the world (Schellnhuber 2008). It is also expected to lead to more disputes over scarce resources such as energy, water, arable land and fish stocks, as well as over the treatment of migrants. As Professor Ole Waever has reported in the *Security Implications of Climate Change* (Richardson et al. 2009 p17), "Factoring security into the climate change equation runs the risk of escalating vicious circles. In the parts of the world where health and well-being are most negatively impacted by climate change, the likelihood of conflict will increase most, and these conflicts will further reduce living standards. More privileged parts of the world are likely to first feel the spill-over effects from these conflicts, such as refugees and diseases, and at higher temperature increases see their own security agenda re-organised around climate change."

Climate Change and Culture

The 2004 UN Human Development Report titled *Cultural Liberty* states, "culture is not a frozen set of values and practices. It is constantly recreated as people question, adapt and redefine their values and practices to changing realities and exchanges of ideas." This, but not the need to protect it from the impacts of climate change, is reinforced in the vision articulated by Françoise Rivière, Assistant Director-General for Culture, UNESCO (UNESCO 2009p). "Culture, has always served as the inspiration and matrix for all transformations within human societies" she says. "The time seems particularly right to focus on culture's capacity for permanent renewal, owing to the creativity of individuals, peoples and societies, and on the capacity for devising alternative models of development rooted in each country's rich cultural diversity. The time has come to rethink our approaches to development if we want to ensure a sustainable future for the generations of tomorrow. Since it is dynamic by nature, culture provides various well-suited opportunities. In the context of the current global crisis, might not culture, given its rich diversity, be part of the solution for sustainable and more equitable development? Should we not move culture to the forefront of our thinking on models for development and for international cooperation?" (UNESCO 2009a).

Ms Rivière could well have added 'might not culture be part of informing and influencing climate change' for culture will also offer the best chances, the 'various well-suited opportunities' for societal survival under climate change stress. The more our diversity of culture and heritage is protected the greater our resilience and chance of achieving a sustainable future.

Tangible Cultural Heritage and Human Impacts of Climate Change

'Tangible' cultural heritage can be moveable, for example the dragon boats in the Intangible Cultural Heritage Listed Chinese *Dragon Boat Festival* (UNESCO 2009k), or immoveable for example the World Heritage Listed *Ancient Maya City of Calakmul, Campeche* in Mexico (UNESCO 2002f). Nearly all of the research into the effects of climate change on tangible heritage has dealt with biophysical impacts. The climate impacted tangible heritage examples shown in Insert 6.1 are all World Heritage sites primarily understood as prone to physical impacts. Very little research has investigated the effect climate change might have on the custodianship of tangible heritage. If the custodians are lost through health impacts or forced displacement, how will the tangible heritage be maintained? Will the cultural memory be lost?

The UN General Assembly is considering a *Draft Resolution on Return (or) Restitution of Cultural Property* asking Member States to actively cooperate in returning displaced cultural artefacts to their rightful home countries (UN 2009a). The UN recognises the international community's growing awareness of the trauma that people and their governments suffer when their cultural properties are stolen or trafficked. This trafficking is increasing despite the 1970 UNESCO *Convention on the Means of Prohibiting and Preventing the Illicit Import, Export and Transfer of Ownership of Cultural Property* and the 1995 UNIDROIT *Convention on Stolen or Illegally Exported Cultural Objects*. While this illicit trade is not the focus of this book it is mentioned here to emphasise that nations suffering from the impacts of climate change, particularly developing ones, are more likely to have their tangible culture threatened by theft (Galla 2003).

Intangible Cultural Heritage and Human Impacts of Climate Change

The threat of climate change to intangible cultural heritage has yet to be fully recognised and the threat it poses to communities' practices, customs, performances, festivals, crafts and traditional knowledge remains mostly unknown. The most vital examples of climate affected intangible heritage are reported as personal stories in reports such as the Friends of the Earth's *Climate Change: voices from communities affected by climate change*, (see Insert 7.1) (FOE et al. 2007) and on the UNESCO internet site *On the Frontlines of Climate Change* (UNESCO 2010j).

'Traditional knowledge' is the aspect of intangible heritage that has attracted the most discussion because of its vulnerability to climate change and its role in adaptation to climate change. Indigenous people, like all people, gain knowledge from the other communities and cultures with whom they mix. They retain culturally significant elements of their traditional way of life. This they combine with contemporary ways that maintain and enhance their identity while allowing their society and economy to evolve and adapt to changing circumstances. Valuing and practising traditional knowledge has become a significant means by which indigenous groups in many parts of the world regain control over their own cultural information. Reclaiming this knowledge has become a major strategy for revitalisation movements (Ecologic 2010).

Insert 6.1 presents some of the few stories and examples of climate impacted intangible heritage accessible in the literature. The most inspirational project is from drought ridden Mali. The entire Mali drought adaption program is built on the baseline of indigenous knowledge and coping capacity (see Insert 6.3).

Mali is a drought-ridden, semi-arid Western African nation with high agricultural dependence — a country highly vulnerable to climate change. Home to some 12 million people, agriculture and livestock generate approximately 40 percent of GDP in Mali and employ over 50 percent of the workforce. Despite being one of Africa's major cotton exporters, Mali is one of the world's poorest nations as approximately two-thirds of its population lives below the poverty line.

In the 1970s Mali's economy was severely impacted by frequent droughts resulting in heavy agricultural losses and negative impacts on livestock and human health. In response, a regional program was launched to combat the effects of drought by enhancing weather data collection, local training and telecommunication capacity. Weather, crop and water data is collected, processed and used to create weather bulletins which are then disseminated in the community via radio, newspapers and television. These notices relay early warning information on rainfall, pestilences and disease, along with advice on optimal timing for planting, crop selection, fertilizer application, etc.

The results have been hugely successful as food security has increased substantially. Since this program's inception, yield has increased by approximately 20 percent and replanting has decreased from 40 percent to 5 percent. The cotton industry saved millions of dollars from advanced warning of out-of season rainfall in 2002 alone. Additional social benefits include reduced rural to urban migration and more effective use of pesticides. Adaptive capacity has thus been tremendously strengthened by combining indigenous knowledge and coping capacity with weather information, technology and training. The Anatomy of a Silent Crisis. Global Humanitarian Forum 2009:Climate Change (Konate 2004) (GHF 2009a)

Insert 6.3 Mali — Building adaptive capacity brings hope to a vulnerable drought-ridden nation

Implications for the Protection of Cultural Heritage

Enough is now known about the physical and biological impacts of climate change to test the efficacy of the two UNESCO Conventions in protecting world cultural heritage. More challenging are the evolving methodological approaches to heritage protection which may demand capacities of the Conventions that were not anticipated when the Conventions were written. Most challenging is the void in the research addressing possible climate induced impacts such as the damage, deterioration, fragmentation, abandonment and loss of core tangible and especially intangible cultural heritage practices, expressions, beliefs and knowledge that give meaning to life. As anthropologists Rosita Henry and William Jeffery argue in their paper *Waterworld: The Heritage Dimensions of 'Climate Change' in the Pacific*, "research on the 'human dimensions' of climate change must be expanded from its current socio-economic emphasis to include baseline local level ethnographic studies that enable understanding of nonmarket impacts and the complex transformative relationship between cosmologies of climate change and heritage values" (Henry and Jeffery 2008 p2). "In other words, we need an expanded notion of what is meant by 'human dimensions' that includes consideration of social practices at the micro-level and how people produce the heritage values that inform everyday practice and that demonstrate capacity to respond to change" (Henry and Jeffery 2008 p16).

Studies in culture related areas such as education and knowledge sharing, disaster preparedness, resilience, advocacy (especially legal), and climate justice are just starting to appear. As "we are clearly dealing with the single greatest emerging humanitarian threat ever encountered" (UNDP 2007b; GHF 2009b) it is not surprising that the role of culture has been eclipsed by studies of vulnerability and adaptation aimed to meet practical needs and to raise awareness.

Concerned thinkers from many walks of life are seeking universal answers to the looming global crisis of climate change. Some as we have seen, are converging on a Gaian/Mother Earth perspective. The question remains as to how the transition to a sustainable future will come about. Judy Ford in *The role of culture in climate change policy making: Appealing to universal motivators to address a universal crisis* explains "addressing climate change requires the engagement of millions of individuals performing multiple steps to reach a more obscure goal. In the Western world, this will likely include modifications of the consuming and mobile lifestyles to which we are accustomed. Successful engagement at this level requires personal, mass commitment to deeper, universal principals, such as those already found in mythology, art, religion, commerce, and human rights." Ford poses that 'universal policy' as well as the concepts of culture and cultural difference, can deliver the transition through its instruments namely - the **Universal Declaration of Human Rights** (UN 1948) and **Vienna Declaration** (UN 1993). She explains that the process has not been successful in the past because it has been effectively subverted for powerful political reasons by the deliberate misinterpretation of culture and cultural differences (Ford 2009a).

Nobel Prize-winning economist, Amartya Sen (Sen 2006 p.xviii) emphasizes, "the prospects of peace in the contemporary world may well lie in the recognition of the plurality of our affiliations and in the use of reasoning as common inhabitants of a wide world." Despite wide differences in customs, there are fundamental human values common to the diverse cultures of the world (Parsons and Vogt 1962 p141). Ultimately, "there does not need to be any trade-off between respect for cultural difference and human rights and development" (United Nations, 2004, p4). This moves the discourse on a universal climate change policy into the discourse on human rights. It moves the responsibility of protecting cultural heritage from the impact of climate change into the domain of the WHC and ICHC, both being instruments of human rights.

The global motivation that can be invoked to bring about the transition to a sustainable future is declared in the UN 1993 **Bangkok Declaration** which contains the human rights aspirations of the Asian region: Paragraph 8 "Recognize that while human rights are universal in nature, they must be considered in the context of a dynamic and evolving process of international norm-setting, bearing in mind the significance of national and regional particularities and various historical, cultural and religious backgrounds." This declaration paragraph brings together human rights, cultural diversity and cultural heritage under the umbrella of universalism. The motivation for change could be seen as 'the evolving process of international norm-setting'. This relates to what Diana Liverman in *The Importance of Behavioural Change* writes, "Mental models vary across different groups in society and affect how people perceive the climate change issue; they are hard to change and can create barriers to communication and action. Thus, a critical challenge to dealing effectively with climate change is to build consensus across society on the nature of the climate change threat and the overall strategy to deal with it. In effect, a single, high-level mental model – or perspective – needs to be achieved. Without it, effective climate and policy action will be unlikely" (Richardson et al. 2009 p34), Could the dynamic and evolving process of international norm-setting be harnessed to achieve the single, high-level perspective? Protecting culture and cultural heritage can be seen as ensuring the richest context for this dynamic process.

Chapter 7
Climate Change, Displacement and Migration

A farmer in the Mekong Delta who sells his daughter to a human trafficker for the money to buy new seed to replace what was washed away by flood waters...These stories show the real cost of climate change and are typical of many who see migration as the only way out. Micheline Calmy-Rey. Global Humanitarian Forum, 2009 Geneva.

The displacement of people is the most demonstrable effect of climate change and the most problematic with regard to protecting cultural heritage. Today about 26 million are 'climate displaced people', displaced temporarily or permanently through drought, desertification, sea level rise and weather related disasters (GHF 2009b p48). This number is expected to reach in the order of 150-200 million by 2050 (Stern 2006; Garnaut 2008) with BAU. Although not the principal driver climate change impacts will contribute to the movement of other displaced people namely refugees, migrants and internally displaced people (UN 2009). (See Appendix 2 for definitions). Displacement is and will continue to be made worse by poverty, employment options, development projects and conflict. The UN has identified five mega-trends - population growth, urbanization, climate change, migration and food, water and energy insecurity which will make contemporary forms of displacement increasingly complex and cause global crises to multiply and deepen (UN 2009). The 2009 Human Development Report *Overcoming barriers: Human mobility and development* calls for lowering the barriers to movement and improving the treatment of those who move, for migration is both an adaptation to, as well as a consequence of, climate change (UNDP 2009).

The Carteret Islanders of Papua New Guinea are the first widely recognised climate driven internally displaced people (Gallery 2009). All are being relocated to neighbouring Bougainville after being advised that their ancestral homeland will likely be under water by 2015 (O'Reilly 2010). Tuvalu and Kiribati will probably be the first climate displaced nations. The people of Tuvalu have reluctantly accepted the idea of relocation, and have started moving to New Zealand under the terms of a negotiated migration scheme (Insert 7.1) (EACH-FOR 2007). Australia is investing in training programmes on Kiribati which will facilitate migration. In a recent development the Kiribati Cabinet has endorsed a plan to buy nearly 3,000 hectares in Fiji in anticipation of the need, over time, to relocate the entire population of 103,000 (Perry 2012). In 2009 Ross Garnaut went on record as saying "The South Pacific countries will end up having their populations relocated to Australia or New Zealand and the rest of the world expects that and in the end, we're likely to accommodate that so there's a solution there" (Garnaut 2009).

Insert 7.1 Living with Climate Change

Statement from Annie Homasi, coordinator, Tuvalu Association of Non-Governmental Organisations

On local impacts: The weather changes and heat affects people, but also sea-level rise. My own experience is that during spring tides in March, my house concrete foundation is now half in the water. This is what I have seen and based on my own markings of the water level at my house.

On global politics of climate change:

The Australian government has not been willing to consider environmental refugees, and is not very friendly. New Zealand has been more flexible and a work scheme has been negotiated between New Zealand and Tuvalu. People in Tuvalu are thinking that they will need to make a move because of global warming. People living in Melbourne, Australia, who have moved there 30 or 40 years ago are very concerned about where people of Tuvalu will be able to go. Moving away from Tuvalu is not good for our culture and values. Where we live now, we know how to behave and live within our means. It will not be comfortable to live in another place. We want to live in our own land, our home and where our forefathers have lived. Tuvaluan people don't like to be called refugees (FOE et al. 2007 p33). Quoted with permission.

Total displacement has international implications. These nations will be stateless but retain seats in the United Nations. How do such nations maintain their cultural heritage? (Insert 7.2 and 7.3). A year later, a confidential US diplomatic cable released by Wikileaks stated "that the Australian government was encouraging Pacific nations expected to be inundated by rising sea levels to think "incrementally", despite the likelihood their citizens might be eventually forced to evacuate" (Dorling and Baker 2010). This is an example of the disturbing situation relatively powerless developing nations may face when forced to relocate.

Insert 7.2 Living with climate change

> Statement from Kilifi O'Brien, Tuvalu Government spokesman. If we lose our land we risk losing our identity. We know if the worst comes to the worst, we would have to relocate. But we would be looking at taking one sovereign country to another — we would want to keep our economic exclusion zone, our United Nations seat and so on. We would want to keep our identity as Tuvalu, in another location. (OXFAM 2009 p37). Quoted with permission.

Ross Garnaut anticipates his Government's concern about the impact of climate change on Asia when he comments, "Much more worrying is (the impact of climate change on) low-lying populations in some of the large and densely populated parts of the world, especially the great river valleys of Asia" (Garnaut 2009). Further diplomatic cable leaks in December, 2010 report Australia's top intelligence agency warning that a "cascade of economic, social and political consequences" in south-east Asia by 2030 would result from scarcity of water from retreating Himalayan glaciers. The agency predicted "Internal migrations in multi-ethnic countries may cause more problems than cross-border migration," (Dorling and Baker 2010).

> "...As you can see by what the map portrays, (of the Torres Strait) we have a lot of reefs up there and that is what we refer to as our supermarket. That is where our lifestyle evolved. This is our world I am looking at. This is my world, my people..."
>
> It never crosses our mind to relocate. Relocation is the last avenue for us. You have to understand who we are. I mentioned that this is our world...
>
> We are keenly aware of the challenges that face us; however, we are also fearful of the loss of our homes—our family homes. Each individual island has its own unique attributes. As an Iama Island person, I cannot live on Saibai, because I will not fit in. We identify with our area. I do not know if you understand, but that is where our identity and everything are derived from. So it would be the last resort for us to leave, because our roots are there.
>
> For generations we have had embedded in our sense of pride that unique identity in our island home. We have found ways to hold onto our traditional practices and our unique culture in this modern day and age. We also have embraced challenges and have adapted to changes in order to protect our island. We have taken whatever steps are needed to ensure our sustainability. We have a traditional saying in the Torres Strait which originated in 1970 during the PNG push for independence: 'Not for one teaspoon of saltwater, not one grain of sand, will we surrender. Border not change.' This determination has ensured a continued existence for each community so far, and I have no doubt that it will do so into the future.
>
> Our region is the frontline in many ways—significantly so due to rising sea levels. We do recognise the urgent need to address climate change and find long-term solutions. Our people are very much aware of the social issues we have—overcrowding, disease and damages and our traditional fishing practices—and we welcome the chance to become involved in a long-term strategy to ensure the protection of our beautiful islands.

> In the community of Warraber back in the 1990s, they had to take into their own hands the building of a seawall because the tides were taking skeletal remains from the cemeteries out onto the reefs. They said to themselves, 'We're not going to sit here and wait for research and studies; we've got to take some action; we've got to do something'—and that is what they did. Even with the sea level today the seawall does its job, and it was built 20 or so years ago (Standing Committee on Climate Change 2009p 110).

Insert 7.3 Living with Climate Change. Evidence to the Committee from Mr Walter Mackie, Member for Iama Island and Portfolio Member for Health and Environment, Torres Strait Regional Authority

The 2009 Human Development Report's migration analysis finds that women and children are disproportionately represented among people displaced by extreme weather events and other climate shocks although there is evidence that migration can empower some traditionally disadvantaged groups, in particular women (UNDP 2009). The Report advocates a human development approach as a means to redress some of the underlying issues that erode the potential benefits of mobility and/or forced migration (UNDP 2009). In the first authoritative overview of the relationship between climate change and migration, the 2011 UNESCO/Cambridge Press publication *Migration and climate change*, authors acknowledge that tropical cyclones, heavy rains and floods, drought and desertification, and sea-level rise are increasingly influencing migration. They emphasise the multicausality of migration, warn of the heavy politicisation of the issue and conclude that the failure, to date, of international negotiations means that "it will be too late for mitigation strategies to prevent or even slow down imminent changes." Again the case is made for the big polluters "to work together globally to provide financial, scientific and logistical support for developing adaptation." A number of options are suggested such as the diversification of economic activity; changes in government attitudes to rural-urban and cross-border migration by abandoning restriction and criminalization, and helping people to move in conditions of safety and dignity" (UNESCOPRESS 2011).

The most seriously affected countries include island states, several African nations, China, India, Bangladesh, Egypt and the delta areas and coastal zones of South East Asian countries. The movement of people, voluntary and involuntary, impacts the work of the two UNESCO Conventions that are under examination, in seeking to protect tangible and intangible cultural heritage.

A Ladakhi drapery shop owner from Leh proudly wearing a traditional goncha. A photograph of the young Dalai Lama, draped with prayer scarves, hangs behind him

Chapter 8

Climate Change, Health and Cultural Heritage

"Solastalgia"- a new type of sadness. People are feeling displaced. They're suffering symptoms eerily similar to those of indigenous populations that are forcibly removed from their traditional homelands. But nobody is being relocated; they haven't moved anywhere. It's just that the familiar markers of their area, the physical and sensory signals that define home, are vanishing. Their environment is moving away from them, and they miss it terribly. Clive Thompson (Thompson 2007).

Epidemiologist Rodolfo Saracchi has expanded the WHO definition of health to "a condition of well being, free of disease or infirmity, and a basic and universal human right" (Saracci 1997). Many indigenous peoples would identify with the widely accepted definition by Australian Aboriginal peoples as health not just meaning the physical well-being of the individual but referring to the social, emotional, spiritual and cultural well-being of the whole community. This is a whole of life view and includes the cyclical concept of life-death-life (Jackson and Ward 1999 p437).

Climate change is the greatest health threat of the 21st century (Costello 2009; Fussel 2008d). In 2000, it was estimated to have been responsible for around 150,000 fatalities (WHO, 2008; Schellnhuber 2008). Now it is causing 300,000 deaths a year and rising (GHF 2009a p30). Without adaptation measures, even a 40 cm rise in sea levels would dramatically increase the storm surge risks for over a hundred million people (IPCC, 2007b).

Climate change has direct health effects such as causing injury and death in cyclones, storms, wild-fires, flooding and heatwaves, see insert 8.1. Such risks as a consequence of extreme weather events will increase in any event, but particularly with business as usual. For example the occurrence of asthma, stroke and circulatory disorders brought on by heatwaves will increase. The elderly, the very young and those with existing respiratory and heart disease are the most vulnerable.

Climate change also has indirect health impacts (see Insert 6.1 and 8.1). The warming climate intensifies epidemic susceptibility because it causes changes in the geographic range and seasonality of some climate-sensitive infectious diseases such as dengue fever and malaria.

"More significantly climate change is harming water flows, plant growth patterns, the biological cycles of animals and insects, and whole ecosystems which support healthy life on earth" (DEA 2009). As a result the threats posed by common poverty-related diseases such as malaria, diarrhoea, malnourishment, cardiovascular and respiratory diseases will be increased by unclean water, contaminated poor quality food and population stresses. The poor, women, children, elderly and migrants will be most at risk. The degraded environment is already causing world food yields to fall, both agricultural crops and fish stocks, with immense repercussions for world food security and health.

Mental health, until recently, has taken a minor place when considering climate change impacts. The 2009 Lancet report *Managing the Health effects of Climate Change* makes mention of anxiety related to extreme events (Costello 2009). However **Doctors for the Environment Australia** has a broader perspective. Within Australia climate induced drought and subsequent falling food yields has caused "stress, social disruption and depression" and for some the abandonment of their farms (DEA 2009). Globally, the millions of climate displaced people will suffer immense mental and physical stress.

The impact of these climate related diseases and conditions on the cultural heritage of a nation or community is yet to be studied. The WHO estimates that environmental factors are contributing to or responsible for 40% of global illness. With health impacts of such scale it is undeniable that climate related health issues must have a significant impact on the viability of cultural heritage, in particular, their custodians. Statelessness, separation from country and deterioration of culture must pose immediate and long term mental anguish that threaten cultural survival. There has been recent recognition of a mental state of mind caused by a large scale loss within a patient's environment. The loss, such as a landscape in drought, could well be climate caused. The condition has been termed 'solastalgia' (Pollard. 2007; Pereira 2008) and is distinct from 'broken heart' disease caused by acute loss, which may also be climate related.

Insert 8.1 Climate Change Impacts on Human Health

Heatwaves	Heatwaves are not only increasing in frequency, intensity and duration, but their nature is changing. Warmer night time temps and higher humidity (7% more for each 1°C warming) that raises heat indices and makes heatwaves all the more lethal.
Asthma and allergies	Asthma prevalence has more than doubled in the U.S. since 1980 and several exacerbating factors stem from burning fossil fuels. Increased CO_2 and warming boost pollen production from fast growing trees in the spring and ragweed in the fall (the allergenic proteins also increase). Particulates help deliver pollen and mould spores deep into the lung sacs. Ground-level ozone which increases during heatwaves primes the allergic response. Climate change has extended the allergy and asthma season two-four weeks in the Northern Hemisphere (depending on latitude) since 1970. Increased CO_2 stimulates growth of poison ivy and a chemical in it (uruschiol) that causes contact dermatitis.
Infectious disease spread	The spread of infectious diseases is influenced by climate change in two ways: warming expands the geographic and temporal conditions conducive to transmission of vector-borne diseases (VBDs), while floods can leave "clusters" of mosquito-, water – and rodent-borne diseases (and spread toxins). With the ocean the repository for global warming and the atmosphere holding more water vapour, rain is increasing in intensity with multiple implications for health, crops and nutrition. Tick-borne Lyme disease (LD) is the most important VBD in the U.S. LD case reports rose 8-fold in New Hampshire in the past decade and 10-fold (and now include all of its 16 counties). Warmer winters and disproportionate warming toward the poles mean that the changes in range are occurring faster than models based on changes in average temperatures project. Biological responses of vectors (and plants) to warming are, in general, underestimated and may be seen as leading indicators of warming due to the disproportionate winter and high latitude warming.
Pests and disease spread across taxa: forests, crops and marine life	Pests and diseases of forests, crops and marine life are favoured in a warming world. Bark beetles are overwintering (absent sustained killing frosts) and expanding their range, and getting in more generations, while droughts in the West dry the resin that drowns the beetles as they try to drive through the bark. (Warming emboldens the pests while extremes weaken the hosts.) Forest health is also threatened in the Northeast U.S. (Asian Long-horned beetle and woolly adelgid of hemlock trees), setting the stage for increased wildfires with injury, death and air pollution, loss of carbon stores, and damage to oxygen and water supplies. In sum, forest pests threaten basic life support systems that underlie human health. Crop pests and diseases are also encouraged by warming and extremes. Warming increases their potential range, while floods foster fungal growth and droughts favour whiteflies, aphid and locust. Higher CO_2 also stimulates growth of agricultural weeds. More pesticides, herbicides and fungicides (where available) pose other threats to human health. Crop pests take up to 40% of yield annually. Marine diseases (e.g., coral, sea urchin die-offs, and others), harmful algal blooms (from excess nutrients, loss of filtering wetlands,

	warmer seas and extreme weather events that trigger HABs by flushing nutrients into estuaries and coastal waters), plus the over 350 "dead zones" globally affect fisheries, thus nutrition and health.
Winter weather anomalies	Increasing winter weather anomalies is a trend to be monitored. More winter precipitation is falling as rain rather than snow in the Northern Hemisphere, increasing the chances for ice storms, while greater atmospheric moisture increases the chances of heavy snowfalls. Both affect ambulatory health (orthopaedics), motor vehicle accidents, cardiac disease and power outages with accompanying health effects.
Drought	Droughts are increasing in frequency, intensity, duration, and geographic extent. Drought and water stress are major killers in developing nations, are associated with disease outbreaks (water-borne cholera, mosquito-borne dengue fever (mosquitoes breed in stored water containers)), and drought and higher CO_2 increase the cyanide content of cassava, a staple food in Africa, leading to neurological disabilities and death.
Food insecurity	Food insecurity is a major problem worldwide. Demand for meat, fuel prices, displacement of food crops with those grown for biofuels all contribute. But extreme weather events today are the acute driver. Russia's extensive 2010 summer heat-wave (over six standard deviations from the norm, killing over 50,000) reduced wheat production -40%; Pakistan and Australian floods in 2010 also affected wheat and other grains; and drought in China and the U.S. Southwest are boosting grain prices and causing shortages in many nations. Food riots are occurring in Uganda and Burkino Faso, and the food and fuel hikes may be contributing to the uprisings in North Africa and the Middle East. Food shortages and price hikes contribute to malnutrition that underlies much of poor health and vulnerability to infectious diseases. Food insecurity also leads to political instability, conflict and war (Hansen et al. 2011 p18).

Chapter 9

Climate Change Vulnerability and Cultural Heritage

The potential for some Pacific islands to become uninhabitable due to climate change is a very real one. Consequently some in our region have raised the issue of their citizens becoming environment refugees ... potential evacuation of island populations raises grave concerns over sovereign rights as well as the unthinkable possibility of entire cultures being damaged or obliterated. Pacific region environment ministers, July 2007 (OXFAM 2009)

Vulnerability to climate change is the degree to which systems e.g. socio-economic, biological etc. are susceptible to, and unable to cope with, the adverse impacts of climate change (IPCC. p19.2.2). The term vulnerability may therefore refer to the vulnerable system itself, e.g. mountain villages dependent on glacial melt; the impact of this on people, e.g., forced migration; or the mechanism causing these impacts, e.g., the melting and retreat of Himalayan glaciers.

The IPCC identifies seven criteria that may be used to identify key vulnerabilities. They are listed here to indicate the demanding nature of making vulnerability assessments. The criteria are: magnitude of impacts, timing of impacts, persistence and reversibility of impacts, likelihood (estimates of uncertainty) of impacts and vulnerabilities, and confidence in those estimates, potential for adaptation, distributional aspects of impacts and vulnerabilities and importance of the system(s) at risk.

Assessing vulnerability is necessary to addressing the physical and biological impacts of climate change and their consequential human impacts. Cultural heritage workers should understand the basis of vulnerability assessments as climate change-related policy and funding decisions often hinge on the quality of such assessments. The situation is currently complicated by a substantial body of partly conflicting vulnerability theory, vulnerability indices and vulnerability assessment practice (Liverman 2008). Fussel has reviewed the evolution of vulnerability theory which is characterized by the progressive addition of non-climatic determinants of vulnerability to climate change, including adaptive capacity, and the shift from estimating expected damages to attempting to reduce them (Fussel and Klein 2006; Füssel 2009a). He explains why "The proposition that those countries most vulnerable to climate change should receive priority assistance for adaptation, applying 'the principle of distributive justice' is ambiguous and its naïve implementation can produce controversial outcomes." He cautions, "The consideration of factors determining the adaptability of countries in decisions on international adaptation assistance will be particularly controversial because they can have contrary effects on the magnitude of national allocations for adaptation, depending on the applied principle of distributive justice." For example "Everything else being equal, poor performance of national governments generally increases vulnerability to climate change, suggesting that a country should receive priority assistance for adaptation. At the same time, poor governance generally decreases aid effectiveness, suggesting that a country should not receive priority assistance for adaptation" (Füssel 2009a p19). As an example, the **International Union for the Conservation of Nature (IUCN)** strategy presented to a World Heritage Working Group (which was developing climate change policy) is summarised in Insert 9.1.

Lessons Learned by IUCN when Supporting Climate Change Adaptation

Vulnerability is a function of ***Exposure*** and ***Sensitivity*** and ***Adaptive Capacity.***

The greater the exposure and/or sensitivity, the greater is the vulnerability. But adaptive capacity is inversely related to vulnerability. So, the greater the adaptive capacity, the lesser is the vulnerability.

- Decrease sensitivity – e.g. reducing other stressors
- Increase adaptive capacity – e.g. knowledge
- Implement early warning systems
- Improve adaptive management & management effectiveness (share best practices)
- Involve local stakeholders and traditional knowledge
- Incorporate uncertainty into decision-making
- Thereby increase resistance and resilience

Insert 9.1 Strategy used by IUCN to assess vulnerability to climate change (Bomhard 2006)

A number of vulnerability indices are easily accessed and commonly referred to in vulnerability assessments. Their usefulness varies as their development is subjective relying on political and scientific choices. The reviewers, Fussel and Klein, (Fussel and Klein 2006) find that all existing indices show substantial weaknesses and many are inappropriate for prioritizing adaptation funding. As a result, there is little agreement regarding the most vulnerable countries. Clarity over the primary purpose is the critical factor that determines the quality and usefulness of any vulnerability index, or set of indicators. Surprisingly Fussel makes a strong case for simply using *current temperature* to assess the generic sensitivity of a country to climate change (other than sea level rise). Almost all least developed countries are located in a climate zone where additional warming has negative effects on all main economic sectors (Füssel 2009a).

Vulnerability to the impacts of climate change varies widely around the world, with ethics and justice issues emerging as key factors in adaptation approaches (Richardson et al. 2009 p22). Of the studies that have been done, such as the UNFCCC commissioned *Vulnerability and Adaptation to Climate Change in Small Island Developing States 2007* (Sem/UNFCCC 2007), cultural heritage is by and large not considered. The exception is the critical intangible heritage of traditional knowledge. The role of traditional knowledge under climate change impacts and its potential in adaptative interventions is now being more fully recognised (Konate 2004). There does not appear to be any applied research on the role of culture and cultural heritage in resilience to climate change. Cultural heritage workers and community leaders coping with climate change impacts would benefit from knowing what cultural elements offer potentially more or less protection. How do communities choose what to abandon?

The Nobel Prize winning economist Elinor Ostrom has described in her book, *Governing the Commons,* how some societies have managed to sustainably manage common pool resources and not overexploit them. She explains that her current work concerns understanding the certain conditions that allow people to self organise to govern themselves. Her second area of interest is developing a general framework of understanding social ecological systems where resources and humans interact. From her research she has determined the certain conditions that favour the truly sustainable management of the commons, those resources such as grazing lands and irrigation systems, which are common pool resources, Insert 9.2.

- Defined boundaries, exclude outsiders
- Participation in decisions
- Mutually agreed rules about who can do what
- Range of Penalties for transgressors
- Effective monitoring of the resource
- Mechanism to resolve conflict
- Self-determination of community recognised by authorities. (Ostrom 1990)

Insert 9.2 Elinor Ostrom's 'Design Principles' for successful management of 'commons' e.g. forests, irrigation systems, grazing lands, fisheries

She sees her work as being particularly important given global warming. Her design principles have relevance to a future where our commons - the atmosphere, our lands and water resources, our treasured cultural heritage will very much depend on successful shared management. A Swedish study on implementation of climate change policy has found that nested organisation can facilitate successful shared management, see Chapter 12, Insert 12.1.

Bestselling author and scientist Jared Diamond, in his book *Collapse* (Diamond 2005), gives a reasoned account of how societies have failed in the past with telling details of the power of culture for survival. His studies of present and past societies lead him to propose a five-point framework of possible contributing factors to a society's collapse, Insert 9.3.

Jared Diamond's Five-point Framework of possible contributing factors to a society's collapse:

- self inflicted environmental damage
- climate change
- hostile neighbours
- decreased support from friendly trading neighbours
- **society's responses to developing problems that depend on political, economic, social institutions and on its cultural values.**

(Diamond 2005 p11-15)

Insert 9.3

By far the most defining point is the last, society's response to developing problems, a response, he explains, which depends on the group's culture as well as on its political, social and economic institutions. Writer and ethicist Clive Hamilton has written bravely in his book *Requiem for a species* about the maladaptative coping strategies of the West, such as denial in all its forms and of our disconnection from nature, cultural attributes that are inhibiting a viable future (2010).

Chapter 10
Climate Change, Mitigation and Adaptation

Traditional owners are dying. No one is interested. They have a name to any wind from any direction of the globe, like the TO'ELAU, LA'I, LA'ILUA, the TUA'OLOA and many others. During this TUA'OLOA season also, my old chief used to tell me the animals would look nice and beautiful and the human beings become not too good looking, and it was true. Much of the oratorical language of Samoa comes from the art of fishing especially bonito and catching sharks the traditional way. And I am further saddened by the fact my grand children will only speak of words of which they will never understand how these words came about or come to be part of the oratorical language In my opinion it is about time the small islands should stand up to these bullying developed countries who are the main cause of these climate changes. They do not care how these have affected us, but they only need our votes for their selfish agendas.
Iteli Tiatia from Samoa (UNESCO 2010)

Mitigation

With respect to climate change, the **Intergovernmental Panel on Climate Change (IPCC)** has defined mitigation as the implementation of policies to reduce greenhouse gas (GHG) emissions and enhance carbon sinks.**The United Nations Framework Convention on Climate Change (UNFCCC)** is the international environmental treaty that came into force in 1994 to stabilize greenhouse gas concentrations in the atmosphere below a level that would prevent dangerous man-made interference with the climate system. There is universal country membership of the UNFCCC. Although the treaty is non-binding in setting limits on GHG emissions for individual countries, its subsidiary protocols do set mandatory emission limits. The principal update is the **Kyoto Protocol** established in 1997 to reduce developed country emissions and slow emissions growth in developing countries.Many developed countries, (the US being the major exception) have now accepted legally-binding emissions targets in the Kyoto Protocol but, as James Hansen reports, it "has been so ineffective that the rate of global emissions has since accelerated to almost 3% per year, compared to 1.5% per year in the preceding two decades" (Hansen 2012a p4)

The IPCC was set up in 1988 by the **United Nations Environment Programme (UNEP)** and the **World Meteorological Organisation (WMO)** to assess the state of knowledge on the various aspects of climate change including science, environmental and socio-economic impacts and response strategies. Through their Assessment Reports, the IPCC, despite a number of serious shortcomings (explained in Chapter 3), is recognised as the most authoritative scientific and technical voice on these issues. Its latest Report was the work of government and IPCC selected people from over 130 countries including more than 2500 scientific expert reviewers, 800 contributing authors, and 450 lead authors. It and the UNFCCC have led the attempt to achieve an effective international global agreement to reduce GHG emissions. Regrettably this is proving too difficult to achieve, thus enabling massive carbon dioxide pollution by industrialised nations to continue (WBGU 2009). Climate change mitigation will require a general rejection of over-consumptive lifestyles and a re-acquaintance with local sources, a prospect that is proving beyond political acceptability.

Yet this lethal lethargy to mitigate is now being confronted by powerfully worded government commissioned (though not fully heeded) reports which are conspicuous for their lack of political correctness. For example in Australia, *The Garnaut review 2011. Australia in the Global Response to Climate Change* quotes the author Ross Garnaut, "However, when we took account of the value of Australians' lives beyond the 21st century, the value of our natural and social heritage, health and other things that weren't measured in the economic modelling, and the value of insuring against calamitous change, strong mitigation was clearly in the national interest" (Garnaut 2011 px).

The climate change report, *The Critical Decade* by Will Steffen, of the (Australian) Climate Commission states: 'This is the critical decade. Decisions we make from now to 2020 will determine the severity of climate change our children and grandchildren experience' (Steffen 2011). He explains:

- Without strong and rapid action there is a significant risk that climate change will undermine our society's prosperity, health, stability and way of life (see Insert 10.1)
- To minimise this risk, we must decarbonise our economy and move to clean energy sources by 2050.
- That means carbon emissions must peak within the next few years and then strongly decline.
- The longer we wait to start reducing carbon emissions, the more difficult and costly those reductions become.

'Unless effective action is taken, the global climate may be so irreversibly altered we will struggle to maintain our present way of life. The choices we make this decade will shape the long-term climate future for our children and grandchildren' (Steffen 2011).

The same urgent message came from the 2011, 3rd Nobel Laureate Symposium on Global Sustainability, *Tipping the Scales towards Sustainability*. Their **Stockholm Memorandum** calls for a **Mind-shift for a Great Transformation**. "We cannot continue on our current path. The time for procrastination is over. We cannot afford the luxury of denial. We must respond rationally, equipped with scientific evidence. Our predicament can only be redressed by reconnecting human development and global sustainability, moving away from the false dichotomy that places them in opposition. In an interconnected and constrained world, in which we have a symbiotic relationship with the planet, environmental sustainability is a precondition for poverty eradication, economic development, and social justice. Our call is for *fundamental transformation and innovation in all spheres* and at all scales in order to stop and reverse global environmental change and move toward fair and lasting prosperity for present and future generations"(3rd Nobel Laureate Symposium 2011a).

The few studies on the protection of cultural heritage from climate change that make reference to mitigation do so by advocating local actions, such as reducing emissions by being energy efficient (Cassar 2005; McIntyre-Tamwoy 2008) (Cassar and Hawkings, 2007). Policies designed to combat the human impacts of climate change centre on strategies integrating adaptation, mitigation, development and disaster risk reduction approaches which, at best, can be mutually reinforcing. In this context Amareswar Galla explains that adaptation, mitigation, humanitarian assistance and development aid underpin each other, but are supported by different sets of institutions, knowledge centres, policy frameworks and funding mechanisms (Galla 2009b). Policy success depends on these different sets having strong linkages, a matter which to date has received inadequate attention (GHF 2009b).

Statement by Maritza Arévalo Amador, a 58-year-old single mother of five, Tegucigalpa, Honduras

On environmental changes: There's been a change in the climate and in the seasons, because before you knew when it was winter or summer. But human beings have made these changes with deforestation, cutting down trees. That has been the worst for our environment, since the deforestation has caused the lack of water in our communities or our country.

On the impacts of these changes: The impacts that we have received from these changes are: hot weather, many skin illnesses in people, lack of water and the pollution of the environment. The destruction of our soil, as well as the mining exploitation in our country that pollutes the air, the water and human beings. Also the children and old people suffer from skin and lung problems.

On working for change: I have had a lot of experience in my life because we have fought for the environment. The struggles have been hard, mainly in the communities where I work planting trees ... We work to improve the environment with talks about environmental health ... You have to plant trees to "breathe" a better environment. Also we have learned to recycle garbage. We prepare compost for our gardens that we have in our homes. We classify the garbage, and use the waste for organic compost, and in that way we have changed our way of life.

My message to other communities affected by climate change: First, organize; second, fight for just causes; third, have the will and spirit to work; fourth, educate yourself and have a vision for the future of our grandchildren, great-randchildren and great great- grandchildren, so that in the future they are well educated, and so that they can have a better environment and a better country (FOE et al. 2007 p11). Quoted with permission.

Insert 10.1 'Living with Climate Change'

Adaptation

Adaptation to climate change refers to "individual or governmental actions to reduce adverse effects or future risks associated with climate change" (The Climate Institute 2010). The IPCC defines adaptation as the "adjustment in natural or human systems in response to actual or expected climatic stimuli or their effects, which moderates harm or exploits beneficial opportunities" but adds "adaptation alone is not expected to cope with all the projected effects of climate change" (IPCC. 2007 p19.1.2). Adaptation means to change behaviour, strategies, programs, investments or individual behaviour based on our knowledge about climate change risks (Smit and Wandel 2006). It is the only option to reduce climate impacts by reducing the risks from present, and from unavoidable future climate change (GHF 2009b).

Adaptation has acquired an aura of being the 'silver bullet' answer to climate change whereas it is more akin to a 'biting the bullet' survival approach. It has severe limitations and is no more than the best preparation (and temporary solution) for a continually worsening situation. Without mitigation, adaptation, as we know it today, and especially for the poor, is a continual process of trying to survive in a deteriorating world. Without mitigation many adaptations may prove short lived. As Archbishop Tutu has said "No community with a sense of justice, compassion or respect for basic human rights should accept the current pattern of adaptation. Leaving the world's poor to sink or swim with their own meagre resources in the face of the threat posed by climate change is morally wrong" (UNDP 2007b p166). Unfortunately, as the Human Development Report 2007/2008 powerfully demonstrates, this is precisely what is happening. We are drifting into a world of 'adaptation apartheid'. Tilman Thomas, President of Grenada, and former chairman of the **Alliance of Small Island States (AOSIS)** says "What I'm saying is that those who are really concerned about humanity and about survival, would they just sit back and permit countries to disappear? It is really an ethical question we are faced with now. A failure to act is sort of really a benign genocide in a sense" (Mottram 2009).

Even more disturbing is Tim Flannery's reporting that some see adaptation as slow genocide with the demise of the vulnerable multitudes and most flora and fauna, at the expense of the select rich who have retreated to safe refuges. He explains, "English environmental politician Aubrey Meyer pointed out how this matter is being discussed at the highest levels. Economists who participated in the IPCC discussions stated that doing anything serious about climate change was too expensive to be worthwhile, leading in Meyer's view to 'the effective murder of members of the world's poorest populations', and whose lives by the economists' estimates were worth only a fifteenth that of a rich person" (Flannery 2008 p208).

Adaptation and Cultural Heritage

Very few of the many theoretical papers and project reports on adaptation to climate change consider cultural heritage (Huq 2004, Fussel 2008d). This reflects the formative stage of this fundamental area of inquiry and indicates how much needs to be done to put culture and cultural heritage central both in theory and practice particularly in relation to its international protection through the UNESCO Conventions.

Adaptation is an involved process. It includes well established practices from disaster risk management (DRM), coastal management, water resource management, spatial planning, urban planning, public health, and agricultural outreach. It is relevant for all climate-sensitive domains, driven by diverse climate hazards and climatic changes of varying predictability. It uses a broad range of technical, institutional, legal, educational, and behavioural measures (Schmitt 2009). Its planning requires close collaboration of climate change scientists, practitioners from the affected sectors, decision-makers, policy analysts and other stakeholders, (Fussel 2008a). Assessing, planning and implementing adaptation measures use different methodological approaches in order to produce knowledge that is relevant in a given decision context. Adaptation is highly context-specific as it depends on the climatic, environmental, social, and political conditions relevant to the particular project. There is nothing more important than targeting adaptation assessment at the specific circumstances of the particular situation. Nevertheless its effectiveness will be inherently uncertain due to the unpredictably of the rates of climate change in general and more specifically in relation to the frequency and intensity of extreme weather events. Across the plethora of theory, guidelines and advice, the role of culture in the adaptive process is not explored.

As has been discussed, the impacts of climate change burden those least responsible for climate change and those with the least material resources to adapt. This is in stark contrast to the polluter-pays principle. Kiribati with a population of 96,000 and an average elevation above sea level of 1 metre has a 95% probability of being inundated by sea rise within the century no matter what adaptations are made (EACH-FOR 2007). Relocation, at the expense of their immovable cultural heritage, becomes their final adaptive strategy. Yet the people of Kiribati rank 163rd in the world for GHG emissions per capita contributing 0.01% of CO_2 global emissions (World Resources Institute 2005).

Adaptation assessment is not only complex but also continually evolving. It has become more inclusive over time, integrating future climate change with current climate risks. The overriding problem for policy makers is that there is no straightforward process in determining the best adaptation option. The process is necessarily complex and time consuming. It must consider global benefits, costs, and limits of adaptation; prioritisation according to common but differentiated responsibilities (CBDR), the capacity to adapt, equity issues, and planning and policy making skills (Füssel 2009a).

Several recent studies have highlighted failures of past adaptation actions and obstacles for future adaptation. Weaknesses often revolve around failure to understand indigenous cultures and failure to recognise weak governance (Füssel 2008c). There are many success stories, one of the most impressive being how preparedness measures and early warning systems helped reduce mortality in Bangladesh during Cyclone Sidr in 2007 which killed approximately forty times fewer people than a similar scale cyclone in 1991 (GHF 2009a p60).

An example of adaptation planning in practice is the ambitious 5 year **Nairobi Work Programme on Impacts, Vulnerability and Adaptation to Climate Change** (2005-2010) developed by UNFCCC for implementation in developing countries, including the **Least Developed Countries (LDCs) and Small Island Developing States (SIDS).** This program aims to improve the understanding and assessment of impacts, vulnerability and adaptation to climate change to enable communities to make informed decisions on practical adaptation actions and measures to respond to climate change on a sound scientific, technical and socio-economic basis, taking into account current and future climate change and variability. The programme operates by giving interactive workshops and providing back-up documentation. Community members learn about climate data and weather observations, climate modelling, scenarios and downscaling climate related risks and extreme events, socio-economic information, adaptation and economic diversification (UNFCCC 2010b). One of the innovations of this extensive programme has been the establishment of the online UNFCCC Database on local coping strategies. Cultural factors are not given priority in this programme.

Australia has committed through the 7 year **Pacific Vulnerability and Adaptation Initiative** to support practical adaptation initiatives to improve water security and coastal zone management in Tuvalu, Tonga, Vanuatu, Samoa, Fiji and the Solomon Islands (Partnerships for Sustainable Development 2006). The single biggest and most important regional adaptation to climate change project that spans 13 countries is the **Pacific Adaptation to Climate Change Project (PACC)** funded by the **Global Environment Facility (GEF)** with the United Nations Development Program (UNDP) as its implementing agency and the **Secretariat of the Pacific Regional Environment Program (SPREP)** as implementing partner. Adaptation projects aim to strengthen water resource management, food production, food security and coastal management capacity (SPREP and UNDP 2009). UNDP has recently launched a *Toolkit for Designing Climate Change Adaptation Initiatives*. It is a hands-on step by step guide useful at national, sub-national and community levels. Although based on lessons that have emerged over the last 4 years from UNDP's support to countries, consideration of culture is not included except for reference to that "wealth of traditional knowledge (which) provides a basis for the design of adaptation measures" (UNDP 2010 p44).

Those adaptive heritage preservation practices that are being initiated in relation to climate change are directed at both tangible heritage (moveable and immoveable) and intangible heritage.

Tangible heritage adaptive preservation practices, as discussed in Chapter 5, are concerned with the built environment, archaeological sites and to a lesser extent cultural landscapes. To date the effort in regard to these domains is generally focussed on physical protection measures (the focus of a growing number of conferences and planning manuals), high technology knowledge banks and investigations of online disaster forecasting. For example **The International Committee of the Blue Shield's (ICBS)** international conference on the theme of *Protecting the World's Cultural Heritage in Times of Change: Emergency, Preparedness and Response* in 2011 emphasised the vital role of geospatial information (ICBS 2011). At the other end of the heritage spectrum the **World Heritage Earthen Architecture Programme (WHEAP)** at their conference in 2012 will focus on *Conservation of Earthen Architectural Heritage* in particular in regards to Risk Preparedness for Natural Disasters and Climate Change.

To date the need for intangible heritage adaptive preservation practices has by and large only serendipitously emerged. The recognition has arisen from community-focussed case studies and regional studies in the climate change frontlines – the Arctic, SIDS, Africa and the Amazon. These studies are primarily practical projects aimed at addressing climate change induced welfare problems and do not yet concern themselves with climate change induced threats to intangible heritage. However what comes out of these studies is the need to incorporate the community's time honoured intangible cultural heritage to ensure ultimate sustainable change. Thus far 'indigenous knowledge' is being identified as salient to project acceptance and success. As for the rest of the spectrum of unprotected intangible cultural heritage there is a lack of any evidence of any safeguarding adaptive practices.

Traditional knowledge is the only specific intangible cultural reference in the UN's report on climate change programmes (UN 2007a) including the large scale Nairobi Work Programme (UNFCCC 2010b). Smaller studies such as that of Amazon farmers experiencing climate change flooding, describe just how intricate is the cultural knowledge about the processes that mediate perceptions of environmental change. The project managers in this case learnt that this knowledge had to be well understood before successful collaborative initiatives to alleviate flooding could be introduced (Moran 2008).

Oxfam's Pacific report, *The future is here: climate change in the Pacific,* recommends that adaptive programs be developed to assist communities threatened by rising sea level, high tides and storm surges. The only references to culture are to local knowledge and local history. This intangible heritage is valued for its usefulness in the adaptive programmes and, despite the lack of specific safeguarding actions, its viability and transmission is ongoing. Importantly the report recommends that women and men participate equitably in all decision-making about climate change and that their differentiated needs be reflected in adaptation efforts (OXFAM 2009).

In the coastal village on Vanua Levu, Fiji, the philosophy of 'vanua' (which refers to the connection of people with the land through their ancestors and guardian spirits) has served as a guiding principle for the management and sustainable use of the rainforest, mangrove forest, coral reefs, and village gardens. This knowledge now serves as an important management framework in developing the skills for adaptive capacity (OXFAM 2009).

Dancers at the Leh Festival

Chapter 11

Culturally Sustainable Development and Climate Change

We are the first generation with the knowledge of how our activities influence the Earth as a system, and thus the first generation with the power and the responsibility to change our relationship with the planet. 3rd *Nobel Laureate Symposium on Global Sustainability, 2011*

Culturally Sustainable Development in a Climate-Changing World

The integration of the economy and environment:

economic decisions to have regard to their environmental consequences. Development promotes economic activity, sustainable investment, community safety, sustainable tourism.

Intergenerational obligation:

current decisions and practices to take account of their effect on future generations. Development owners take a holistic perspective to develop long term ecological practices to leave the world a better place for the next generation and beyond.

Social justice:

all people to have the equal right to an environment in which they can flourish (or have their basic human needs met). Development incorporates social justice, promotes good health and distribution of economic resources to all.

Environmental protection:

resources to be conserved and the non-human world protected. Development monitors and manages environment, develops sustainable forestry use, soil quality and stability, protects biodiversity, provides sustainable transport, avoids wasteful consumption of goods, reduces unnecessary energy and water use, fossil fuel use and waste; minimises air and water pollution, recycles and limits carrying capacity.

Quality of life:

a wider definition needed for human well-being beyond narrowly defined economic prosperity.

Participation:

institutions to be restructured to allow all voices to be heard in decision-making (procedural justice). Development promotes a self-learning society through community awareness and participation.

Culture:

all people and groups to be encouraged to express their cultural identity with pride and dignity. Development encourages cultural diversity and enriching cultural heritage leisure activities.

Climate Change:

Policy and management decisions to consider climate impact and climate change resilience. Development aims to be carbon neutral, to build best *ecological footprint*, adaptative organisations and climate change resilience.

Insert 11.1 Culturally sustainable development in a climate-changing world. Adapted and altered from two sources with permission: Jacobs M (1995) 'Sustainable development: Assumptions, contradictions, progresses. In Lovenduski J, Stanyer J (eds.) Contemporary political studies: Proceedings of the Annual Conference of the Political Studies Association (London: PSA). and appearing in (Simms, Johnson, and Chowla 2010 p19) and FOCUS, 1997, Sustainable Development Updating the Theory, for National Capital Authority, Canberra (FOCUS 1997)

Sustainable Development and Population Growth

The UN defines sustainable development as 'development which meets the needs of the present without compromising the ability of future generations to meet their own needs' (UN 1987). Since 1987 when this term was adopted, much more has been learnt about the growing ecological crisis of which climate change is a major part. Many environmentalists would simplify this definition to 'development which meets the needs of all people, all species and all generations'. Overall, sustainable development is understood to improve the total quality of life, both now and in the future, in a way that maintains the ecological processes on which life depends (Australia's National Strategy for Ecologically Sustainable Development 1992). The concept has become conventionally accepted as encompassing economic, ecological and social values. These are expressed as 'the triple bottom line' when assessing any sustainable development project. Furthermore there is now recognition of the importance of cultural factors in sustainable development (Insert 11.1) – hence the fourth pillar - cultural vitality and the authors proposes a fifth pillar, that of 'resilience to climate change impacts' (Insert 11.2).

Sustainable community development				
environmental responsibility	**social equity**	**economic health**	**cultural vitality**	**resilience**
ecological balance	justice engagement cohesion	material prosperity	diversity innovation	to climate change impacts

Insert 11.2 The proposed five pillars of sustainability

Regrettably the term sustainable development is widely used as 'greenwash' by developers to defend and promote projects that are anything but sustainable. The term is even applied to coal mining! Thus the growth in use of the term across many disciplines has been accompanied by a loss in meaning. Now at best it is an aspirational term, a term for a goal that will rarely be achieved. A preferable and more realistic term might be 'ecologically or environmentally managed' with these latter two terms defined holistically.

Climate change threatens sustainable development by undermining its fundamental principles of protecting resources and the interests of future generations. Similarly the eight **Millennium Development Goals (MDGs)** (Insert 11.3) which aim to encourage development by improving social and economic conditions in the world's poorest countries are also threatened.

1. Eradicate extreme hunger and poverty
2. Achieve universal primary education
3. Promote gender equality and empower women
4. Reduce child mortality
5. Improve maternal health
6. Combat HIV/AIDS, malaria and other diseases
7. Ensure environmental sustainability. (Integrate the principles of sustainable development into country policies and programs and reverse the loss of environmental resources.) Reduce biodiversity loss.
8. Develop a global partnership for development

Insert 11.3 Millennium Development Goals (MDGs)

It is argued that since climate change will only intensify, resilience in the face of climate change must be added as an additional pillar of sustainable development. Development must not only be sustainable, but also climate-proof. Sustainable development is becoming closely linked to adaptation, vulnerability and mitigation. There is a need to look at all the dimensions of sustainability in determining an appropriate response to issues affecting cultural heritage protection. "Strong sustainability" will require, "radical changes in our relation with the non-human natural world, and in our mode of social and political life" (Simms, Johnson, and Chowla 2010 p19). Insert 11.1 summarises the components of culturally sustainable and climate proof or climate resilient development.

The question of population growth has been intensely debated since Paul and Anne Erhlich's book *The Population Bomb* confronted the public in 1968 with dire predictions of widespread starvation by the 1970s and 1980s from overpopulation. Although their predictions did not eventuate the book began a debate that is even more relevant today. The world population is 7 billion and expected to reach 9 billion by 2045, climate change permitting. The *Living Planet Report 2010* using 2007 data reveals that it now takes the Earth one year and six months to regenerate what we use in a year. "Turning resources into waste faster than waste can be turned back into resources puts us in global ecological overshoot, depleting the very resources on which human life and biodiversity depend" (WWF 2010). It raises profound questions as to whether the planet can reach, let alone sustain, a population of 9 billion.

Population stabilization and the onset of a decline is predicted in the second half of this century (Lutz and Samir 2010). This prediction relies on a number of factors including the 'development' of many 'developing' nations and the education and empowerment of women through the availability of cheap contraception. Such empowerment has allowed a "profound enigma" to surface – this being that "both death rates and birth-rates decline as a country becomes more affluent and educated" (Flannery 2010 p207).

Population growth remains a threat to sustainability, one that will only resolve in a non-catastrophic way over time if the standard of living improves in the poorer nations.

Sewing machine repairman from Leh. His shop was behind the Jamia Masjid, Sunni mosque (founded about 1666) (#19 on the map at insert 29.1) near to the tailors' shops

Chapter 12
Knowledge Sharing, Cultural Heritage and Climate Change

Nowadays we have a lot of means to communicate; everybody in this room has access to facebook and twitter. But that is the way literate people communicate. What about poor and illiterate people living in remote zones? They do not have access to internet, but they all have a radio. In many parts of the world it is better to choose a very easy and simple way to communicate with the local population. I'm from Benin, and I don't understand English that well. Sometimes, when people clap, I clap also, although I did not understand why we are clapping. For this reason, it is very important to bring to the people in information about West Africa climate change in their local languages! If we manage to bring to people living in remote areas information about climate change in their local language, then they will understand that climate change is not happening because gods or ancestors are angry with them, but they will understand what the real problem is and how they can cope with it. Folachade Bello aged 23 from Benin, Global Humanitarian Youth Forum 2009

Information Technology, Knowledge Transfer and Networks

Climate change is such a complicated problem it requires a multidisciplinary approach strongly supported by data, information and knowledge sharing on an unparalleled scale (Galla 2009b). As knowledge about potential impacts grows, it is evident that policy makers and heritage managers need a global system to assemble and assimilate the vast amount of scientific information to assist with the development of good management practices (Heath 2008 p19). Despite the obvious overlaps, the mainstream climate change literature and debate has, until very recently, given little or no attention to human rights concerns. Information is far more detailed for those areas likely to experience lesser impacts than for those where the consequences will be most devastating (ICHRP 2008 p4). Skills and knowledge about getting the message across also need to be shared. Institutional arrangements that are 'nested' from policy maker to policy recipient have proved successful in modern and traditional societies when dealing with common pool resources, see Insert 12.1.

Climate change deals with 'common pool resources' (often known as 'The Commons') such as the atmosphere, the oceans, fresh water. Top down implementation of 'common pool resource' policy has a history of failure.

Climate change policy implementation may benefit from institutional arrangements that are **nested** in layers.

The nested layers should embrace a **mixture of institutional types** in order to employ a **variety of decision rules** and to induce compliance.

Analytic deliberation among involved parties and the general public is essential in order to improve information, increase trust and hopefully produce consensus on governance rules.

This organisational structure works in the sustainable management of the commons in some traditional societies (Ostrom 1990).

In 2007 it contributed to successful implementation of climate change policy in Sweden because it reduced social uncertainty - everyone knows about others and what they are doing at different levels thus increasing trust. Dialogue and cooperation are the keys to success (Lundqvist 2007).

Insert 12.1 Making Climate Change Policy work with 'nested' organisation

Stakeholders in climate change negotiations relating to culture may include indigenous communities, **non government organisations (NGOs), community based organisations (CBOs),** international organisations such as the UNFCCC, the **United Nations High Commissioner for Refugees (UNHCR),** technical and professional experts and government public servants. Increasingly their work demands access to the knowledge of the other stakeholders. Knowledge sharing, 'cross-fertilisation of ideas' and networking are vital for sharing experiences, developing response strategies and building capacity in climate change science, policy and management. Cultural heritage workers need a system for reporting and monitoring climate change effects on heritage that is accessible to both stakeholders and managers (Heath 2008 p19).The internet has revolutionised information sharing, processing and dissemination, and permitted the more sustainable practice of online conferencing.

In a partial response to similar 'needs', the UNFCCC has made ambitious advances in the information sharing realm. Although cultural heritage issues are not addressed by the UNFCCC, it has employed techniques with cross-cultural and interdisciplinary potential to address climate change that are worth considering. The **Climate Change Information Network, CC: iNet** is UNFCCC's web portal which serves as a clearinghouse for climate change information. It assists users to link to potential partners to gain rapid and easy access to ideas, strategies, contacts, experts and materials that can be used to motivate and empower people to take effective action on climate change. During the current prototype phase the focus is on identifying sources in the fields of education and public awareness.

Insert 12.2 A tent home of the nomadic herders, the Changpas, equipped with solar cell for lighting. Ladakh, being in the rain shadow of the Himalayan Range, has ideal fine and sunny conditions for photovoltaic power generation

Broader based is the online **Technology Transfer clearing house, (TT: CLEAR).** This portal has a searchable database that accepts and integrates multiple input variables to provide a multifaceted response about a) environmentally sound technologies and know-how, methods, models and tools; b) projects including links to case studies of successful technology transfer (eg. insert 12.2, 12.3) (UNFCCC 2011a); c) funding opportunities; d) publications; e) events; f) organisations and experts and g) internet sites... all of which could potentially contribute to providing an accounting framework for technology transfer activities.

In the Ladakh region of India, government funded photovoltaic (PV) systems are used for rural electrification. The system capital costs are covered by the government; homeowners must pay for maintenance costs.

Impacts

Approximately 5,000 PV systems are installed in Ladakh; 25% of the population is receiving electricity from solar energy. However, 70% of these installations have occurred in the last three years, and lack of maintenance may prove problematic (UNFCCC 2011a).

Insert 12.3 Rural Electrification Using Photovoltaics in Ladakh, India

The UNFCCC has established within TT:CLEAR a**Technology Transfer Framework** which facilitates the transfer of environmentally sound technologies and knowledge to developing countries and a specialised **Technology Mechanism** to facilitate "enhanced action on technology development and transfer to support action on mitigation and adaptation to climate change" (UNFCCC 2011b). The ambitious Technology Mechanism which will become fully operational in 2012 is led by a high level **Technology Executive Committee** of 20 experts. The Committee will concentrate on the specific technologies available to it, recommend actions and guidance on policies and programmes especially for the least developed member Parties and facilitate collaboration, address barriers and promote coherence and co-operation. A second component of the Technology Mechanism is a **Climate Technology Centre and Network** which, at the request of a developing State Party, will provide information, advice, assistance and training. As well, the Centre will encourage the development and transfer of existing and emerging environmentally sound technologies to developing Parties along with opportunities for North–South, South–South and triangular technology cooperation. It will facilitate, a) national, regional, sectoral and international technology centres, networks, organization and initiatives for enhancing co-operation and b) partnerships to accelerate the innovation and diffusion of environmentally sound technologies. TT:CLEAR also has an **Implementation Support Program** to finance transfer projects.

The innovations and efforts made by UNFCCC to embrace the less well resourced Parties of the UNFCCC community are to be applauded. However these may well fall short of that needed by a number of the SIDS, the very existence of which are threatened by rising sea levels and worsening storm surges. The Alliance of Small Island States (AOSIS) member's message to the Technology Mechanism team at their 2011 meeting was that their future survival and viability depended on a "global transition to low carbon energy sectors" and "the availability of low carbon and carbon negative technological systems, within the next couple of decades to keep average temperature increases below 1.5 C", as well as adaptation (Binger 2011). Their 20th AOSIS Anniversary logo is ***1.5 to stay alive.***

Barley fields, Ladakh. Barley is the staple crop as it is adapted to the high altitude and the short growing season. It thrives in these centuries-old carefully tendered composted fields that are irrigated under strict village regulation with glacial meltwater. This system is suffering under climate change. The water supply is already less predictable in continuity and volume. As glaciers further retreat, the water supply will further diminish with serious consequences on crop yields

Museums

Museums have a long and authoritative history of information sharing. Along with libraries, they rank higher in trustworthiness than all other information sources, on or offline, among adults of all ages, educational levels, races and ethnicities (Johnson 2009 p13); (Trautmann 2007; CAMD and Directors 2011). Museums are also accepted as secure and safe places of congregation, unencumbered places to learn. The **Council of Australasian Museum Directors (CAMD)** explain "Museum exhibitions and programs, both actual and online, inspire and educate the community by fostering confidence, critical thinking, creativity and problem solving abilities" (CAMD and Directors 2011 p11) . Museums are also a trusted source of information on contested issues such as climate change as "they bring together a multitude of viewpoints on significant issues, interpreting complex science, promoting community understanding and engaging on and offline users in the search for solutions" (CAMD and Directors 2011 p4). Teachers prefer authoritative sources, such as museums, for online education purposes as they are known to offer high level information, unlike the opinion dominated, crowd sourced and unvetted information flooding the blogosphere. Natural history museums make a fundamental contribution to the public's climate change knowledge through research based on their collections and through the sharing of this information through display and online. It is common practice in many museums to use online social networking tools like **Facebook** and **Flikr** to engage a wider audience.

Chapter 13
Climate Change Justice and Heritage

Climate Change is the worst case of intergenerational abuse ever perpetrated.
Rev Tim Costello, social activist and head of World Vision (Aust), 2011,
Climate Change rally, Brisbane, 2011.

Climate Justice, Human Rights and Intergenerational Justice

Climate justice brings together human rights, collective rights, equity and ethics to address climate change "by looking at who is hurt, how they will be hurt and who is responsible" (FOE and Walker 2007).

Mary Robinson, President of *Realising Rights: The Ethical Globalisation Initiative* has said "Human Rights law is relevant because climate change causes human rights violations" (ICHRP 2008 piii). These violations may be understood as stemming from our 'carbon footprint', that is the amount of carbon dioxide generated by our lifestyles. An average person weighing 68 kg puts his/her weight in carbon dioxide into the atmosphere every 6 days (Henson 2008 p34). However the average is not the norm. The industrialised world is putting in up to 200 times more carbon dioxide per person than the developing world (see Insert 13.1). These footprints go deep as they are linked to the history of industrial development and "reflect the large 'carbon debt' accumulated by rich countries—a debt rooted in the over-exploitation of the Earth's atmosphere" (UNDP 2008). The carbon footprint of the world's most powerful nations is starkly obvious, with the injustice perpetuated, unequivocal (GHF 2009e). GHG emissions from any one country affect the climate of all countries. The effect of this amount of pollution means irreversible changes to life are happening (see Insert 13.2).

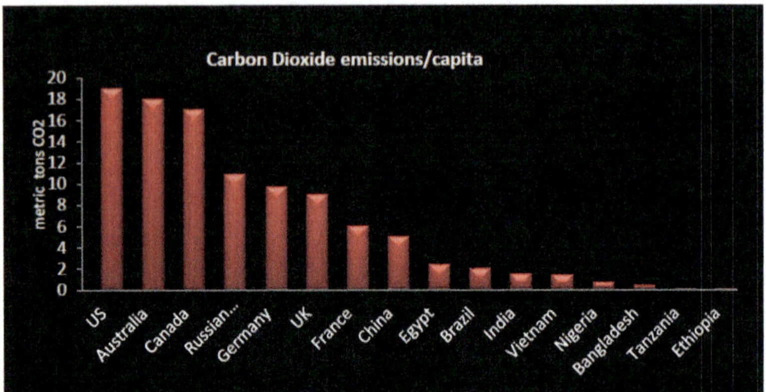

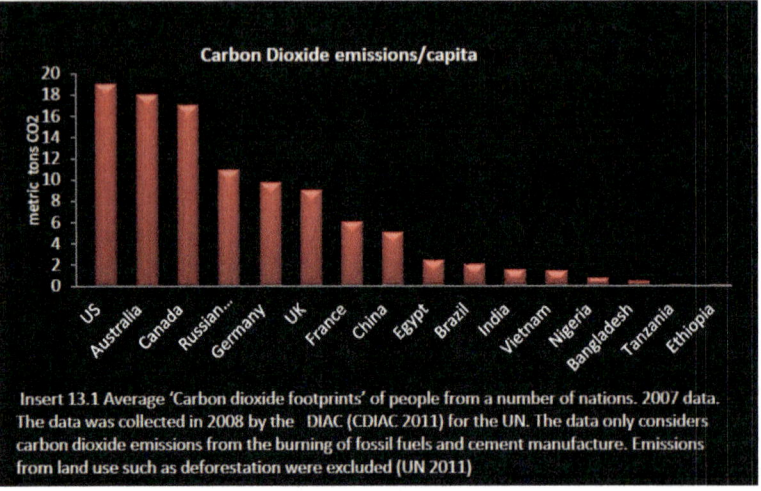

Insert 13.1 Average 'Carbon dioxide footprints' of people from a number of nations. 2007 data. The data was collected in 2008 by the DIAC (CDIAC 2011) for the UN. The data only considers carbon dioxide emissions from the burning of fossil fuels and cement manufacture. Emissions from land use such as deforestation were excluded (UN 2011)

The 2010 Pakistan floods killed 1400 and changed the lives of 3 million displaced people. Climate change intensifies such naturally occurring flooding events. Ninety-nine percent of all human climate casualties, are and will continue to be those from minority groups, the materially poor, the indigenous - particularly women, children and the elderly from these groups - living in developing countries. At greatest disadvantage are those living in vulnerable rural areas and urban slums.

As touched upon in Chapter 1, systematic assessment of the international equity implications of climate change shows that those countries least responsible for climate change generally have the least capacity to deal with its adverse impacts. Their lower standards of living mean that their people have the least resources to adapt to gradational change and capacity to withstand climate shocks. This does not imply that they are ignorant of what is happening. Although lacking knowledge of the science of climate change due to poor education, many understand firsthand the impacts of climate change and its links with the activities of industrialised nations. Unsurprisingly, it is these very people who are the least represented in the halls of power. As a result questions are emerging about international equity and climate change (Füssel 2008b, 2009b; HREOC 2008; ICHRP 2008).

Statement by Make Nhleko, traditional council elder, Zombodze Emuva, Shiselweni region, Swaziland, South Africa.

On extreme weather damage: This has been a very bad year as I have cultivated very little. The hail storm in December 2006 made things even worse. The roof of the supermarket behind us was totally blown off during the storm and crops and houses were destroyed. I now have to buy maize and beans which I used to plant. But at least I can afford to buy maize. There are many that cannot and it is much worse for them.

Time-worn traditions wearing out: In the past, the chief would call the people to weed or harvest his fields. This was a way of unifying the people of the area. After weeding the fields the people would gather at the chief's *kraal* [livestock enclosure] and issues affecting the community would be discussed. The chief would slaughter a cow for the people and food harvested from the chief's fields would be used to feed those people in the community who had nothing and could not afford to feed themselves. In this way everyone had something to eat. But nowhere is anything to harvest so even the chief cannot help those that have nothing to eat.

How livestock are affected: I also keep goats and cattle. But there is very little grass for them to feed on. In the past the grass was always lush and plentiful. The cattle would survive the winter through eating whatever was left on the fields after harvest. But nowhere is nothing left. There has always been stock theft but now it is worse because our cattle have to travel long distances in search of water. This pains me because for us our cattle are gold.

On water shortages: Water is a big problem. Our rivers and wells have dried up. Even some boreholes provided by the Canadian government back in 1997 have since dried up. The community is now digging some trenches for water pipes which we hope will carry water from an old borehole. At the moment we must fetch from the rivers and wells that have not yet dried up. We share the water with livestock. Diseases like cholera and diarrhoea are very common because the water is always dirty (FOE et al. 2007 p29). Quoted with permission.

Insert 13.2 Living with climate change

The poorest billion live in the semi-arid dry-land belt countries from the Sahara to the Middle East and Central Asia, as well as in Sub-Saharan Africa, the south Asian deltas and in **Small Island Developing States (SIDS)** (Füssel 2009b, 2008b; GHF 2009e). These people and their cultural heritage will be the first and worst affected by the impacts of climate change. The numbers affected are expected to more than double within 20 years (GHF 2009a p12). As has been acknowledged, climate change is a 'multiplier' of disadvantage. Anders Kompass speaking on behalf of the UN Human Rights Office explained at the *UN 2010 A Social Forum- Climate change and human rights*, "A human rights-based approach brings into focus how climate change-related threats affect individuals and groups differently. Climate change impacts exacerbate existing vulnerabilities, which in turn, are rooted in discrimination, disparate health status and imbalances in access to knowledge and information" (OHCHR 2010).

Other international organisations are arguing that human rights principles, protecting the access right of each person to life's essentials, such as food, water, shelter and security, be put at the heart of international climate change policy (Mearns 2009). Their members are advocating a dual approach, of aggressive mitigation and pro-poor adaptation, mitigation by rich countries as an emergency and help for developing countries struggling to adapt (UNDP 2007b; GHF 2009e; OXFAM 2009; OHCHR 2009).

The national differences in carbon footprint per person point to an inverse relationship between climate change risk and responsibility. The situation is not that direct. Industrialised countries bear the most responsibility for climate change, but there is an increasing number of cases where low and middle income countries with high economic growth rates and rich natural resources, also contribute significantly to climate change. Often they find it difficult to achieve sustainable policies as they do not always have access to appropriate and affordable technologies. In these countries climate justice issues surrounding deforestation, black soot from diesel engines and domestic stoves and the growth of the biofuel industry are becoming particularly difficult but also sensitive — the countries being both large emitters and highly vulnerable to climate change (GHF 2009a p64).

However the equity principle of **common but differentiated responsibilities (CBDR),** enshrined in Article 3.1 of the UNFCCC has been widely accepted (UN, 1992). It proposes "States shall cooperate in a spirit of global partnership to conserve, protect and restore the health and integrity of the Earth's ecosystem. In view of the different contributions to global environmental degradation, States have common but differentiated responsibilities" (Thorson 2008 p20). "Accordingly, the developed country Parties should take the lead in combating climate change and the adverse effects thereof." This acknowledges that long term industrialized countries are responsible for far more greenhouse gas emissions than developing nations. Wealthy nations face the biggest liability and task to address climate change and therefore must support developing nations to adopt a sustainable approach, and avoid the polluting path to development through technology transfer, capacity building and financial support.

This is not to dismiss the technological, economic and health benefits gained from the developed world, (Smith 2007) but to recognise the overall pollution contributions from the mostly village-based lives of those in the developing world are miniscule compared to the emissions of people from the developed economies. Only one percent of global emissions are attributable to some 50 of the **least developed countries (LDCs).**

The Bottom line: "The essence of the potential climate tragedy is its intergenerational injustice, and the fact that effective approaches that could avert the tragedy make sense for many reasons" (Hansen 2011a).

James Hansen is widely regarded as the world's foremost climate change scientist. He has advised the US government, testified to Congress and most famously appeared in Al Gore's documentary *An Inconvenient Truth*. Extreme frustration has forced him to combine his research life with going public about what he describes as "The predominant moral issue of the 21st century..." (Hansen 2010a)

Clearly out of his comfort zone in his media role, Hansen has striven to use every approach open to him to bring his message to world leaders and the public. As well as being the author of many of the most authoritative papers on the science of climate change, he has written to world leaders, advised popular global environmental movements such as '350. Org', given evidence supporting climate activists who have been charged, and has himself been arrested for civil disobedience or 'peaceful civil resistance' (his term) for speaking out about the dangers of coal mining at a mining site. He has taught and lectured internationally, appeared on the media, proposed an alternative scheme for national emissions reduction by pricing carbon called a carbon 'fee and check' approach to replace the officially favoured 'cap and trade' schemes he considers terminally flawed. In 2009 James Hansen wrote *Storms of my Grandchildren* explaining the urgency, the science and the politics of the problem, revealing again the power of the fossil fuel lobby and other "special interest" groups have over government. He has become increasingly outspoken as he has watched the leaders of the fossil fuel industry choose to ignore climate change science facts and "to support activities intended to keep the public ill-informed. These kingpins are guilty of high crimes against humanity and nature. It is little consolation that the world will eventually convict them in the court of public opinion or even, unlikely as it is, that they may be forced to stand trial in the future before an international court of justice."(Hansen 2012a p8). Angrily Hansen explains his current approach because "Begging Congress to be responsible does not work. Exhorting the president to be Churchillian does not work." He is preparing to file suit on behalf of his grandchildren. The suit -"Sophie, Connor, Jake and Lauren versus Obama and the United States Congress", is for breaking the Declaration of Independence and the Fourteenth Amendment of the Constitution of the United States of America... "Obligations to young people, it seems to me, are already clear in the second sentence of the Declaration of Independence, "We hold these truths to be self-evident, that all men are created equal, that they are endowed by their Creator with certain unalienable Rights, that among these are Life, Liberty and the pursuit of Happiness." This basic tenet leads directly to the right to equal protection of the laws."

The Fourteenth Amendment of the Constitution declares: "No State shall make or enforce any law which shall abridge the privileges or immunities of citizens of the United States; nor shall any State deprive any person of life, liberty, or property, without due process of law; nor deny to any person within its jurisdiction the equal protection of the laws." Over time the courts ruled that "any person" includes minorities and women, for example, and equal protection provided the principal basis for extension of civil rights to minorities" (The young and unborn).

> "Human-made climate change now raises a moral issue as momentous as any that the courts have considered in the past. Today's adults are reaping the benefits of burning fossil fuels while leaving the consequences to be borne by young people and future generations" (Hansen 2011b p293-4).

Insert 13.3 Climate change and Intergenerational Injustice

For many, climate impacts are a final step of deprivation. Garvey terms it a coming slow motion catastrophe (Garvey 2008). Allowing this human tragedy-in-the-making to evolve is a political failure. It precludes the human rights of future generations (see Insert 13.3). The **Universal Declaration of Human Rights (UDHR),** Article 3, establishes that "everyone has a right to life, liberty and personal security" (ICHRP 2008). It is a travesty that our knowledge-rich developed world has given voice to climate change sceptics and contrarians and to so many that see themselves as 'exceptions' to making modifications in their lives to reduce their carbon footprint. Those living on the frontlines (Inserts 13.4) of climate change have no such illusions or doubts (FOE et al. 2007). Interestingly an American study (Adams 2010) has shown that significantly more Hispanic Americans, despite their social and educational disadvantages, understand climate change than do non-Hispanic Americans. This is probably because Hispanics have had closer ties to small scale farming and have a grass roots understanding of our dependency on the environment.

Inaction to the threat posed by climate change represents a violation of the UDHR. Because of their equal entitlement we should not discount the human rights of future generations by passing on to them a natural environment that we have degraded.

> Statement from Babaga Island resident, Solomon Islands.
> They talk about us moving. But we are tied to this land. Will we take our cemeteries with us?
> For we are nothing without our land and our ancestors. (OXFAM 2009 p36).
> Quoted with permission.

Insert 13.4 Living with climate change

Discounting the Future

This raises the controversial topic of **discounting the future**. Discounting the future is "taking short term gain even though doing so might cost us immensely in the longer term" (Flannery 2010 p211). According future generations a low discount rate means that you value future generations and their human and environmental rights as you value your own. According future generations a high discount rate means that you do not value future generations and their rights. Those that hold the latter view may do so because they see the future as bleak, offering no opportunities. This view often arises in those who are disadvantaged by poverty, discrimination, inequity and war. They might express this as "I may as well go for it now as there is no future". High discounters would be more likely to 'trash' the environment. Discounting the future is a significant threat to sustainability.

If BAU continues the world will suffer massively from climate change, therefore finding the correct social discount rate is very important. Economist Nicholas Stern explains his ethically based low discount rate of 0.1% in his *Stern Review* by giving the following example. "Selecting a 2 percent rate of pure time preference would halve the ethical weight given to somebody born in 2043 relative to somebody born in 2008" (Stern and Taylor. 2007) and quoted in the (UNDP 2007b p63). In fact, the only justifiable reason for discounting the welfare of future generations, according to Stern (Stern 2006 Chapter 2, p43) was the possibility of extinction (Stern 2006 chapter 6 p 160). Stern was in no doubt that the benefits of strong, early action on climate change outweighed the costs.

The hypocrisy of understanding the ethical implications of climate change yet maintaining a BAU trajectory raises significant concerns and difficult questions regarding international equity, global justice and human rights both within and across generations and goes to the core of climate change justice. Humanity's response to this challenge will reveal the moral stature of mankind. Dangerous climate change is a threat, not a pre-ordained fact of life. We can choose to confront and minimise the threat, or we can choose to let it evolve into a fully fledged crisis for future generations (UNDP 2007b). Tackling climate change should be seen as integral to the broader goals of eradicating poverty in the developing world. These are the twin challenges of the 21st century and should and can be tackled together (Richardson et al. 2009 p25). As more people worldwide are raising these issues (Ford 2009a) UNESCO takes another stand for a better future by proposing the preparation of a draft **Universal Declaration of Ethical Principles in relation to Climate Change** to be considered later in 2011 (UNESCO 2010s, 2009n).

Climate Change Justice and Cultural Heritage

The question of cultural rights of those displaced by climate to life in another culture is still largely unexamined. Indigenous researchers in Papua New Guinea (Leahy 2009a) and The Solomon Islands (Leafasia 2010) have written and spoken about their island coastal communities facing relocation through **sea level rise (SLR).** These observers report that the fear of cultural loss is so profound that some villagers claim they would rather disappear with their land than relocate. As Leahy has recently documented, the Tami Islanders in Morobe Province, Papua New Guinea are seeing their land and crops eroded, swamped and contaminated with salt water. Uniquely, the Islanders have experience of what relocation to the mainland is like.

In the 1930s, overpopulation forced some Tami Islander families to relocate to land "provided by Yabem people in view that they were 'helping out' their brothers and sisters in need, a deed long invested in them by early German missionaries and through their traditional trade" (Leahy 2009a p51). Since then, those displaced have become tense and anxious through their spiritual separation from their ancestral land. Disputes over land with their new neighbours have broken out. Their traditional language has been partly lost. And most seriously, there has been the appropriation of Tami carving designs by the neighbouring Yabem. This has been made worse by the Tami peoples' emerging dependency on the Yabem for suitable timber to carve. This is a consequence of a scarcity of suitable trees on the Tami Islands. The distinctive Tami carvings are highly valued both within and outside the Tami community. Their carvings are world renown, collectable and an important source of income. Traditionally, the copying of Tami designs by the Yabem was not allowed without a proper public agreement (with accompanying feast) to share design ownership for the duration of the 'loan'. Leahy records Tami Island Community Leader Stephen Mesa's words, "Tami Islanders are already watching helplessly while their carvings, mats, baskets, show up at local markets with strangers" (Leahy 2009a p66). Second to the loss of ancestral land, the devaluing of their intangible cultural heritage (through lack of intellectual property rights and copyright) is seen by the Tami community as the greatest disadvantage to relocating.

It is possible that the Yabem feel justified in their cultural acquisitions on the basis of their earlier actions in providing land to the Tami Islanders. A lack of understanding and acceptance between currently merging groups may not bode well for the climate change displaced of the future.

People from developing and indigenous communities can have strong coping strategies nurtured by their cultural heritage. It is ironic that while their tangible cultural heritage may be more vulnerable to climate change than that of the industrialised nations (which safeguard many cultural treasures in museums and galleries), their intangible cultural heritage may be less vulnerable (UNESCO 2009). A great many traditional communities have potentially strong adaptative traditional knowledge, customs that connect to their environments and faiths which see environmental care as self-evident and wholly in keeping with centuries of theological practice.

The Ladakhis, for instance, who live in the mountains just south of Tibet, are subsistence farmers with only hand tools and a low average income. First hand experiences have found these Buddhists' culturally rich village lives are largely free of poverty, crime and social breakdown. Their communities are resourceful, with time for vibrant ceremonies and religious practice. Their cultural heritage contributes to a community resilience that may, under stress of climate change impacts, be more robust than that experienced by industrialised societies suffering from similar stress. Island communities present another well recognised case of traditional strength in the face of adversity. Sonia Smallcombe writes of Pacific Islanders adapting by merging indigenous and atoll technologies to "improve post-crisis resilience" (Smallacombe 2008 p77). Western societies have become more ecologically unaware and disconnected from their environment as has their intangible heritage, making them potentially less prepared for protracted or extreme climate induced events.

Autumn at a village in Zanskar, Ladakh. Note the treeless lunar landscape that characterises this rain shadowed high altitude region. Hillsides are terraced to enable the cropping of barley and goats graze on the sparsely vegetated slopes

Part III

Can the UNESCO Conventions Protect and Safeguard Cultural Heritage from Climate Change Impacts?

Chapter 14

What do the Statistics Indicate About the Protection of World Cultural Heritage Against Damage Caused by Climate Change?

You really do have to wonder whether a few years from now we'll look back at the first decade of the 21st century — when food prices spiked, energy prices soared, world population surged, tornados ploughed through cities, floods and droughts set records, populations were displaced and governments were threatened by the confluence of it all — and ask ourselves: What were we thinking? How did we not panic when the evidence was so obvious that we'd crossed some growth/climate/natural resource/population redlines all at once? By T. L Friedman (Friedman 2011)

What do the statistics concerning the World Heritage Convention and the Intangible Cultural Heritage Convention indicate about the state of protection offered to world cultural heritage against damage caused by climate change?

The World Heritage Convention

By: Type, Region, Zone, Latitude, LCDs, SIDS and Inscription Status

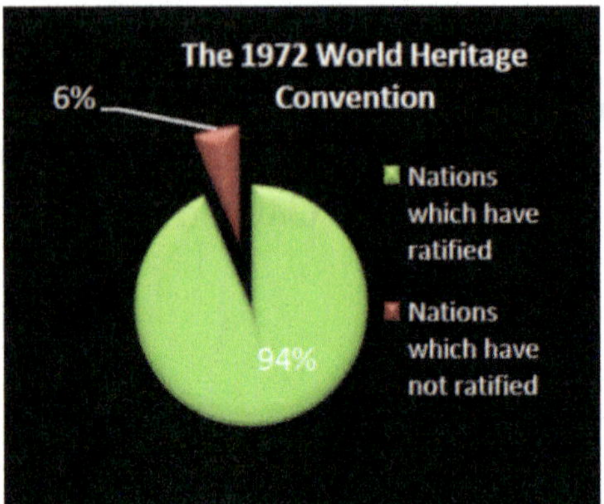

Insert 14.1 WHC ratification status, 2012

The **World Heritage Convention (WHC)** has succeeded in attracting 188 out of 201 nations to become Member States over 38 years (see Insert 14.1). The **World Heritage List** holds 936 inscriptions of properties judged as meeting criteria for **Outstanding Universal Value (OUV)**. Of these, 725 are cultural properties, 183 natural properties and 28 mixed properties (see Insert 14.2). **(From here on, in this study, the 'mixed' properties have been included with the 'cultural' properties).** The imbalance between the two major 'types' has been of ongoing concern to the **World Heritage Committee** and UNESCO. The WHC legally ties **States Parties** to protect inscribed properties and to implement national, regional, local (including traditional), protective measures to ensure this (See Chapter 2 for more details. For the WHC text, see Appendix 3). The WHC is a success story having attracted almost universal ratification to become the foremost international legal tool in support of the conservation of the world's cultural and natural heritage. The WHC, as mentioned, has no enforcement agency but carries out its work with encouragements, assistance and requests. Of the 936 listed properties 35 are on The **List of World Heritage in Danger** (See Insert 14.3).

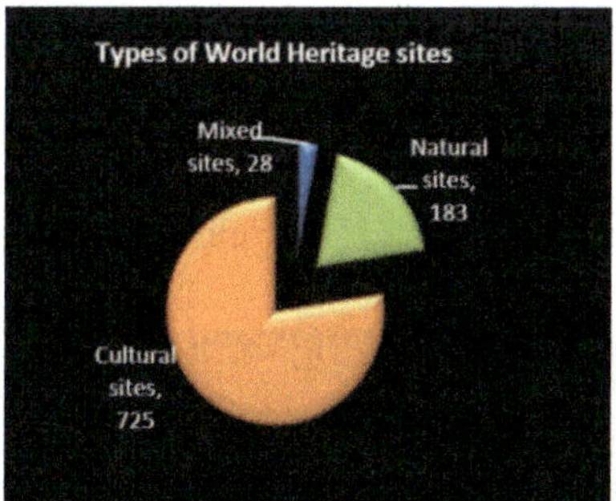

Insert 14.2 Composition of WH sites by type 2012

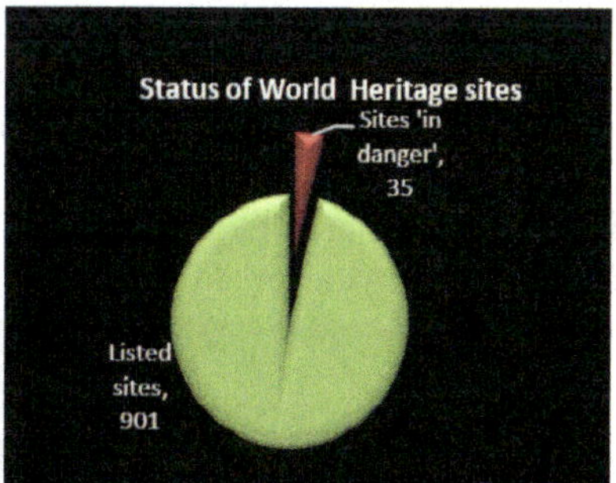

Insert 14.3 Composition of WH sites by status, 2012

Member States are invited to submit a **Tentative List** of World Heritage properties to the World Heritage Committee for publication but not evaluation. Out of 188 States Parties, 167 have complied. The preparation of this List spurs national effort, draws attention to newly listed properties, and introduces the World Heritage 'processes' of identification, inventory making, nomination and cooperation. It particularly encourages those 35 States Parties which have ratified the WHC but have not, as yet, had any sites inscribed (see Insert 14.4 and 14.5).

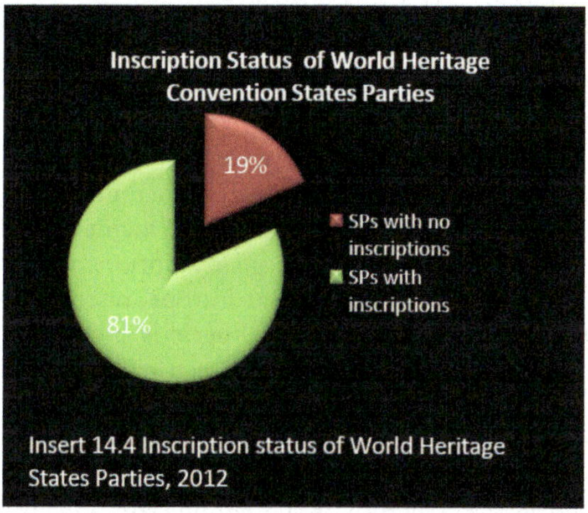

Insert 14.4 Inscription status of World Heritage States Parties, 2012

A breakdown of the latter and reference to Inserts 14.6 and 14.7 identifies the States Parties of Africa, Asia and the Pacific, including the LDCs and particularly the SIDS, as in need of support in the nomination process (See **Appendix 1** for inscription status of all States Parties linked to Human Development, latitude, climate change vulnerability, population, area and whether States Parties are also SIDS, LDCs or both.)

The WHC's great strength is that it is instrumental in implementing protection for properties from most of the world's nations. As most of these properties are cultural, the WHC can be seen as being a protector of a significant breadth of global cultural heritage and therefore of being in an influential position to raise awareness of, and help address, climate change threats to this widespread heritage.

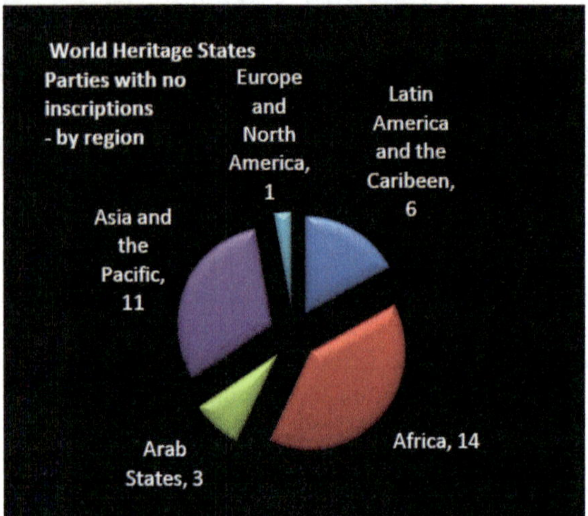

Insert 14.5 Regional distribution of the 35 WH States Parties with no inscriptions, 2012

The **Global World Heritage Strategy for a Representative, Balanced and Credible World Heritage List** was launched in 1994 to identify and fill major gaps in the World Heritage List and to maintain a reasonable balance between cultural and natural heritage inscriptions and geographic representation. Research had "revealed that Europe, historic towns and religious monuments, Christianity, historical periods and 'elitist' architecture (in relation to vernacular) were all over-represented on the World Heritage List; whereas, all living cultures, and especially 'traditional cultures', were underrepresented" (UNESCO 2011k).

The reasons for the gaps in the World Heritage List have been identified as being related to the structure of the nomination process itself; to how cultural properties are to be managed and protected; and to the qualitative information needed for property identification and assessment. By adopting the Global Strategy, "the World Heritage Committee wanted to broaden the definition of World Heritage and to provide a comprehensive framework and operational methodology for implementing the WHC." Its implementation was by exerting "all efforts" to encourage more countries to become States Parties and to support the development of Tentative Lists and the preparations of nominations of properties from categories and regions currently not well-represented on the List. The Global Strategy is both a conceptual framework and a pragmatic and operational methodology (Pedersen 2002 p17). It brought a new vision of World Heritage sites as "outstanding demonstrations of human coexistence with the land as well as human interactions, cultural coexistence, spirituality and creative expression". New categories have resulted such as the categories of cultural landscapes, itineraries and industrial heritage.

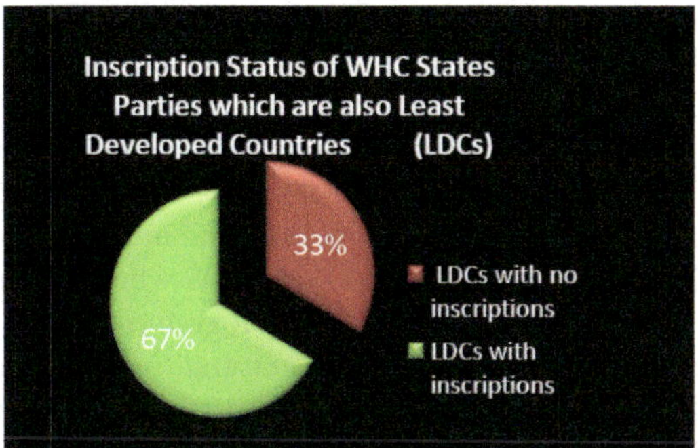

Insert 14.6 Inscription status 2012

"Important conferences and thematic studies aimed at implementing the Global Strategy have been held in Africa, the Pacific and Andean sub-regions, the Arab and Caribbean regions, central Asia and south-east Asia. These well-focused studies have become important guides for the implementation of the WHC in these regions" (UNESCO 2011k). For example the World Heritage Committee recognised that despite its extraordinary cultural and biological diversity and richness, the Pacific was the most under-represented sub-region on the World Heritage List.

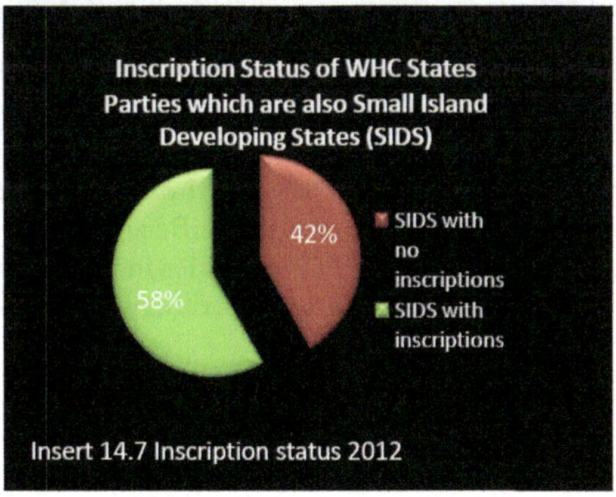

Insert 14.7 Inscription status 2012

To address this imbalance, Global World Heritage Strategy meetings were held in the Pacific - in Fiji in July 1997 and in Vanuatu in August 1999. As a result, many Pacific Island countries joined the WHC. However, by 2003 there was still an imbalance, with only one property in the Pacific Sub-Region on the World Heritage List (East Rennell in the Solomon Islands). The World Heritage Committee responded by launching the 5 year **World Heritage - Pacific 2009 programme** in 2003 (UNESCO 2009w). To further the work of the **Mauritius Strategy** (the Further Implementation of the Programme of Action for the Sustainable Development of SIDS), UNESCO and the World Heritage Centre designed the **World Heritage's Intersectoral Platform on Small Island Developing States** (UNESCO 2008h). This Programme coordinates and develops World Heritage activities on islands of the Caribbean Sea and the Atlantic, Indian and Pacific Oceans, raising awareness, building capacity of staff and institutions (UNESCO 2009v) and educating to increase the number of nations ratifying the WHC. It concentrates efforts on SIDS which remain under-represented on the World Heritage List. The Programme helps develop integrated heritage policies covering natural, cultural, moveable and intangible heritage. Assistance for the conservation and management of sites with a view to 'sustainable development' is available after inscription on the List.

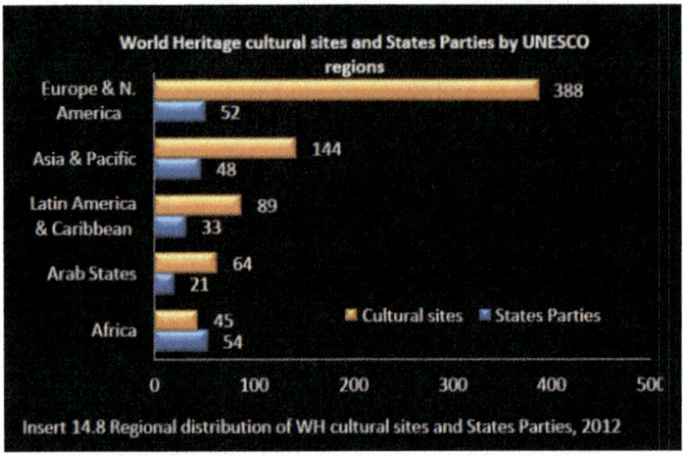

Insert 14.8 Regional distribution of WH cultural sites and States Parties, 2012

In the Pacific region, the **Action Plan for the Implementation of the World Heritage Pacific 2 Programme, 'Pacific 2009'**, provides the overall framework for development of field activities including support for the widely offered **World Heritage National Strategy Workshops.**

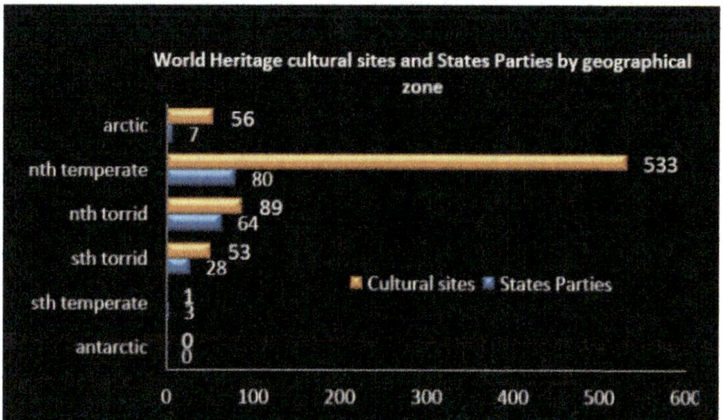

Insert 14.9 Geographic zonal distribution of WH cultural sites and WH States Parties, 2012

Although these strategies have led to significantly more nations ratifying the WHC, more States Parties developing Tentative Lists and nominating sites, there remains a geographic bias as shown in Insert 14.8 and representative bias as shown in Insert 14.2. An examination of Insert 14.9 shows that the ratios of sites per State Party per zone are 8.0:1 for the Arctic, 6.7:1 for the north temperate, 1.5:1 for the tropics (North Torrid and South Torrid zones) and 0.3:1 for the south temperate. Insert 14.10 lists the cultural sites held by States Parties plotted from north to south (the northernmost point of each State Party being used as the reference point). The graph reveals an overall relationship skewed in favour of the northern, often wealthier States Parties with generally less sites per State Party as the equator is approached.

The wide variation in numbers of sites per State Party between each geographic zone is apparent from Insert 14.9. To further improve the chances of nominations from underrepresented regions and of natural types of sites succeeding there is now a limit of two nominations per round per State Party.

The World Heritage Committee works with its three advisory bodies: **International Council on Monuments and Sites (ICOMOS), International Union for Conservation of Nature (IUCN)** and **International Centre for the Study of the Preservation and Restoration of Cultural Property (ICCROM).** These three professional bodies plus member States have a track record of continuing commitment for developing better ways of diversifying the World Heritage List and of making it truly balanced and representative. The World Heritage Committee's leadership has over 40 years demonstrated how the WHC has the capacity to be responsive to change and to address entrenched and emerging issues. The statistics show the success of World Heritage strategies in attracting ratifications and in its ongoing response to improve geographic and zonal representativeness and inscription equity. Global ratification is close.

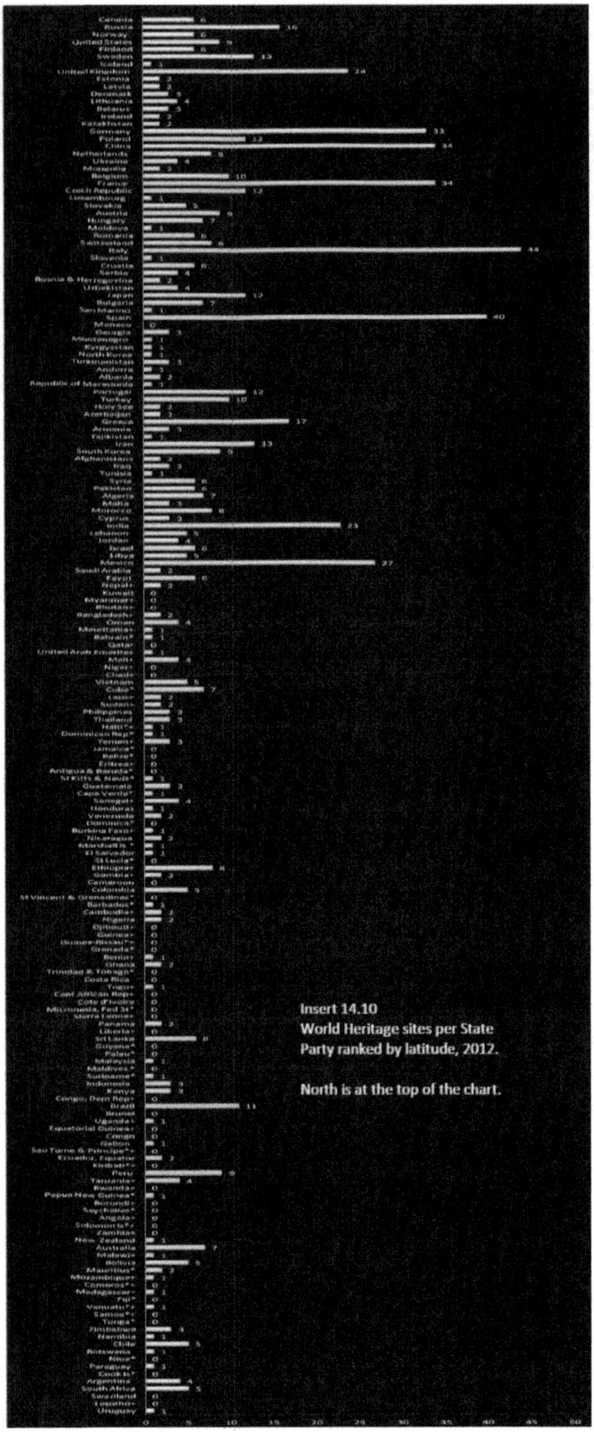
Insert 14.10
World Heritage sites per State Party ranked by latitude, 2012.
North is at the top of the chart.

Recommendations

14.1 Ensure that the World Heritage Centre continues its analyses of statistics thus supporting the development of evidence based World Heritage protection programmes that implement the proposed strengthened and broadened *Policy document on the impacts of climate change on World Heritage properties*. Refer to recommendation 15.2.

14.2 Ensure that the World Heritage Committee continues to encourage, assist and give preference to nominations from States Parties with zero inscriptions as such States Parties are likely to be the most disadvantaged and hence benefit most from World Heritage protection.

14.3 Support the World Heritage Committee's continued facilitation of nominations. In 2009 Cherif Khaznadar, Chairperson of the General Assembly to the ICHC said, "I have been recommending for some time, tying the nomination of an element from a country of the North for inclusion in the Representative List to that country's sponsorship of the inclusion of an element from a country of the South on the Urgent Safeguarding List" (Khaznadar 2009).Although these remarks were directed at Intangible Cultural Heritage nominations, it is recommended that Cherif Khaznadar's suggestion for partnerships be also applied to World Heritage nominations, taking into account the similar views of UNESCO experts (see under Chapters 12 and 23) about increasing the nominations from underrepresented States Parties, directed assistance to SIDS and LDCs and the pivotal need for information exchange and cooperation in the face of climate change.

14.4 Encourage World Heritage authorities to use their statistics to support the proposed worldwide **heritage site vulnerability campaign** (see recommendations 15.4).

The Intangible Cultural Heritage Convention

By: Domain, Region, Zone, Latitude, LDCs, SIDS and Inscription Status

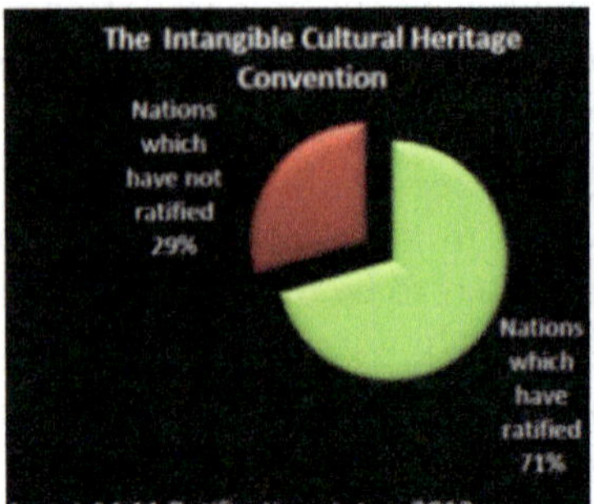

Insert 14.11 Ratification status, 2012

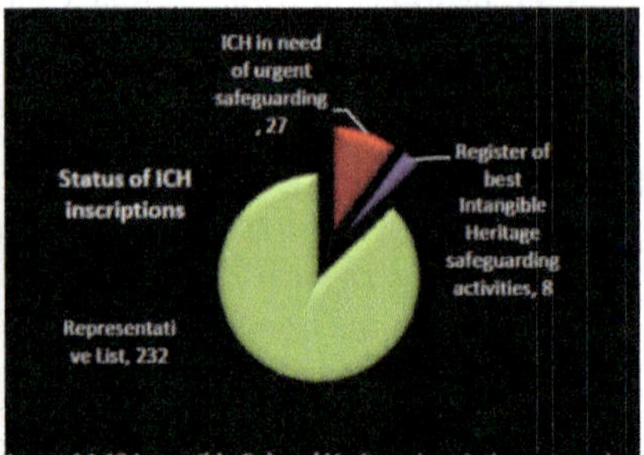

Insert 14.12 Intangible Cultural Heritage inscription categories, 2012

Since 2003, the **Intangible Cultural Heritage Convention (ICHC)** has attracted 142 out of 201 nations to become Member States (see Insert 14.11). Ratification is significant as it is the first step in a commitment of a State Party to safeguard its **intangible cultural heritage** and, when resources permit, the intangible cultural heritage of other States Parties. Ratification brings a complex of obligations, legal, scientific and economic; reciprocal relationships; networks of support, and benefits from the intangible cultural heritage community membership. The **Representative List** holds 232 intangible heritage inscriptions of elements judged as meeting inscription criteria listed in Chapter 2. Inscription show-cases intangible cultural heritage safeguarding. It brings national pride, widespread appreciation and respect and contributes to global civilisation. The **List of Intangible Cultural Heritage in Need of Urgent Safeguarding** has 27 inscriptions of elements that have met very specific criteria, also listed in Chapter 2 (Insert 14.12). The inscriptions in the latter List do not have to derive from the former as is the case under the WHC where its sites 'in Danger' derive from the major List. See Insert 14.13 for a breakdown of the inscriptions on these two lists into the various 'domains'. The third list is the **Register of best Intangible Heritage safeguarding activities**. It has 8 programmes inscribed, representing 8 States Parties. An amendment to Paragraph 20 of the ***Operational Directives*** suggests that States Parties be allowed to propose up to 3 nominations for inclusion in any category per annual cycle. A limit of 100 nominations overall will be accepted for assessment per round by the **Intergovernmental Committee for the Safeguarding of Intangible Cultural Heritage** (hereafter referred to as the **Intangible Cultural Heritage Committee**). Of the 142 States Parties, 87 have inscriptions, leaving 55 yet to have successful nominations, see Insert 14.17 and 14.20.

The ICHC, 32 years junior to the WHC, has had unexpected outcomes. Mr Koïchiro Matsuura, Director-General of UNESCO in 2009 summed up the reactions in this formative period when addressing the 4th session of the Intangible Cultural Heritage Committee, "I want to be frank and express my dismay about the marked imbalance between the Urgent Safeguarding List and the Representative List. This is particularly surprising because during the elaboration of the Convention and in adopting the *Operational Directives* in June 2008, many countries repeatedly emphasized that the primary aim was to safeguard living heritage facing threats of deterioration, disappearance and destruction. To that end, overriding importance was placed on the Urgent Safeguarding List. On several occasions, many countries have reiterated that the Urgent Safeguarding List is the more important of the two because it can catalyse international assistance and cooperation. As we all know, developing countries can receive financial assistance for the preparation of nominations to the Urgent Safeguarding List, and endangered elements are eligible for financial assistance for the implementation of safeguarding action plans. One of the main purposes of the Intangible Heritage Fund is to assist endangered intangible heritage, and the Secretariat is available if countries require any help in completing assistance request or nomination files....... Looked at objectively, this crucial list is hardly 'representative' from a global perspective. The indications are that the marked geographical imbalance of these first nominations is likely to deteriorate in the future. We must find ways of ensuring that the Representative List is truly representative on a global scale" (Matsuura 2009b p3).

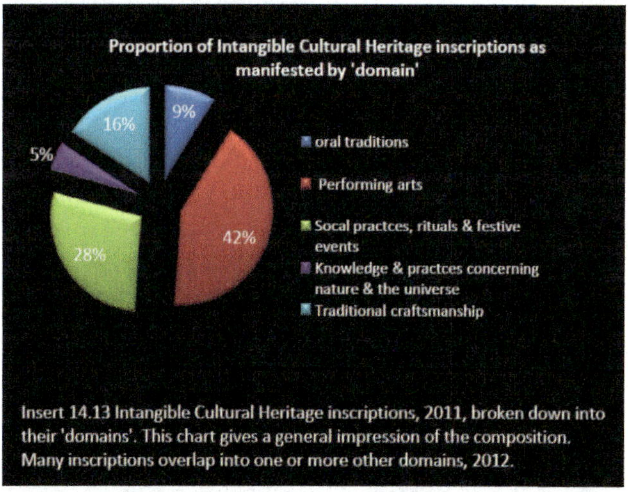

Insert 14.13 Intangible Cultural Heritage inscriptions, 2011, broken down into their 'domains'. This chart gives a general impression of the composition. Many inscriptions overlap into one or more other domains, 2012.

The ICHC has no formal equivalent to the The Global Strategy for a Représentative, Balanced and Credible List as is the case with the WHC. The Intangible Cultural Heritage Committee is responsible for examining the nominations to the 3 Lists. It has underlined that its workload was very heavy and proposed amendments to the *Operational Directives* so as to have a more manageable workload in the future and to more equitably handle nominations, especially first nominations. The response is being worked through taking legal advice to decide further on Decision 4 COM 19 and Decision 4.COM 1.SUB/6.

As well as trying to right the imbalanceby restricting the number of nominations, the Intangible Cultural Heritage Committee is embarking on information sharing initiatives, one of which is the recent establishment of the **Intangible Cultural Heritage Centre for Asia and the Pacific** in the Republic of Korea, a UNESCO Category 2 Centre (Galla 2010). The third method of rectifying the imbalance is by education through dedicated programmes such as UNESCO's **Priority Africa** and **SIDS.** Although efforts are mainly directed to capacity-building, one of the end aims is to safeguard heritage in African and SIDS by inscription. Insert 14.14 shows that the world's intangible cultural heritage is better represented across all regions under the ICHC, than is tangible cultural heritage under the WHC.

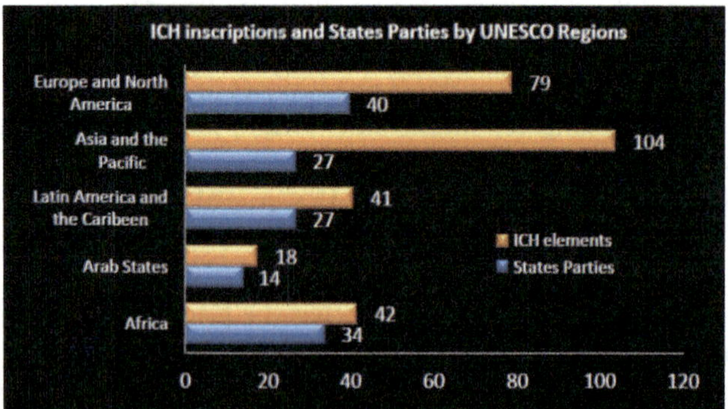

Insert 14.14 Regional distribution of ICHC inscriptions and States Parties, 2012

By holding regional workshops the Intangible Cultural Heritage Committee is endeavouring to encourage ratifications from nations not yet signatories to the ICHC and to encourage nominations from the high proportion of States Parties, particularly those that are also LDCs, which have not proposed successful nominations (see Insert 14.17, 14.18 and 14.19).

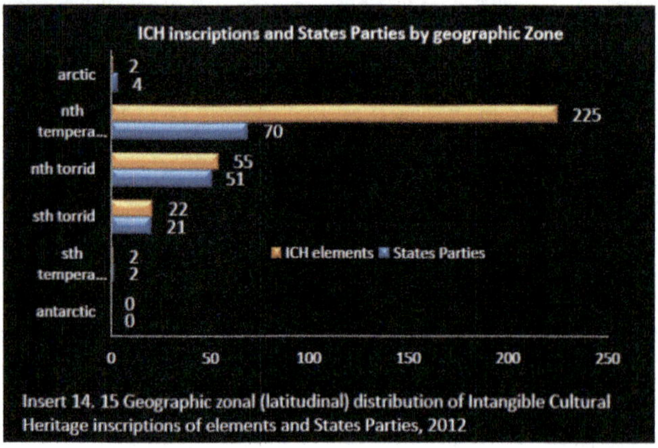

Insert 14. 15 Geographic zonal (latitudinal) distribution of Intangible Cultural Heritage inscriptions of elements and States Parties, 2012

The imbalance between the 'Urgent Safeguarding' and 'Representative' lists is a more challenging dilemma. The Intangible Cultural Heritage Committee is earnestly disseminating the intentions of 'Listing' (Galla 2009a, 2009b), especially of the primacy of the 'Urgent Safeguarding' List through regional meetings. Under the ICHC, a State Party may decide that an inscribed element should be transferred from one list to another. It does this by renominating the element for the other List. When the WH Committee transfers a listed property from the 'Representative' to the 'in Danger' List it is a public acknowledgement that the State Party responsible has not been able to protect its WH. The ICH Committee is at pains to dispel the perception that this is the case when an element is listed on the 'Urgent Safeguarding' List. There should be no suggestion of failure associated with States Parties that nominate elements to the 'Urgent Safeguarding' List. Another factor dissuading nomination for the 'Urgent Safeguarding' List or transfer from the Representative List to the 'Urgent Safeguarding' List may be the high status associated with the Representative List and the economic benefits in the form of the tourist market that might be associated. This regrettable mindset overlooks the priority of safeguarding fragile intangible heritage and could lead to further financial disadvantage in the least developed Member States.

Insert 14.14 indicates that there is geographic over-representativeness, with most inscriptions originating from the Asian Pacific region. On examination of the data tables (Appendix 1) and the latitudinal distribution graph, see Insert 14.16 it becomes apparent that one of the reasons for this is that two States Parties have majorly large numbers of inscriptions. China has 36 inscriptions and Japan 20. A group of States Parties consisting of the Republic of Korea with 14 inscriptions, Croatia with 12, Spain with 11, France with 10, and Turkey, Mongolia, Belgium and Iran with 9 each hold the 'middle ground' for inscriptions, with the majority of States Parties having a very low number of elements inscribed. The Pacific is under represented as is evident from the listings in Appendix 1 and the rankings in Inserts 14.16, 19.7, 19.8, 19.9 and 19.10.

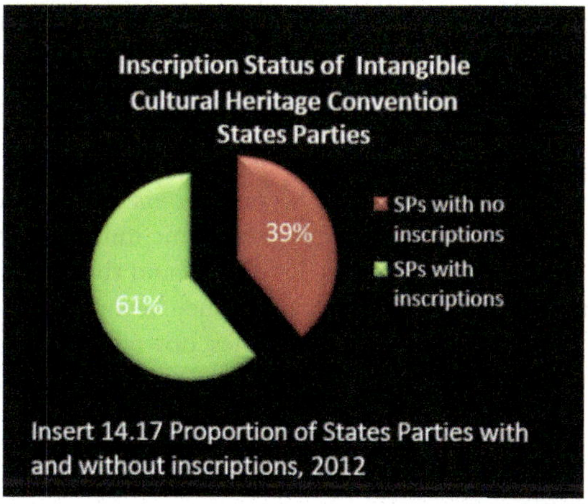

Insert 14.17 Proportion of States Parties with and without inscriptions, 2012

Inserts 14.16, 14.17, 14.18, 14.19, 14.20, and 19.7, 19.8, 19.9 and 19.10 provide information on the States Parties with zero inscriptions, their SIDS, LCD status and distribution by UNESCO regions, latitude, area, population, climate change vulnerability and human development. The similarity between Inserts 14.17, 14.18 and 14.19 indicates a degree of equity in achievement not mirrored by the WHC figures. In fact the proportion of LDCs that are States Parties and have no inscriptions is less than that of the average State Party. It suggests that the work of the Convention promoters and the follow up has been successfully directed at those in most need. Chart, Insert 14.15 shows the dominance of the inscription-holding States Parties is in the northern temperate latitudinal zone. All other zones have similar low numbers of average holdings per State Party. The next chart, insert 14.16 has more information. Every State Party is represented. The States Parties are ranked in order of their most northerly points by latitude, north to south, top to bottom.

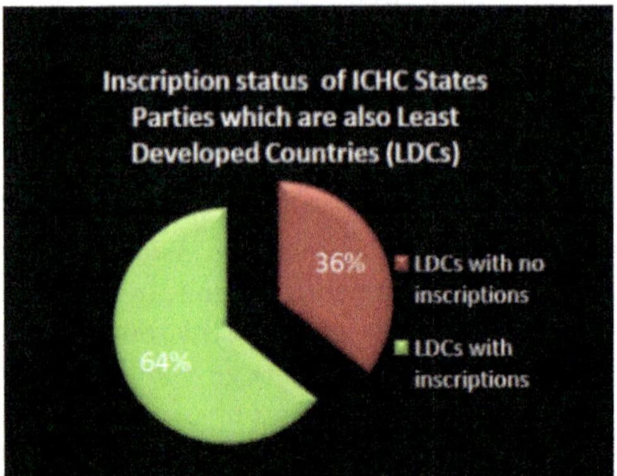

Insert 14.18 Proportion of States Parties (which are also LDCs) with and without inscriptions, 2011

Insert 14.16 shows a scatter of States Parties with zero inscriptions, and a large group of nations holding one inscription. The next group of inscription holders is biased to the northerly latitudes with the two stand-out States Parties being Japan and China. The reason for the 'north–south' relationship proposed by Cherif Khaznadar is evident. This is what prompted him to suggest that the well endowed and experienced States Parties from the north should assist their southerly neighbours in nominating their intangible heritage for inscription. This geographic imbalance is consistently and persistently being addressed by UNESCO and the Intangible Cultural Heritage Committee and every year the balance becomes more equitable.

Sixty one nations (30% of world countries, down from 37% in 2009) have not ratified the ICHC. 39% of the States Parties, (down from 44% in 2009) that have ratified the Convention have yet to hold an inscription, see Insert 14.17.

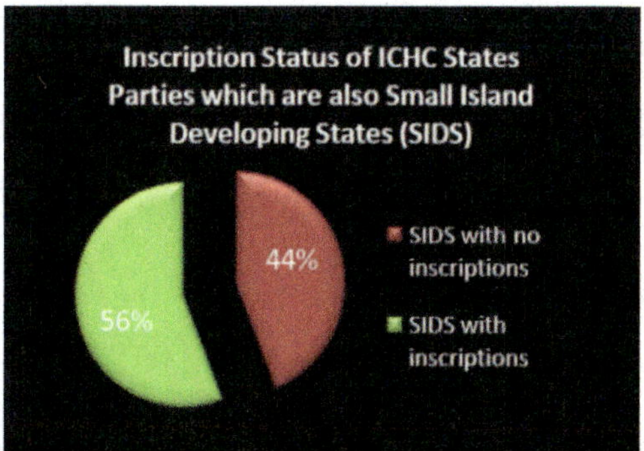

Insert 14.19 Proportion of States Parties (which are also SIDS) with and without inscriptions, 2012

States Parties making up the 39% is a significant proportion, to some extent reflecting the youth of the Convention. States Parties are ratifying at a steady rate, but there is a lag time, for some, before they nominate. Insert 14.20 shows that the States Parties with no inscriptions are to be found in every region. The possible reticence of less developed States Parties to nominate elements under the ICHC in its formative period may be related to unfamiliarity with procedures, hesitancy to request assistance in preparing nominations and lack of resources with which to implement the required conditions (UNESCO 2009bb). The 'lag' period may have more to do with pressing national economic priorities than with anything else. However, the 53 States Parties without inscriptions may have started to implement their safeguarding policies and practices nationally, so some protection of intangible heritage may be in place.

This lively and colourful Convention, judging from its ICH-inspired management style, is working to meet the challenges that the initial three rounds of inscriptions have raised. The ICHC is growing in popularity as far as the Representative List reflects but is yet to overcome the resistance to listing on the Urgent Safeguarding List. Perhaps it is a matter of national pride, not wanting to see one's heritage struggling? Of relevance here is that whatever climate protection the ICHC may provide will directly assist the current practice of customs and ceremony, performers, dancers, musicians, singers, craftspeople, narrators, knowledge holders and ritual practitioners of 81 nations.

States Parties are invited to submit **Proposals of Programmes, Projects and Activities as Best Reflecting the Principles and Objectives of the Convention** 'taking into account the special needs of developing countries' for inscription. This third List is an ICHC design innovation. It provides for the wide dissemination of ideas and practical examples to inspire and encourage further safeguarding by nomination and a potential means by which climate change responsible programmes can be recognised and copied.

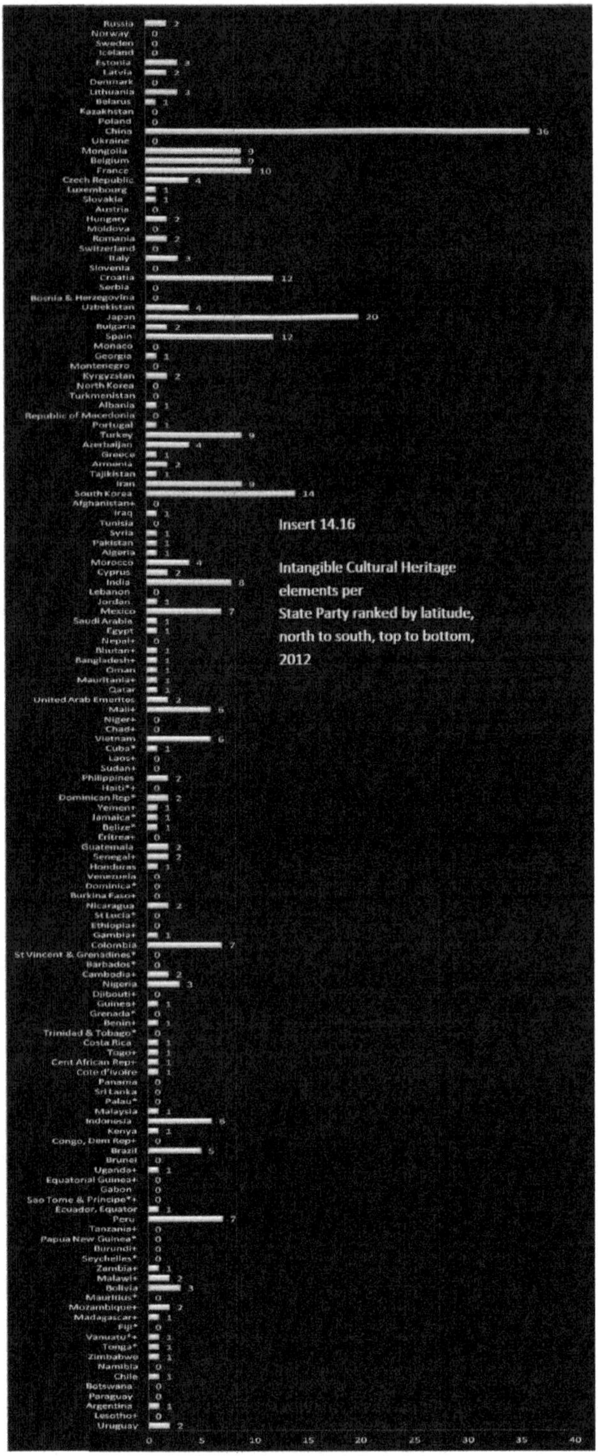
Russia 2
Norway 0
Sweden 0
Iceland 0
Estonia 3
Latvia 2
Denmark 0
Lithuania 3
Belarus 1
Kazakhstan 0
Poland 0
China 36
Ukraine 0
Mongolia 9
Belgium 9
France 10
Czech Republic 4
Luxembourg 1
Slovakia 1
Austria 0
Hungary 2
Moldova 0
Romania 2
Switzerland 0
Italy 3
Slovenia 0
Croatia 12
Serbia 0
Bosnia & Herzegovina 0
Uzbekistan 4
Japan 20
Bulgaria 2
Spain 12
Monaco 0
Georgia 1
Montenegro 0
Kyrgyzstan 2
North Korea 0
Turkmenistan 0
Albania 1
Republic of Macedonia 0
Portugal 1
Turkey 9
Azerbaijan 4
Greece 1
Armenia 2
Tajikistan 1
Iran 9
South Korea 14
Afghanistan+ 0
Iraq 1
Tunisia 0
Syria 1
Pakistan 1
Algeria 1
Morocco 4
Cyprus 2
India 8
Lebanon 0
Jordan 1
Mexico 7
Saudi Arabia 1
Egypt 1
Nepal+ 0
Bhutan+ 1
Bangladesh+ 1
Oman 1
Mauritania+ 1
Qatar 1
United Arab Emerites 2
Mali+ 6
Niger+ 0
Chad+ 0
Vietnam 6
Cuba* 1
Laos+ 0
Sudan+ 0
Philippines 2
Haiti*+ 0
Dominican Rep* 2
Yemen+ 1
Jamaica* 1
Belize* 1
Eritrea+ 0
Guatemala 2
Senegal+ 2
Honduras 1
Venezuela 0
Dominica* 0
Burkina Faso+ 0
Nicaragua 2
St Lucia* 0
Ethiopia+ 0
Gambia+ 1
Colombia 7
St Vincent & Grenadines* 0
Barbados* 0
Cambodia+ 2
Nigeria 3
Djibouti+ 0
Guinea+ 1
Grenada* 0
Benin+ 1
Trinidad & Tobago* 0
Costa Rica 1
Togo+ 1
Cent African Rep+ 1
Cote d'Ivoire 1
Panama 0
Sri Lanka 0
Palau* 0
Malaysia 1
Indonesia 6
Kenya 1
Congo, Dem Rep+ 0
Brazil 5
Brunei 0
Uganda+ 1
Equatorial Guinea+ 0
Gabon 0
Sao Tome & Principe*+ 0
Ecuador, Equator 1
Peru 7
Tanzania+ 0
Papua New Guinea* 0
Burundi+ 0
Seychelles* 0
Zambia+ 1
Malawi+ 2
Bolivia 3
Mauritius* 0
Mozambique+ 2
Madagascar+ 1
Fiji* 0
Vanuatu*+ 1
Tonga* 1
Zimbabwe 1
Namibia 0
Chile 1
Botswana 0
Paraguay 0
Argentina 1
Lesotho+ 0
Uruguay 2
0 5 10 15 20 25 30 35 40
Insert 14.16
Intangible Cultural Heritage elements per
State Party ranked by latitude, north to south, top to bottom, 2012

The statistics have been analysed to present the status of global safeguarding coverage given by the ICHC. This informs the potential extent of safeguarding under climate change impacts on intangible cultural heritage of inscribed elements and also on the national inventories of Intangible Cultural Heritage that member states are committed to compile under the Convention.

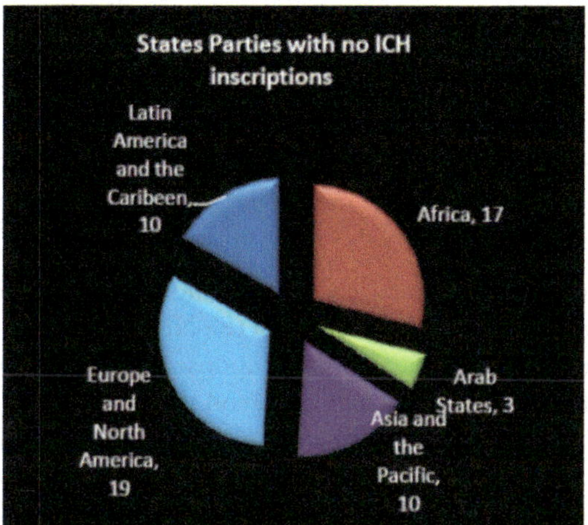

Insert 14.20 Regional distribution of States Parties with no inscriptions, 2012

In summary, the statistics reveal that the Convention is safeguarding a growing body of intangible cultural heritage from an improving geographic representation of the world's nations but that there exists the possibility that the most at risk heritage in the tropical poorer nations of Africa and the Pacific is yet to be internationally safeguarded.

Recommendations

14.5 Continue to initiate programmes that encourage ratification of the ICHC; nominations from 'developing' States Parties which have one inscription or no inscriptions; and promotion of the opportunities for multinational, serial and trans-boundary inscriptions ....as a 'race against time'.

14.6 Propose addressing marginality by formalising a strategy to address imbalances within and between the 'Lists'. This should take into consideration human development, climate change vulnerability, size, geographic placement, remoteness, (plays a role in multinational nominations), educational attainment and information technology capacity. See Chapters 15 and 19 for further information.

14.7 Ensure UNESCO continues capacity building programmes for disadvantaged States Parties. This should assist listing which is a proven safeguarding action, and in time may prove to be a powerful protective action against the impacts of climate change.

14.8 Continue to provide technical support and expertise to assist ratification, implementation and translation of the ICHC into national policies and guidelines.

14.9 Use the language(s) native to the State Party in assisting the State Party to implement the ICHC.

14.10 Consider instigating a **'Tentative Safeguarding List'** for intangible heritage to provide easier access to formal listing for inexperienced States Parties.

14.11 Promote Cherif Khaznadar's recommendation of "tying the nomination of an element from a country of the North for inclusion in the Representative List to that country's sponsorship of the inclusion of an element from a country of the South on the Urgent Safeguarding List" (Khaznadar 2009). This 'tied' relationship may lead to greater empathy with and responsibility for the insecurity of intangible cultural heritage of climate change vulnerable States Parties.

14.12 Propose making universal ratification by the 10th anniversary of the ICHC in 2013 a goal.

14.13 Include programmes to encourage the nomination of ICH elements in the UNESCO longer term strategies such as 'Priority Africa', SIDS and Pacific Century Premium Development (PCPD) countries.

14.14 Encourage States Parties to adopt the methodologies from the ICHC's Best Practices List.

The late Syed Ali Shan, photographer and well known identity, outside his shop in Leh. Ali is wearing a Goncha, a traditional Ladakhi men's garment. He sold his own black and white photographs of Leh life. His photographs spanned back decades and provided an invaluable historic resource. His photographic legacy has yet to be safeguarded. With his passing the shop was shut and his photographs became unavailable

Chapter 15

How do the Conventions Address the Physical and Biological Impacts of Climate Change on Cultural Heritage?

We can talk endlessly, but with limited results, about how climate change is severely threatening biological diversity worldwide. However, if we point out that a World Heritage site is under extreme pressure from climate change and may be irreversibly damaged, causing loss in biodiversity at the same time, our message and call to action may be heard more clearly........the World Heritage Convention can contribute a voice that may break through the din when others cannot." Ahmed Djoghlaf, Executive Secretary, Convention on Biological Diversity, April 2008 (ACJP, CANA, and Australia 2008)

To what extent are World Heritage cultural properties and Intangible Cultural Heritage elements protected and safeguarded by the World Heritage Convention and the Intangible Cultural Heritage Convention from the physical and biological impacts of climate change?
The major physical and biological impacts of climate change are listed in column 2 of Insert 3.1 and 6.1 ***'Diagrammatic summary of the relationships between climate change and cultural heritage'****.*
Impacts include temperature rise and increased humidity; extreme weather events (storms, tropical cyclones, floods, storm surges, coastal erosion, drought, desertification, heatwaves, bushfires); fresh water stress (shortage , contamination); poleward shift of climate zones; acidification of oceans; extinctions including coral bleaching.

The World Heritage Convention

Assessing the Problems

The **World Heritage Convention (WHC)** text makes no mention of climate change. Article 11.4 states "changes in water level, floods and tidal waves" are "serious and specific dangers" that would threaten World Heritage properties and qualify them for special assistance. Such properties are eligible for entry on the **List of World Heritage in Danger**. Although the "serious and specific danger" may be ultimately caused in part by anthropogenic climate change, the site would not be listed as linked with this cause. Prior to 2004 UNESCO had received warnings about climate change induced damage to World Heritage properties from the **Intergovernmental Panel on Climate Change (IPCC)** and from traditional owners (UNESCO 1999). In 2004 (LAW 2004), 2006, 2007, (CANA et al. 2007) 2008 (ACIP 2008) and 2010 (ACJP and Earthjustice 2009; Climate Justice 2011) UNESCO received, in all, 8 petitions from over 40 groups and individuals expressing concern about the threats to World Heritage sites from climate change. Chapter 26 gives a more detailed legal history of these and the sequence of responses and contested repercussions that followed each. Petitioners urged UNESCO to give leadership in offering greater protection to the world's heritage from the physical threats of climate change.

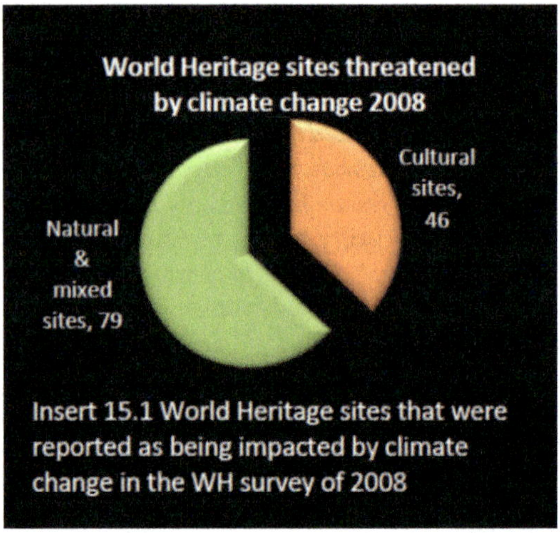

Insert 15.1 World Heritage sites that were reported as being impacted by climate change in the WH survey of 2008

As a result of the initial petitions, the UNESCO World Heritage Committee at its 29th session noted that "the impacts of climate change are affecting many and are likely to affect many more World Heritage properties, both natural and cultural in the years to come". UNESCO initiated a questionnaire, a **Survey of States Parties,** which was sent to States Parties in

2005 to assess the extent and nature of climate change impacts on World Heritage properties and the actions taken. 83 States Parties replied. 72% acknowledged climate change impacts. 125 World Heritage sites were being threatened of which 46 were cultural (Insert 15.1). The nature of the impacts to the cultural properties was physical. See Insert 15.2 for a breakdown.

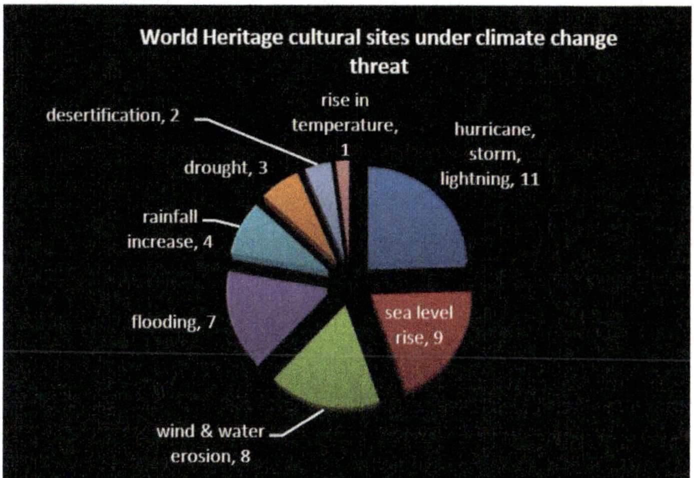

Insert 15.2 Breakdown of climate change impacts as reported in the WH 2005 survey

Strategies and Policy

UNESCO organised 'a meeting of experts' in March 2006. Over 50 representatives from the States Parties, various international organizations, NGOs, the advisory bodies to the World Heritage Committee, academic and scientific experts convened to discuss current and future impacts of climate change on World Heritage sites. The outcome was a clear direction to develop and implement appropriate management responses to protect World Heritage in the face of climate change.

Through the interpretation of Articles 4, 5 and 29, the Convention became a vehicle for climate change action. Article 4 offers the promise of protection as it legally binds each State Party to recognise its duty "of ensuring ...., protection ... and transmission to future generations of the cultural and natural heritage..." "referred to in Articles 1 and 2" and "situated on its territory". A State Party "will do all it can to this end, to the utmost of its own resources and, where appropriate, with any international assistance and co-operation, in particular, financial, artistic, scientific and technical, which it may be able to obtain."

Article 5 binds agreement from a State Party to protect its national heritage, including World Heritage properties, by adopting "a general policy document.", "to integrate the protection of that heritage into comprehensive planning programmes;" "to set up...services... for the protection...", "to develop ... studies and research and to work out such operating methods as will make the State capable of counteracting the dangers that threaten its cultural or natural heritage", "to take the appropriate legal, scientific, technical, administrative and financial measures necessary for the identification, protection, conservation, presentation and rehabilitation of this heritage; and to foster the establishment or development of national or regional centres for training in the protection, conservation and presentation of the cultural and natural heritage and to encourage scientific research in this field."

Article 29 commits States Parties to regularly report on World Heritage listed properties to the World Heritage Committee.

Thus Article 5 enables climate change policy and plans to be introduced and Article 29 enables climate change feedback in the form of the prescribed processes already formulated in WHC's **Operational Guidelines,** namely 'reactive monitoring' and 'periodic World Heritage Reporting'. By using these opportunities UNESCO is able to develop important short and medium term climate change impact protection measures and influence States Parties to introduce similar measures to protect national and regional cultural and natural heritage.

Following the 2006 Meeting of experts, the World Heritage Committee produced 5 publications: a report titled *Predicting and Managing the Effects of Climate Change on World Heritage,* two World Heritage *Strategies*, and later, following a working group meeting, a *Policy Document on the Impacts of Climate Change on World Heritage Properties* and a book – *Case Studies on Climate Change and World Heritage.* They are outlined here to give a sense of the potential of the WHC to respond to a full scale challenge. They are placed in a seven year World Heritage legal context in Chapter 26.

The 55 page report *Predicting and Managing the Effects of Climate Change on World Heritage* (UNESCO 2006a) is a foundation document.It presents a detailed analysis of climate change threats to World Heritage and discusses some of the preventive and corrective actions that are possible and the importance of sharing information and knowledge related to climate. It acknowledges shortcomings especially in the addressing of human impacts of climate change (that is *the impacts of climate change affected humans*) on cultural heritage. The report recognises that little is known about how climate change affects social structures and habitat and how it "could lead to changes in, or even the migration of, societies that are currently sustaining World Heritage sites." Under the heading 'Adaptation,' paragraph 120 reports: "However in the context of enhanced desertification, abandonment of cultural heritage must be anticipated. Although the relative importance of climatic and anthropogenic factors as a cause of desertification remains unresolved, evidence shows that an increase in dust storms would result in damage to settlements and infrastructure, and will affect human *health* and population *migration*. Thus, the impact on cultural heritage could range from erosion of physical structures to the break-up of the societies and communities supporting World Heritage sites or even *abandonment*, with the eventual *loss of cultural memory*", (UNESCO 2006a p53). *Predicting and Managing the Effects of Climate Change on World Heritage* also notes that "more needs to be done on monitoring, research and maintenance of cultural heritage."

A 5 page document titled ***A Strategy to Assist States Parties to Implement Appropriate Management Responses***, (UNESCO 2006g) was published at the same time as *Predicting and Managing the Effects of Climate Change on World Heritage*. The two documents were combined and published as World Heritage Reports 22, *Climate Change and World Heritage* in 2007 (UNESCO 2007c).

A Strategy to Assist States Parties to Implement Appropriate Management Responses, derives from the report *Predicting and Managing the Effects of Climate Change on World Heritage* to which it refers for detailed guidance. It proposes a process of preventative, adaptive and knowledge sharing measures that should be considered when preparing to implement preventive and/or corrective management responses to deal with threats to the *physical and biological* deterioration of World Heritage sites, monuments and landscapes. Reflecting on the gaps in knowledge expressed in the report *Predicting and Managing the Effects of Climate Change on World Heritage*, and the *Strategy* specifies the need to develop coordinated approaches to research, with the help of expert UNESCO and other bodies. It anticipates that climate change impacts on cultural World Heritage will include disruption as a result of "changes in society such as movement of peoples, displacement of communities, their practices, and their relation with their heritage" (UNESCO 2006g p9). These issues are not explored further in the publications.

An allied document the ***Strategy for Reducing Risks from Disasters at World Heritage Sites*** (UNESCO 2006g) was endorsed and adopted in 2006. This World Heritage *Strategy*, backed by Article 11, arose from the same expert meeting as did the report *Predicting and Managing the Effects of Climate Change on World Heritage*. It is a response to the need for a World Heritage climate change strategy for short term, more frequent or severe disasters implicitly integrating concern for climate change impacts in its provisions. It is referred to in the WHC *Operational Guidelines* paragraphs 118, 161, 162, 181, 241 and Annex 7 which deal with disaster management. It is recommended that risk preparedness be an element in all World Heritage site management plans. Importantly, this *Strategy* is applied in the World Heritage Periodic Reporting section II.5 (UNESCO 2010ff) where States Parties must report on its implementation. Cross references to it occur in the publication, *A Strategy to Assist States Parties to Implement Appropriate Management Responses,* and in the report *Predicting and Managing the Effects of Climate Change on World Heritage*.

The *Strategy for Reducing Risks from Disasters at World Heritage Sites* operates through objectives, actions and a 'by whom' approach. The first action it emphasises is the strengthening of support at different levels for reducing risk. It then builds a culture of disaster prevention; identifying, assessing and monitoring disaster risks; reducing underlying risk factors and strengthening disaster preparedness. It is salutary to note that the accelerating nature of climate change was not recognised then, as referred to in Chapter 3, to contribute to the "increasing number of natural disasters affecting World Heritage properties". The World Heritage Committee's request for the preparation of a World Heritage progress report (UNESCO 2009cc) on the implementation of the *Strategy for Reducing Risks from Disasters at World Heritage Sites* was a submission at UNESCO's 34th session in 2010 (UNESCO 2010r). In the same year the World Heritage Centre launched its first manual, a 69 page online publication titled ***Managing Disaster Risks for World Heritage*** authored by the World Heritage Centre, International Centre for the Study of the Preservation and Restoration of Cultural Property (ICCROM), International Council on Monuments and Sites (ICOMOS) and International Union for the Conservation of Nature (IUCN). It was written in response to the need for more concentrated training and capacity development in **disaster risk management (DRM),** a need clearly evident in the feedback data from Periodic Reports and from meetings of experts.

The 12 page World Heritage ***Policy Document on the Impacts of Climate change on World Heritage Properties*** (UNESCO 2008a) was adopted and published in 2007. It provides policy makers with guidance on research needs, legal issues and synergies with international conventions and organisations including the United Nations Framework Convention on Climate Change (UNFCCC), IPCC and various NGOs. It states that "priority in all climate change related actions.....will be given to properties in Africa and Small Island Developing States (SIDS)" (UNESCO 2008a p4). It anticipates some properties being able to sequestrate carbon and adopt carbon offsets. Of these climate related publications, the *Policy Document on the Impacts of Climate Change on World Heritage Properties* has generated the most criticism as detailed in Chapter 26.

The enthusiastically received and well illustrated book ***Case Studies on Climate Change and World Heritage*** was also published in 2007 (Colette 2007). It is now translated into French and Spanish and is in its third English printing. 26 case studies of 18 natural and 8 cultural sites are summarised. Each describes the impacts of climate change that have already been observed and those which can be anticipated in the future. On site, ongoing and planned physical adaptation measures are explained.

Implementation: Operational Guidelines, Reporting and Monitoring

The ***Operational Guidelines*** for the Implementation of the WHC are the means by which the World Heritage Committee carries out the *Policy Document on the Impacts of Climate Change on World Heritage Properties* internationally, and to various extents, nationally and regionally, including climate change management and actions arising from its report *Predicting and Managing the Effects of Climate Change on World Heritage and Strategies.* The 163 page *Operational Guidelines* elaborate the Convention: defining world heritage; describing the criteria for the assessment of **Outstanding Universal Value (OUV)**, integrity and authenticity. As a result of the aforementioned publications, the *Operational Guidelines* were updated to include as part of the nomination process, climate change impact protection, management, and monitoring requirements. All new nominations have to answer questions about climate change in the Nomination Form 4b (ii) under "Factors affecting the property – Environmental pressures (e.g. *climate change*, desertification, pollution)". Boundaries, buffer zones and extensions to a property are factors in protection insurance that are included in the *Operational Guidelines.* These factors may prove of additional value under climate change effects. Of further relevance to implementing climate change protection, the *Operational Guidelines* detail the form of feedback required when reporting on climate change management implementation. The reporting forms are **Reactive Monitoring, Reinforced Monitoring** and **Periodic World Heritage Reporting**.

Reactive Monitoring refers to the World Heritage reporting by the World Heritage Centre, other sectors of UNESCO, and the advisory bodies to the Committee, on the state of conservation of World Heritage properties that are under threat. It draws on specific World Heritage reports and impact studies requested from States Parties each time exceptional circumstances occur or work is undertaken which may have an effect on the state of conservation of the property. It is used when considering inscriptions on the List of World Heritage in Danger and when considering deletions of properties from the List. The 2011 revision of the *Operational Guidelines* further strengthened the protection of World Heritage properties by specifying that the OUV of a property be regularly monitored. Paragraph 96 reads: "Protection and management of World Heritage properties should ensure that their OUV, *including* the conditions of integrity and/or authenticity at the time of inscription are sustained or enhanced *over time. A regular review of the general state of conservation of properties, and thus also their OUV, shall be done within a framework of monitoring processes for World Heritage properties, as specified within the Operational Guidelines*". Reactive monitoring is the referred 'monitoring process". The amendments are in italics.

There are two reactive monitoring report formats, one for properties on the List of World Heritage in Danger, one for inscribed properties on the World Heritage List. Both ask for responses to only three inquiry areas. In 2009 the World Heritage Committee acknowledged "the increasing number of State of conservation Reports" (UNESCO 2009cc p14), (partly driven by climate change related factors) and stated that reviewing these reports was a key tool for ensuring the effective conservation and credibility of World Heritage. The Committee had earlier requested a summary of the trends, changes and threats based on an analytical summary of state of conservation reports received over the previous 5 years. This summary was discussed at the 34th session of the World Heritage Committee in 2010, with a view to making recommendations for prioritizing management efforts in the context of the Global World Heritage Strategy (Decision 33COM 7C) (UNESCO 2009cc). The 2010 decision arising out of the summary document titled ***Reflection on the trends of the state of conservation*** notes the potential of satellite imagery and other remote sensing techniques for providing evidence over time to determine whether some impacts on World Heritage values continue to occur or are being addressed (Decision: 34 COM 7C) (UNESCO 2010r).

The ***Progress Report*** on the implementation of the *Strategy for Reducing Risks from Disasters at World Heritage Sites* (Document WHC-10/34.COM/7.3), as well as the new ***Managing Disaster Risks for World Heritage* manual** also considered the feasibility of improved utilization of remote sensing (UNESCO et al. 2010). These moves are indicative of the World Heritage Committee's openness to adopting new ways of achieving its "global lead responsibilities" (UNESCO 2008c) as manager and guardian of its universal lists of World Heritage and Intangible Cultural Heritage.

Reinforced Monitoring is activated when information is needed on a more frequent, even urgent, systematic and proactive basis to protect the integrity or authenticity of a site. It strengthens existing monitoring mechanisms, bringing required information to the attention of the members of the Committee in the period between two sessions.

Periodic World Heritage Reporting (UNESCO 2010ff) is an intensive and extensive exercise. It refers to the preparation of a World Heritage report for the World Heritage Committee to be written every 6 years by a State Party including on 'the state of conservation' of the World Heritage properties. This analytic Periodic World Heritage Reporting process is the vehicle for climate change assessment, climate change monitoring, regional co-operation, exchange of information and experiences. The format of this report is sophisticated. The 330 short answer; multiple choice, ranking, and 'full-commentary-answer' questions are comprehensive and split into 2 'Sections'. There are extensive explanatory notes accompanying the questions which collect baseline and ongoing data. Section I covers the State Party's legislative and administrative provisions, policy, protection, research, training, education, information, awareness building and cooperation. Section II questions are wide ranging, seeking information on national inventories, OUV, nomination development, local collaboration, threats, monitoring and management. Physical threats are addressed in Question 3 of this Section. This is the section that allows climate change impacts to be reported. Information is sought about 'severe weather events': storms, tornadoes, hurricanes/cyclones, gales, hail damage, lightning strikes; river / stream overflows; extreme tides; flooding; drought; desertification; changes to oceanic waters; changes to water flow and circulation patterns at local, regional or global scale, changes to pH, changes to temperature, and other impacts plus information on the occurrence of fire. Further factors are listed e.g. invasive/alien species-dieback due to pathogens; invasive/alien terrestrial species, invasive / alien freshwater species and invasive / alien marine species. Section I and II develop into self administered management tools (as well as reporting vehicles) for the State Party. Building on given information, the State Party has, by the end of the Periodic Reporting process, formulated tailored priority-driven management action plans with time frames and lead agency involvement. There are 'three major management plans' that result from completion of the Periodic Reporting process. These become the working plans for the State Party, the benchmark against which the next Periodic Report can be measured and a contribution to UNESCO's data bank.

UNESCO CLIMATE CHANGE INITIATIVE:

Core Program:

Climate Change, Culture and Biological Diversity and Cultural Heritage:

Field Observatory

"Besides its impacts on biodiversity and natural heritage, climate change also variously impacts the world's cultural heritage. Archaeological remains and historical buildings can be affected when the hydrological, chemical and biological processes of the soil change in response to climate change. The predicted increase in flooding which may damage building materials, desertification, salt weathering and erosion are already threatening cultural heritage in desert areas" (UNESCO 2009j).

Insert 15.3

As mentioned, UNESCO has requested that the next IPCC Assessment include a chapter on climate change and World Heritage. In 2007 the IPCC4 draft report called on the Australian Government to identify how it can reduce the nation's vulnerability to climate change, especially in 'hot spot' regions like the Murray-Darling Basin, south-east Queensland, Kakadu, the Queensland wet tropics, the Snowy Mountains and the drought-prone south-west of Western Australia. Journalists Marion Wilkinson and Deborah Smith reported that "The draft summary noted that many of these hot spots include World Heritage sites. The Australian Federal Government had this reference cut during the process where every line of the report's summary for policy makers is debated" (Wilkinson and Smith. 2007). Unfortunately this is reflective of how some governments are prepared to censor IPCC reports in order to deflect criticism and shirk responsibility.

Examination of the Periodic World Heritage reports could offer much information for consideration in the proposed IPCC 5th *Assessment World Heritage Report (2014)*.

Undertaking Periodic World Heritage Reporting provides on the job-management training in and an education about protection mechanisms. Even when States Parties experience severe internal problems such as happened in the Solomon Islands in 2003, their partly completed periodic World Heritage report was still useful for setting benchmarks and for providing planning frameworks.

Also pertinent to climate change impact protection for cultural heritage are the *Operational Guidelines* implementation strategies for "encouraging support for the WHC", namely capacity building, awareness-raising and education. Additionally the *Operational Guidelines* explain the role of the **World Heritage Fund** and how to apply for financial and expert advisory support. The World Heritage Committee strongly advocates the adoption and dissemination of its climate change publications to all States Parties and the public at large.

Addressing Climate Change

The most specific climate change related decision made by the World Heritage Committee was **Decision 30COM 7.1** made in 2006 (UNESCO 2006e). It states the World Heritage Committee "Considers that the decisions to include properties on the List of World Heritage in Danger because of threats resulting from climate change are to be made by the World Heritage Committee, on *a case-by-case basis,* in consultation and cooperation with the State Party, taking into account the input from Advisory Bodies and NGOs, and consistent with the *Operational Guidelines* for the Implementation of the WHC." This contentious issue is addressed in Chapter 26. The latest change to *update* the WHC in relation to climate change impacts has been made in the *Operational Guidelines* which were changed to now read: "the Committee may inscribe a property on The List of World Heritage in Danger when certain requirements are met including the potential danger of threatening impacts of climatic... factors." This phrase replaced "gradual changes due to ...climatic... factors" **(Decision 32 COM7A.32)** (UNESCO 2008e). More details are tabled in Chapter 26.

The World Heritage protective measures outlined above do not address climate change as a global challenge, that is, as an emissions driven phenomenon requiring universal action. They offer adaptive and preventive measures to mainly physical and biological threats that can be taken at the local 'World Heritage site' scale. In 2008 UNESCO began incorporating climate change into its **Medium term 2008 -2013 Strategy**, but again only at a site-based level. "Specifically, UNESCO will draw on its extensive list of sites, including World Heritage sites ... for global monitoring of climate change, biodiversity loss and sustainable development" (UNESCO 2008c).

UNESCO's Climate Change Initiative

The following year the **UNESCO Climate Change Initiative** was launched by the Director-General of UNESCO, Ms Irina Bokova to reinforce the scientific, mitigation and adaptation capacities of countries and communities that are most vulnerable to the effects of climate change, effects that are clearly stated: "Climate change is increasingly recognized as the driver of biodiversity change with the most rapid increase in impacts and related cascading effects on human livelihoods" (see Insert 15.3) (UNESCO 2009j).

The Initiative brings together – science, education, ecology and ethics to address climate change. Through one of the four Core Programs, **Climate Change, Culture and Biological Diversity and Cultural Heritage, a Field Observatory** has been set up to use UNESCO "World Heritage Sites ... as priority reference sites for understanding the impacts of climate change on human societies and cultural diversity, biodiversity and ecosystems services, the world's natural and cultural heritage and the possible adaptation and mitigation strategies" (UNESCO 2009j) see Insert 15.4.

UNESCO CLIMATE CHANGE INITIATIVE: Core Program:
Climate Change, Culture and Biological Diversity and Cultural Heritage:
Development of a Global Climate Change
Field Observatory of UNESCO Sites

The objective of this Observatory is to use UNESCO World Heritage Sites and biosphere reserves as priority reference sites for understanding the impacts of climate change on human societies and cultural diversity, biodiversity and ecosystems services, the world's natural and cultural heritage and the possible adaptation and mitigation strategies.

The main objectives of the Field Observatory are the following:

Promote the recognition and use of biosphere reserves and World Heritage sites as priority sites for implementing the UNESCO Climate Change Initiative and other UN-level climate change initiatives. Collect, document and analyse scientific and traditional and local knowledge of both women and men on impacts of climate change on biodiversity and ecosystem services and human /community responses to adapt to these impacts.

Promote skills and competencies in cultural industries linked to films, computer animation and products, photographic and art festivals and events as well as other public/media events to tell climate change adaptation and mitigation linked stories drawn from science, religion, traditional knowledge, etc (UNESCO 2009j).

Insert 15.4

Of upmost importance and relevance to the World Heritage protective (including potentially climate change protective) juggernaut of a process (namely: ratifying, nominating, complying, implementing, policy-making, assessing, adapting, monitoring, reporting), is that States Parties are encouraged to apply for

- funding from the World Heritage Fund to assist at all stages
- skill and knowledge back-up from experienced experts through UNESCO

Limitations to Protection

Describing this five year long path of climate change recognition in World Heritage operations is worthy. It demonstrates the ongoing responsiveness, though skewed to local physical impacts, and criticised by some as inadequate (see Chapter 26), of this Convention to going part-way in protecting World Heritage against climate change impacts. It allows comparison with the ICHC. The determination of the World Heritage Committee to monitor and upgrade its climate change protection through its *Operational Guidelines* and through its free and promoted publications is evident.

Climate change is accelerating and the 6 year cycle of Periodic World Heritage Reporting may not be adequate for protecting cultural World Heritage sites. For example **The end Grave Site** component of the World Heritage cultural site **Chief Roi Mata's Domain, Vanuatu** (UNESCO 2008g, 2008f) is in direct threat of flooding and erosion from storm surges and sea level rise. The referees for the 2008 nomination, (Decision 32COM 8B.27), comment that climate change was an 'issue' but it was not regarded as dangerous in the short term. The 'threat' graph on the Chief Roi Mata's World Heritage website shows a near zero threat. In the light of current knowledge (see Chapter 3) it is now inevitable that the site will be severely impacted by sea level rise, storm surge and high tides.

In summary, the World Heritage Committee has explored the options offered by the WHC and its *Operational Guidelines* to put in place a foundation process of climate change accountability, albeit mostly to physical and biological impacts of climate change, for World Heritage. Its methodical feedback mechanisms and awareness-raising has led the way in helping States Parties address climate change impacts on World Heritage sites. For instance Australia has taken its first step in developing a national approach to climate change protection for World Heritage properties with the cautiously worded aforementioned 2009 publication *Implications of climate change for Australia's World Heritage properties: a preliminary assessment*. In future the World Heritage Committee's leverage, encouragement and direct assistance with documentation, and opportunities for funding and expert support will be crucial for ensuring States Parties develop *national* climate change protection policies for managing, monitoring and reporting on physical and biological impacts. Its latest support comes in the form of another online manual, ***Preparing World Heritage Nominations***, 2010. As the World Heritage List grows and evolves so does the World Heritage Committee's responses to emerging critical needs for guidance for States Parties on the implementation of the Convention.

The World Heritage Committee has been criticised for being prey to political lobbying, not prioritising expert advice over national sensitivities, diplomacy in regards to placing sites on and off the 'in Danger' list, lack of transparency in that some meetings are closed to the public, and for lacking robust means to implement policy. The most pertinent criticism relates to the Convention's lack of commitment to a strong mitigation policy at the national and international level. The resultant difficulty is that all World Heritage properties will differentially suffer the impacts of climate change which is overwhelmingly caused by powerful GHG polluting States Parties which are reluctant to take mitigating action and for which no amount of local adaptive management can ameliorate to ensure the long-term protection of the OUV. The World Heritage Committee has to tread a careful path to be effective within its resources and capitalise on its strengths within the environment of persuasion and leverage while lacking enforcement mechanisms.

The 2008 journal article *Climate Change: How Should the World Heritage Convention Respond?* by Greg Terrill shows a disappointing lack of support for needed action. The author, an Australian federal government public servant and one of the most senior international climate change negotiators, advises "While the Convention can illustrate the impacts of climate change upon World Heritage, it must avoid being seen as an interest group and taking sides on debates that properly occur under other conventions. On grounds of expertise alone, the World Heritage Convention is well advised to avoid moving from describing problems and possible adaptation responses at World Heritage Sites to advocating particular mitigation levels or approaches" (Terrill 2008 p397). The WHC is judged as working upon a presumption of relative stability and manageable change and is not suited to addressing broad, interdisciplinary matters beyond world heritage. Nor does he advise the World Heritage Committee to support climate change research as, he observes, it lacks enough research-based and monetary resources.

- Understanding the vulnerability of materials (indoor, outdoor, buried) to climate variables (for example, particularly too much or little moisture effects).
- Understanding how traditional materials and practices need to adapt to extreme weather events and a changing climate.
- Development of fail-safe methods and technologies for monitoring the impact of climate change at properties.
- Modelling and projecting climate behaviour,
- Managing cultural heritage
- Preventing damage.
- Understanding climate change impacts causing changes in society i.e. movement of peoples, displacement of communities, their practices, livelihoods and their relation with their heritage (UNESCO 2008a p10).

Insert 15.5 Climate Change and Cultural Sites Research Priorities

Terrill recognises the development of the *Policy Document on the Impacts of Climate Change on World Heritage Properties* on climate change but is concerned about the continuing lack of a declared role for the WHC on climate change and the lack of climate change knowledge possessed by participants of World Heritage working and expert advisory groups. He emphasises the global lack of examination of the impacts of climate change upon built heritage, overlooking the advocacy by the International Council on Monuments and Sites (ICOMOS) in Australia. The author concludes that "The Convention has an interest in seeing that emissions mitigation occurs, but it has no mandate or expertise to become involved in decisions about how, where, when or what sort of mitigation. In relation to mitigation, its role does not extend beyond registering its interest. The mandate and the expertise for any further level of detail reside in other conventions, particularly the UNFCCC. Instead, the key area where the World Heritage Convention needs to be active is in ensuring that sites are as well equipped as possible to adapt to the inevitable impacts of climate change" (Terrill 2008 p400).

It is indicative of the divisive stance that this issue has generated that the preceding 2007 report *Impacts of climate change on Australia's World Heritage properties and their values* by Lance Heath (referred to in Chapter 5) specifically states that we must *mitigate or lose* World Heritage sites, specifically stating "Mitigation involves the intervention by humans to reduce atmospheric greenhouse gas concentrations through the creation of carbon sinks (e.g. forest plantations) and by reducing emissions through the use of sustainable technologies and renewable energy sources."

The former Director General of UNESCO, Mr Matsuura has a broad vision, "But aside from these principal physical threats, climate change will also have tremendous impact on social and cultural aspects, with communities changing the way they live, work, worship and socialize in buildings, sites and landscapes, possibly leading to migration and the abandonment of their built heritage altogether. The fact that climate change poses a threat to the OUVs of World Heritage sites has several implications for the implementation and monitoring of the 1972 WHC. Lessons learnt at several sites worldwide show the relevance of designing and implementing appropriate adaptations measures. Research at all levels would also have to be promoted in collaboration with the various bodies involved in climate change work, *especially for cultural heritage* where the level of involvement of the scientific community should be reinforced. The global network of the World Heritage sites is ideally suited to build public and political support through *improved information dissemination* and *effective communication on the subject*, given the high profile nature of these sites. Similarly, the 2003 UNESCO Convention for the Safeguarding of the Intangible Cultural Heritage, while underlining the deep-seated interdependence between the preservation of the world's tangible and intangible heritage, pays close attention to knowledge and practices concerning nature and the universe..." (Matsuura 2006).

The present Director-General of UNESCO, Irina Bokova seemed to have retreated to a more conservative stance as regards the potential of World Heritage for multilevel broad scale climate change advocacy. She said in Copenhagen in 2009 "Thanks to its interdisciplinary capacities, UNESCO can render a unique contribution to mitigation and adaptation to climate change through distinct action in education, the sciences, culture, communication and information. ....UNESCO will assist Member States to harness the iconic values of World Heritage sites and Biosphere Reserves to showcase adaptation and conservation measures" (UNESCO 2009n). However, in the lead up to Rio+20, the UN's Conference on sustainable development, UNESCO has taken a determined position that humanity must break with past development practices. While not describing any radical protective initiatives for climate change threatened World Heritage properties, Ms Bokova asserts in the report ***From green economies to green societies. UNESCO's commitment to sustainable development*** that "We need a new way forward"; that "science holds many of the answers..." and so must drive the transition to a global green society. Further she claims "We need a change of culture to tackle climate change" (UNESCO 2011e) heralding perhaps a rethinking that could involve empowering the WHC and ICHC anew as protective legal instruments.

Recommendations

15.1 Urgently adopt **Resolution 16 GA 10 A**. This resolution "Requests the World Heritage Committee to institute a mechanism for the World Heritage Centre and the Advisory Bodies to periodically review and update the *Policy document on the impacts of climate change on World Heritage properties*, and other related documents, so as to make available ***the most current knowledge and technology*** on the subject to guide the decisions and actions of the World Heritage community" (UNESCO 2007d). This adoption should facilitate the imperative of making climate change action a major factor at all levels of World Heritage management.

15.2 **Strengthen and broaden the scope of the *Policy Document on the Impacts of Climate Change on World Heritage Properties*** by adopting Resolution 16 GA 10A. The policy should address predicted climate change impacts on World Heritage over the short, medium and long term and include positions on mitigation, adaption, climate justice and ethics. It should also advise on amendments to the Organisational Guidelines in order to implement the 'revised' Policy document especially in relation to the criteria for listing of properties on the **List of World Heritage in Danger**.

15.3 Propose developing a **climate change vulnerability scale** with a matching set of **climate change crisis symbols**. Each symbol should indicate a level of threat on a climate change vulnerability scale. The climate change crisis symbol could be displayed alongside the World Heritage symbol at each heritage site. In each case the climate change crisis symbol could be accompanied by an explanatory statement about the nature and magnitude of the climate change risk threatening that particular site. Included in the statement could be predictive estimates of the likely condition of the site at set intervals into the future under varying greenhouse gas emission scenarios. The aim would be to raise awareness of the link between ineffective greenhouse gas emissions control and the survival of the OUV of World Heritage Sites

15.4 Propose launching a worldwide campaign – a **heritage site vulnerability campaign** employing the climate change vulnerability scale and climate change crisis symbols - to introduce the 'reviewed and updated' **World Heritage position** (refer 15.1) on climate change.

15.5 Propose developing an on-line facility, a **World Heritage and Intangible Cultural Heritage Climate Change Database and Network, (WH:ICH:CC Database & Network)** something analogous to a combination of the UNFCCC's CC:iNet and TT:CLEAR (Chapter 12), which like CC:iNet could serves as a clearinghouse for **climate change information** but in relation to both natural and cultural heritage (tangible and intangible) and be inclusive of impacts on custodians. Included should be regional and international topics, such as mitigation and adaptation and links to related information sources from organisations such as the **UNESCO Climate Change Adaptation Forum (UCCAF)**, UNFCCC, IPCC, ICOMOS, ICCROM and **United Nations Development Programme (UNDP).** Also as with CC:iNet it could assist users to link to potential partners to gain rapid and easy access to ideas, strategies, contacts, experts and materials that can be used to facilitate effective action on climate change in relation to both natural and cultural heritage. Included it should also have, like TT:CLEAR, a **searchable database that accepts and integrates multiple input variables to provide a multifaceted response concerning more demanding enquiries.**

Ideally the network should link to World Heritage site-specific information collected through reporting procedures and thus have access to all available information on all aspects of every site including that relating to climate change such as sea level rise and frequency of extreme weather events. It should attempt to satisfy all or most of the information needs of site managers, policy makers, scientists, NGOs, CBOs, volunteers and the public.

Innovations may include a searchable and interactive web page access structured like Wikipedia. The website should allow online contributions from those with an interest in the heritage site such as advisers and traditional owners. Site specific information could include climate change vulnerabilities, projected regional climate and weather expectations related to the site, current and planned mitigation and adaptation measure as well as sustainable development, particularly sustainable tourism statistics.

The Network could link to the UNFCCC web services **Technology Transfer Clearing House (TT: CLEAR)** and to the **Climate Change Information Network (CC: iNet)** as addressed further in Chapter 23.

15.6 Review the Nairobi Work Programme's 'coping strategies' database portal to assess the applicability of such an approach being incorporated into the **WH:ICH:CC Database & Network.**

15.7 Consider revising the six year Periodic Reporting regime to a four year simplified process including **mandatory site specific climate change reporting**. This may help facilitate the monitoring of sites such as the threatened Chief Roi Mata's Domain and other threatened cultural sites and give the plight of such sites regular publicity.

15.8 Encourage the World Heritage Committee to **review** and support the specific **research priorities** for cultural heritage as described in the *Policy Document on the Impacts of Climate Change on World Heritage Properties* and listed in insert 15.5.

An elderly Amchi from Leh. Amchis practice traditional Tibetan medicine. They co-operate well with medical practitioners trained in Western medicine and play a vital role in healthcare particularly in the more remote areas of Ladakh

The Intangible Cultural Heritage Convention

An Emerging Problem

UNESO staff understands the multidisciplinary nature of climate change impacts. The **Climate Change Initiative** (UNESCO 2009n, 2009j) is far reaching and brings a world-view perspective to climate change. The threat of losing intangible cultural heritage is acknowledged (see Insert 15.4). It aims to address this potential loss through the **Global Climate Change Field Observatory** of UNESCO Sites (see Insert 15.3) the objectives of which include collecting, documenting and analysing intangible heritage, including "traditional and local knowledge of both women and men on the impacts of climate change on biodiversity and ecosystem services and human/community responses to adapt to these impacts" (UNESCO 2009c). The other reference to intangible heritage is the Observatory's intent to "promote skills and competencies in cultural industries linked to films, computer animation and products, photographic and art festivals and events as well as other public/media events to tell climate change adaptation and mitigation linked stories drawn from science, religion, traditional knowledge, etc." The Initiative provides an overarching perspective which will materialise in new directives for the inter-collaborative work of many UNESCO bodies and trickle down into policy and plans for climate change safeguarding of intangible heritage.

The ICHC does not specifically recognise the physical impacts or the 'human impacts' of climate change on a community's intangible cultural heritage. The Convention's 'safeguarding' measures do not address climate change impacts as such. The Convention's implementation tools – the ***Operational Directives*** and **Reports** (*Report on the Status of an Element inscribed on the List of Intangible Cultural Heritage in need of Urgent Safeguarding* and *Report on the Implementation of the Convention and on the Status of Elements inscribed on the Representative List of the Intangible Cultural Heritage of Humanity*) do not specify climate change induced physical or human impacts nor request any data about such threats or potential threats.

The ICHC is pivotal as the leading instrument together with the WHC, for protecting the culture and heritage of humankind from harm and damage. Presently the world's tangible heritage, through the WHC, has specific climate change proactive and reactive processes in place. The ICHC does not.

By interpretation, the ICHC's most responsive capacity for safeguarding against climate change resides in its ability to encourage States Parties, which have identified their intangible cultural heritage "whose viability is endangered despite the efforts of the community", to nominate for inscription on the **List of Intangible Cultural Heritage in Need of Urgent Safeguarding**. The Nomination has 28 sections to be answered. The nominated 'element' must meet *all* of the criteria U1 –U6 as outlined in the *Operational Directives* (See Chapter 2). The criteria require the State Party to describe how the viability of the intangible cultural heritage is "at risk", or, in a more urgent situation "is in extremely urgent need ...because it is facing grave threats." "The necessary safeguarding measures that may enable the community ...to continue the practice and its transmission...." have to be explained. The nomination has to have evidence of the widest possible participation of the community and be already included in a national inventory.

The **Intergovernmental Committee for the Safeguarding of the Intangible Cultural Heritage (the Intangible Cultural Heritage Committee)** can be approached for funding and assistance in the nomination process (Form Intangible Cultural Heritage-05, 14 sections to be completed) and for funds to assist with safeguarding measures (Form Intangible Cultural Heritage-04, 27 sections to be completed). 27 elements have been listed for urgent safeguarding. None have described climate change as a threatening factor although factors such as migration, pressures on land resources, low water level and drift to urban areas may well reflect an underlying climate change contribution.

The nomination, assistance and funding application processes are too challenging for a State Party that lacks a relatively sophisticated governmental structure and is struggling economically. Fourteen States Parties have nominated the 27 inscriptions that are presently on the 'in need of Urgent Safeguarding' List. Four have very high human development, four have high human development, four have medium human development and two have low human development, the point being, that States Parties need to be able to invest considerable resources into completing successful nominations, resources that are too scarce in those States Parties that need the nominations the most. The Urgent Safeguarding List is the most important list, but as has been mentioned, it has not met with the response that clearly was anticipated, and for which the Convention was primarily designed.

The ICHC is not proactive in making prescriptive, dedicated and precautionary provisions for the impacts of climate change on intangible cultural heritage at the community or national level. Theoretically, the Convention could facilitate the trans-inscription of elements from the Representative List to the Urgent List to effect more protection to an element undergoing originally unforseen impacts of climate change.

In 1999 (pre-convention times), the Second UNESCO World Heritage Global Strategy Meeting for the Pacific Islands region in association with the Pacific Islands Museums Association (PIMA) and the Vanuatu Cultural Centre suggested far-sightedly that considering global warming "the WHC should ensure protection of the intangible heritage .....recognition and respect for..... language, oratory, ritual, song and dance" (UNESCO 1999).

Because of the geographic and 'human development' imbalance in the Representative List and a lack of interest in the Urgent List, UNESCO has made intangible cultural heritage capacity building in Africa and SIDS a Medium-Term Strategy priority for 2008-2013 (UNESCO 2008c). As a consequence people from these States Parties have been attending ICHC meetings and workshop which teach how to develop integrated heritage policies and to improve safeguarding measures. Despite this their intangible cultural heritage remains largely unprotected by the Convention through a lack of ratifications and/or nominations and unprotected through States Parties' underprivileged conditions. 39 out of the 55 States Parties with no inscriptions, have a high to severe to acute vulnerability to climate change factor and 36 of those have either a 'low' or 'medium' human development. Of the 59 nations that have not ratified the Convention, 12 lack human development data, 24 have 'low' or 'medium' human development, and of those 24, all except 1, are 'acutely ' to 'severely' to 'highly' vulnerable to climate change impacts (UNDP 2011), (DARA and Forum 2010).

In 2006, the head of UNESCO, Mr Matsuura, when giving an address about UNESCO's response to global warming, linked the ICHC with the changing climate, when echoing Article 2d, he said that climate change is connected to the ICHC as intangible cultural heritage is manifested in "knowledge and practices concerning nature and the universe" (Matsuura 2006). This does not anticipate the emerging need for the Convention to be made *more responsible* for actual policy that brings intangible cultural heritage face to face with the tsunami of threats it will suffer with climate change. What rights do communities have when their cultural links are severed as ceremonial places or raw materials for local woodcarvers are submerged in the sea or sand? How do communities look after their intangible cultural heritage when they are fragmented by displacement and disease outbreaks? How do communities retain mental resilience when their intangible cultural heritage – their monastic spaces – are now eroded by unprecedented torrential rain, not blanketed, as was the case, by snow?

Recent Intangible Cultural Heritage inscriptions have acknowledged the impact of climate change. ***The Mexican Places of memory and living traditions of the people of Tolimán: the Peña de Bernal, guardian of a sacred territory*** (UNESCO 2010kk) was inscribed, with one of the selection criteria being the adoption of safeguarding measures respecting nature in the *context of climate change.*

A second example is the urgently listed ***The Sanké Mon: collective fishing rite of the Sanké*** from Mali. The examiner's report stated "There are no physical memorials for it (the rite) and the pond or lake is the only necessary physical element that endures after the festival and even this cannot be said to have an assured existence *given climate change* and its potential impact on the water levels of the pond" (UNESCO 2009z).

> Papua New Guinea. Mali Voi. I am originally from Papua New Guinea but I now live in Samoa, South Pacific. The debates so far on climate change are centred around weather and how it is affecting plants and animals and the environment. These are very important issues and they ought to be brought to this forum. In the Pacific the issue that is not so obvious is the effect of **loss of land** due to the sea level rise. Land is very important to and for humanity without which humanity, or for that matter, all other living things cannot conduct the business of sustaining their lives. In the Pacific land is not a commodity that one buys and sells or leases for money. Not only Land is for sustaining life but it is associated with spirituality of the people. Land is the people's life and soul. A person without land or land connection is soulless. Moreover there is also loss of cultural heritage. Now land loss is happening through sea level rise. For example the case of Cartright or Cartret Island in Papua New Guinea where the people are being re-settled in Bougainville. There are other low lying atolls such as Tuvalu, Tokelau, Kiribati and others in this category. Land loss has psychological and sociological dilemma and is devastating to the affected people. It has adverse effect on self-esteem and a lack of motivation and drive to survive.
>
> As regards to loss of land, I have made this preliminary observation from belief systems of the Pacific Islands. As you know the people of the Pacific are closely associated with land.

> In the traditional practice, when one is born, one's placenta is buried on land at which a tree (usually a coconut tree or rosewood tree) was planted.
>
> As the baby grew up so too that tree. Once the baby was old enough she/he was told to look after the tree. The land on which both the tree and the human being became their common reference both physically, sociologically and spiritually,
>
> Physically the tree when matured it provided food in the case of coconut tree. You know coconut tree provides more things such as: broom, shelter, timber for house, fibre for mattresses, oil, medicine, ropes and string, fishing line, posts, but to name some and as well as food.
>
> As both the human being and the tree grew up, they both became interdependent. The human being depended on the coconut tree to provide food direct contribution and the coconut tree depended on the human being to keep it safe from fire.
>
> The entity (the site, coconut and human life) became social reference for this member of the community a place in it to participate. People in the community say, "This is, for example Mali's tree", which means Mali is a full member of the community in which he has every right to participate in the social, cultural, political and economic life of the community.
>
> In the case of the rosewood tree, it signalled the periodic seasonal changes between winter and spring. The second function was when it was big enough it would be felled and made into a canoe. Both functions signalling seasonal changes and transportation i.e. are but symbols of communication.
>
> Land is both life bearing and life giving and parts of it were declared sacred for the places of abode of the dead, on whose guidance for the living to conduct life. Once land is removed or destroyed neither what had been briefly described above nor the sacred abode of the dead exist. The above is a general observation. What could be considered as a possible project is to conduct impact of re-settlement of the Carteret Islanders or similar situations elsewhere. How and what do they feel of loss of land?
>
> I have viewed interviews conducted by TV Samoa screened some two weeks ago on a number of people in Tokelau on the same topic.
>
> Some of them said they will look to their aiga or relatives in Samoa for re-settlement. They still, however, feel the loss of land with "psychological pain." How do we measure such pain is yet another question? What of the other Tokelauans who have no aiga in Samoa? How and what do they feel when they are disposed of land by sea level rise. (UNESCO 2010j)

Insert 15.6

Mali Voi's story from *On the Frontlines of Climate Change* UNESCO's online global forum (UNESCO 2010j) , most dramatically describes the crippling repercussions of climate instability on traditional knowledge relating to farming and loss of cultural identity through loss of land (Insert 15.6).

While Mali's experience and the many other 'internet reported' stories are not candidates for ICHC Lists (yet) they are indicative of the emerging status of vulnerable intangible cultural heritage that the convention is mandated to safeguard.

How can the ICHC Better Fulfil its Role in Relation to Climate Change?

Although not specifically addressing climate change the ICHC is receptive to preventing and/or addressing events that may arise as a consequence of climate change. The Preamble sets the context, "recognizing...the processes of... social transformation ... give rise... to grave threats of deterioration, disappearance and destruction of the intangible cultural heritage, in particular to a lack of resources for safeguarding such heritage" (Paragraph 4). It describes "the far-reaching impact of the activities of UNESCO in establishing normative instruments for the protection of the cultural heritage, in particular the Convention for the Protection of the World Cultural and Natural Heritage of 1972" (WHC). In referring to human rights instruments (paragraph 1); to aforementioned threats to intangible cultural heritage (paragraph 4); to the global influence of the WHC (paragraph 7) and to "bringing human beings closer together and ensuring exchange and understanding among them" (paragraph 13) it offers a good platform for interpreting the ICHC's intention of seeking to safeguard heritage using a 'rights based' approach and for using the WHC as an example for meeting the climate change challenge.

The ICHC 's purpose "to safeguard intangible cultural heritage" and "to provide for international cooperation and assistance" (Article 1) and its definition of ICH as "knowledge and practices concerning nature and the universe" and of safeguarding as meaning "...ensuring the viability...including...transmission... generation to generation ...providing a sense of identity and continuity, thus promoting respect for cultural diversity and human creativity" (Article 2), validates using the opportunities the Convention offers, to begin the process of **addressing climate change impacts.** Article 7 offers the best opportunity for informing and influencing climate change impacts: The Committee functions to "prepare ..." and *Operational Directives* for "the implementation of, this Convention;" to "provide guidance on best practices and make recommendations on measures for the safeguarding of the intangible cultural heritage". It "encourages and monitors" the implementation of the conventions objectives: to safeguard, respect, raising awareness and appreciation of and to provide for international cooperation and assistance. The ICH Committee therefore has the power to introduce strategies, guidelines, processes, plans, projects, expertise, capacity building and financial assistance as means by which climate change impacts on intangible cultural heritage may be addressed.

Articles 11, 12 and 13 "ensure" that each State Party identify, safeguard, update and inform on their intangible cultural heritage at the national and local level as a condition of accepting nominations for intangible cultural heritage inscriptions for international safeguarding. This offers the chance of obliging each State Party to be climate change prepared at national and local levels through *Operational Directives* to provide 'relevant information'.

Each State Party must adopt "a general policy aimed at"... "integrating the safeguarding"... "into planning programmes" and must "foster scientific"... "studies", "research methodologies" to safeguard "in particular the Intangible Cultural Heritage in danger" and "adopt legal, technical, administrative and financial measures aimed at fostering ....training" and "establishing documentation institutions." Article 14 sets the onus on States Parties to provide education, awareness-raising and capacity building programmes and to foster non formal transmission. Significantly, States Parties shall endeavour, by all appropriate means to "keep the public informed of the dangers threatening such heritage and of the activities carried out in pursuance of this Convention." Article 15 reinforces the importance of community participation in managing intangible cultural heritage. Although the Convention lacks empowerment to make and implement these directives at the national and community level it does *exert influence through the pledge* member Parties make on ratification.

At the International level Articles 16, 17 and 18 commit States Parties to nominate intangible cultural heritage for inscription, and to register Intangible Cultural Heritage safeguarding programmes, practices and activities "taking into account the special needs of developing countries" so that "best practices" may be disseminated, thereby presenting *two more opportunities to introduce measures to combat* climate change.

Articles 19, 20 and 21 offer directives for international cooperation to exchange "information and experience," to grant financial assistance for "studies", "provision of experts", "training" "the elaboration of standard- setting and other measures", "infrastructures" and "equipment," to make loans and donations to "safeguard listed heritage, prepare inventories and support safeguarding programmes, projects and activities at national, subregional, and regional levels." The motivating directive to States Parties is to "recognise that the safeguarding of intangible cultural heritage is of general interest to humanity, and to that end they undertake to cooperate at the bilateral, subregional, regional and international levels." The Convention here pledges States Parties to act in the highest interests of safeguarding intangible cultural heritage and, for which, it can provide support. *These articles provide opportunities to inform and influence climate change impacts.*

Article 29 offers a direct way for climate change impact information to be both gathered and climate change impact measures to be monitored. "States Parties shall report on the measures taken by them to implement the Convention" (UNESCO 2010ff).

"Should we worry about change that happens to intangible heritage?" was asked in the *Sub-regional Meeting on the Convention for the Safeguarding of the Intangible Cultural Heritage: Implementation and Inventory-making, Dar es Salaam, Republic Tanzania in 2007*

"Certain kinds of changes are inevitable in a continued enactment of Intangible Cultural Heritage. The important question to ask is, 'is the change damaging to the core significance of the intangible heritage in question?' For example, in talking about traditional dance, it is important to ask,

'What is the valuable part of the dance?' These decisions need to be made by dancers and in cooperation with heritage practitioners. Defining significance of a given intangible heritage is important in considering changes" (UNESCO 2007g p15).

The protective success of the Wold Heritage Convention is largely due to **knowledge-based vigilance.** Knowledge based vigilance may similarly be employed by the ICHC, this forming the basis of the following recommendations.

Recommendations

15.9 Propose writing an ***Intangible Cultural Heritage Climate Change Response and policy Document***. The document should be compatible with UNESCO's Climate Change Initiative. It should adopt a human rights approach and include a climate change strategy, a disaster risk management strategy (supported by the ICHC Article 7b) and policy implementation options.

15.10 Consider the following to achieve this recommendation:

a. **Study the WHC's response to the threat of climate change**. This could involve an appraisal of the WHC's response process, published documents and implementation as described earlier in this chapter and in chapter 26. Specific issues may include:

   - The response of World Heritage States Parties to the WHC's 'case-by-case' approach in determining whether the climate change impact on a World Heritage property is severe enough to warrant inscribing a site on the List of World Heritage in Danger and to be " amenable to correction by human action," (decision 32COM7A.32),
   - The World Heritage position regarding the equity principle 'common but differentiated responsibilities' (CBDR),
   - The WHC's lack of formal recognition of the differential impact of climate change on States Parties due to their geographic position,
   - The appropriateness or otherwise of the WHC's lack of obligatory requirements supporting climate change mitigation.

b. Gain a greater understanding of the **impacts of climate change, especially the human impacts,** on ICH through consideration of reports e.g. *Implications of Climate Change for Australia's World Heritage Properties: A Preliminary Assessment* (ANU 2009), by surveying people who are holders of both listed and unlisted ICH and by reference to academic articles e.g. the **International Journal of Intangible Heritage** (IJIH 2006-2011)).

c. Recognise that the World Heritage Committee was stimulated to act on climate change by a handful of petitioners. Propose initiating **engagement** with Mali Voi and other disaffected ICH owners. Encourage full participation leading to consultation and partnerships in developing the ***Intangible Cultural Heritage Climate Change Response and Policy Document.***

d. Consider initiating a **questionnaire survey of States Parties** for the purpose of formulating a means by which climate change threats may be monitored, prevented, ameliorated or managed. Particular interest should be paid to the added effects of climate change on poverty, ill-health and migration on cultural custodians, their communities, the element's practice, and its transmission and inventorying. Central to the identification of threats will be possible repercussions of climate change on ICH including damage, deterioration and disappearance.

15.11 Publicise the ICHC's role in providing **financial support** (through the ICHC Fund) and in providing **expert assistance** to disadvantaged States Parties to aid in the nomination, implementation, policy-making, monitoring and reporting processes and projects that safeguard ICH from impacts of climate change.

15.12 Consider establishing a Wikipedia style **Global Database of Intangible Cultural Heritage Events** (relying on 'owner' sourced updating) of online timetables for festivals, performances, fairs, exhibitions, recitals, and meetings of practitioners including artists and musicians and online conferences of artisans – to maintain viability, dynamism and transmission and above all to bring advocacy for climate change mitigation to safeguard heritage. This could be developed into a computer application.

15.13 Consider obliging States Parties to adopt, implement and report on **climate change management plans** requiring specific climate change information by amending *Operational Directives*.

15.14 Consider amending *Operational Directives* to include climate change threats when **nominating** ICH for inscription on the Representative List, and especially when nominating ICH on the 'in need of Urgent Safeguarding' List . Consider amending *Operational Directives* in the nominating **examiner's report** where 'environmental transformation' (*Operational Directive* 7) could be elaborated and climate change could be detailed as a risk factor; and in paragraph 30 relating to the transfer of elements from one list to another for safeguarding reasons.

15.15 Consider promoting the inclusion of climate change management in the programmes that are nominated for listing under the Convention's third List: **Register of programmes, projects and activities for the safeguarding of intangible cultural heritage considered**

to best reflect the principles and objectives of the Convention. Climate change impacts will become so ubiquitous that their consideration in every list of best practice is appropriate. Consider adding 'climate change' to the **criteria for selection** on safeguarding (*Operational Directives 52, a*).

15.16 Be mindful of the challenges of **remaining true to the ICHC.** One of the challenges that could face the ICH Committee in developing a response to climate change will be to draw what is appropriate from the World Heritage experience and no more - to develop peculiarly ICH specific measures which are useful without being an antithesis of the living, changing, recreating, revitalising essence of ICH. In the publication *Policy Document on the Impacts of Climate Change on World Heritage Properties* documenters have stated in more than one instance that their advice, guidelines, strategies, policy and management plans relating to cultural heritage are less adequate, less well researched and less understood than those relating to natural heritage.

15.17 Consider the adoptability of the **information gathering opportunities** described in the WHC's *Operational Guidelines*, Annex 7, 11.5 and the World Heritage Periodic Reporting (section II).

15.18 Consider the adoptability of the World Heritage ICH **Reporting** Form at Insert 15.7. In its original form it was used to record ICH from WH properties and has been adopted by the Intangible Cultural Heritage Committee as a possible outline for **inventorying** elements (UNESCO 2011i). It has been amended here to capture further ICH data. It could be considered

- For inventorying ICH to monitor climate change impacts
- For using in ICH climate change vulnerability or risk assessments as the basis for adaption plan decisions.

15.19 Support the development of the proposed **World Heritage and Intangible Cultural Heritage Climate Change database and Network** (WH:ICH:CC Database & Network).

Insert 15.7 World Heritage Periodic ICH Reporting Form with suggested Amendments

1. Identification of the element

1.1 Name of the element, as used by community or group concerned;
1.2 Short, maximally informative title (including indication of domain(s));
1.3 Community (ies) concerned; *describe composition, estimate size, are there restrictions on participating e.g. physically demanding, age related, gender specific, seasonal factors.*
1.4 Physical location(s) of element; *describe essential needs- area, indoor, outside, stage*, full moon, paid for?

1.5 Short description. *Have there been changes to this element e.g. new tangible attributes, security, health issues, times when participation was more or less. Have there been times when participants* moved or were *displaced? What caused these changes e.g. economic – crop changes, fishing changes, severe weather e.g. storm surge, drought, flood, earthquake etc. Please describe.*

2. Characteristics of the element

2.1 Associated tangible elements; *e.g. plants, animals, raw materials, works of art, instruments, costumes,*
2.2.Associated intangible elements; *space, audience, time*
2.3 Language(s), register(s), speech level(s) involved; *by how many?*
2.4 Perceived origin. *Please describe history.*

3. Persons and institutions involved with the element

3.1 How many practitioners(s)/performer(s): name(s), age, gender, social status, and/or professional category, etc.;
3.2. Other participants (e.g., holders/custodians); *how many?*
3.3 Customary practices governing access to the element or to aspects of it;
3.3. Modes of transmission; *have these changed?*
3.4 Concerned organizations (NGOs and others).

4. State of the element: viability

4.1 Climate change threats to the enactment: Other threats to the enactment;
4.2 Threats to the transmission;
4.3 Availability of associated tangible elements and resources;
4.4 Viability of associated tangible and intangible elements;
4.5. Safeguarding measures in place.

5. Data gathering and inventorying

5.1 Consent from and involvement of the community/group in data gathering and inventorying;
5.2 Restrictions, if any, on use of inventoried data;
5.3 Resource persons(s): name and status or affiliation;
5.4 Date and place of data gathering;
5.5 Date of entering data into an inventory;
5.6 The inventory entry compiled by....

6. References to literature, Photography, audiovisual materials, and archives.

(UNESCO 2007g, 2011i, 1999)

This is a controversial proposal in that it challenges the 'dynamic' nature of intangible cultural heritage. However traditional owners may require formal assessments of their intangible heritage in order to make decisions about how and what to safeguard or abandon under pressures such as forced displacement, pandemics, and extreme weather events. For example it may be valuable for intangible cultural heritage holders to assess how many people and of what age and sex are needed for the safeguarding of a tradition such as a 'handmade paper' tradition. If the community was displaced and fragmented, what intangible cultural heritage traditions would be worth safeguarding for the size of the subgroups of the fragmented community? For instance, what seeds would be worth taking to allow the chance of growing traditional foods or for preparing a special dye? If these questions had a **documented basis**, communities may be able to make better safeguarding and adaptive decisions.

Chapter 16

How do the Conventions Address the Human Impacts of Climate Change on Heritage?

In one of the neighbouring districts, the girls have stopped going to school since over 80% of their time is allocated to water sourcing, meaning that some stay awake overnight due to the long queues and by the time they get back home they are so weary and can barely do any other work. To those who are lucky to be near the schools; once they get to the classes they are sleepy throughout the sessions, affecting the rate of concentration in class work. The circle completes when they get back home and travel to fetch the scarce commodity throughout the night again.
Chabari Z.K, Kenya (UNESCO 2010)

To what extents are World Heritage cultural properties protected, and Intangible Cultural Heritage elements safeguarded, by the Conventions from the 'human impacts' of climate change?

The major impacts on humanity of climate change are listed in column 3 of Insert 6.1 Diagrammatic summary of the relationships between climate change and cultural heritage. They include: loss of land; decline in farming, fishing, forestry and tourism leading to livelihood insecurity; shortages of food and fresh water leading to famine, poverty exacerbation; migration; health deterioration; potential insecurity (conflict); social and cultural fragmentation; damage, deterioration and possible loss of culture; loss of social cohesion, dignity, sense of place and identity; and for some the ultimate loss – the loss of nation leading to - 'statelessness.'

THE CONSEQUENCES OF CLIMATE CHANGE IMPACTS ON HUMANITY WILL, IN TURN, AFFECT HUMANITY'S IMPACT ON THE ENVIRONMENT. 'HUMAN IMPACTS' AS DEFINED IN THIS BOOK RELATE TO THE IMPACTS OF CLIMATE CHANGE AFFECTED HUMANS ON CULTURAL HERITAGE

The World Heritage Convention and the Intangible Cultural Heritage Convention

The human impacts of climate change on heritage whether tangible or intangible share common drivers, the effect of climate on custodians and artisans. Because of this commonality of causation both Conventions are discussed together.

The Scale of the Impacts

The **UNESCO Climate Change Initiative** recognises that climate change has a direct effect on our great tangible cultural heritage, namely our World Heritage cultural sites, and through its impacts on humans, has an additional affect on our tangible (and intangible) cultural heritage(Insert 15.4). The **World Heritage** site of **Timbuktu in Mali**, inscribed in 1988, has been threatened by climate change and is likely to be further threatened directly and indirectly. Its three impressive Mosques, those of Djingareyber, Sankoré and Sidi Yahia have been stressed by sand inundation and windblown sand. Their condition deteriorated as desertification of the surrounding areas was exacerbated by drought which in turn led to further deterioration. The site was put on the World Heritage in Danger list in 1990. The site was removed from the 'in Danger' List in 2005 after having been restored by local crafts people who had used time honoured methods (intangible heritage). However with hotter and drier weather predicted, the threat will intensify and likely re-emerge. Desertification causes land degradation which feeds back to amplify the climatic warming, a positive feedback which will make the threat worse. Another climate change related problem is the increase in frequency of the extreme weather event of torrential rainfall. Such unusual downpours are particularly damaging to the adobe Mosques. A third climate change related threat could well come in the future. With the undermining affect of desertification on the local and national economy, the maintenance of the Mosques may deteriorate; local communities may possibly shrink due to migration (Colette 2007). The consequent human impact - lack of maintenance, may be the defining threat to this site.

Lack of Knowledge

The relationships of these direct and indirect impacts with their drivers are summarised in Insert 6.1. The human impacts on heritage have not been studied in depth and have not, as yet, been taken into consideration in World Heritage and ICH protection, although awareness of their impacts is growing, see Insert 16.1. As recognised in the report *Predicting and Managing the Effects of Climate Change on World Heritage* the 'socio-economic changes that will result (from climate change) will have a greater possible impact on the conservation of cultural heritage than climate change alone' (UNESCO 2006g p29). The report's authors found that, as compared with the research on natural heritage sites, more needs to be done on monitoring, research and site maintenance in relation to climate change impacts on cultural heritage (UNESCO 2006a p55). They concluded that little is known about climate change impacts on cultural aspects of the listed properties and *the cultures of the property owners.* There is a *gap in knowledge generally about the human impacts* of climate change on cultural heritage. This gap is evident in climate change related heritage studies (OXFAM 2009; ANU 2009; FOE et al. 2007) although the villagers' stories as related by Leahy in Papua New Guinea and Leafasia in the Solomon Islands (see Chapter 6) and online (Oxfam 2011) poignantly document the mostly deleterious local 'human impacts' on cultural heritage.

The report *Predicting and Managing the Effects of Climate Change on World Heritage* lists "community transformation, disruption, collapse, abandonment, and loss of cultural memory' as possible impacts of climate change on humanity leading to 'human impacts' on cultural heritage (UNESCO 2006a p37). The only opportunities for these types of human impacts on World Heritage sites to be translated into World Heritage protective action occurs through the 6 yearly 'Periodic World Heritage Reporting' (section II) (and in its directives in the Operational Guidelines (Annex 7). The opportunities for reporting on 'human impacts' are to be found under 'Factors affecting the Property'. They are listed as "Changes to identity and social cohesion: changes in livelihoods, migration to or from site and changes in local population and community" (UNESCO 2011a).

UNESCO CLIMATE CHANGE INITIATIVE:

Core Program:

Climate Change, Culture and Biological Diversity and Cultural Heritage

Includes.......

"………Climate change may cause other social and cultural impacts, with communities changing the way they live, work, worship and socialise in buildings, sites and landscapes, *possibly migrating and abandoning their built heritage and losing their intangible cultural heritage.*"

(UNESCO 2009j)

Insert 16.1

The *Policy Document on the Impacts of Climate Change on World Heritage Properties* explains that there are gaps in knowledge, information and understanding of climate change impacts on cultural properties. It outlines research priorities; see Chapter 15, Insert 15.5. These are, especially in developing countries, compounded by a lack of financial support and capacity for research and the application of both at the national level. The *Policy Docu-*

ment on the Impacts of Climate Change on World Heritage Properties proposes welcome changes to the *Operational Guidelines* including incorporation of the precautionary principle, see Insert 16.2.

Climate change case studies are currently being conducted at several World Heritage sites (see Insert 6.1, column 4, and UNESCO's book, *Case Studies on Climate Change and World Heritage*) implementing physical adaptive measures including ongoing monitoring. 'Human impacts' have not specifically been included in these interventions.

The landmark decision to list **Uluru - Kata Tjuta National Park, Australia** in 1994 as a 'mixed site' (UNESCO 2010q) demonstrates the wisdom of the World Heritage Committee in adopting a holistic approach, inclusive of symbolic values. This positive action bodes well for future World Heritage climate change management plans to recognise the significance of human impacts and to be more inclusive of human impacts in protection planning.

Post- 2006, the World Heritage Committee has made around 64 Decisions that have 'climate change' in the wording, refer chapter 26. (There are many more references made to climate change in World Heritage documents including in nominations, State of Conservation Reports and in draft Decisions that are not reflected in the final wording of the resultant decision.) The majority relate to natural sites. It is anticipated that this situation will change as climate change bites, potentially having far greater impacts on cultural sites than natural sites. World Heritage cultural sites are relatively small in area, immoveable, and therefore more fully prone to rising sea level, extreme weather events and the human impacts of deteriorating custodianship and migration leading to neglect and abandonment.

The World Heritage Committee Decisions reveal that it has been diligent in following through with directives for State Parties to integrate the methodology outlined in their publication *A Strategy to Assist States Parties to Implement Appropriate Management Responses* into nominations. Decision 29COM7B.a encourages UNESCO to do "its utmost to ensure that the results about climate change affecting World Heritage properties reach the public at large, in order to mobilize political support for activities against climate change and to safeguard in this way the livelihood of the poorest people of our planet." In decision 33COM 7C (UNESCO 2009cc) the WH Committee's approach is reinforced. A request is made for the World Heritage Centre and Advisory Bodies to adopt a *consistent approach* to World Heritage Reporting on the impact of climate change on World Heritage properties and to ensure that future Decisions in this respect are based on the aforementioned *'Strategy'* publication. Reactive monitoring, periodic World Heritage reporting, international assistance, capacity building and other training programmes are part of the 'consistent' approach.

The World Heritage Committee will consider specifically incorporating reference to the precautionary principle within the Operational Guidelines.

The UNFCCC includes this principle under Article 3 as follows: 'The Parties should take precautionary measures to anticipate, prevent or minimize the causes of climate change and mitigate its adverse effects. Where there are threats of serious or irreversible damage, **lack of full scientific certainty should not be used as a reason for postponing such measures**, taking into account that policies and measures to deal with climate change should be cost-effective so as to ensure global benefits at the lowest possible cost. To achieve this, such policies and measures should take into account different socio-economic contexts, be comprehensive, cover all relevant sources, sinks and reservoirs of greenhouse gases and adaptation, and comprise all economic sectors. Efforts to address climate change may be carried out cooperatively by interested Parties.'

The explicit adoption of the precautionary approach by the World Heritage Committee as a consideration in decision making in general will encourage States Parties and the Advisory Bodies to use the emerging knowledge relating to the implementation of the precautionary approach to deal more actively with risk and uncertainty when making decisions concerning the effects of climate change on World Heritage properties. (UNESCO 2008a)

Insert 16.2 The precautionary approach in World Heritage decision-making in the context of climate change

The 'human impacts' of climate change on all cultural heritage will be far reaching and although the World Heritage Committee has provided some opportunities for their monitoring as part of the assessment and reporting processes, there is, as yet, no body of knowledge on which to draw. There is no adequate protective policy and management strategy addressing the climate change driven 'human impacts' on World Heritage properties.

The issues raised in this discussion relating to migration, health, and security warrant individual consideration in the following chapters.

The ICHC does not have a position on climate change as has been addressed in chapter 15.

It is timely that the 2012 *UNESCO's Input to the Rio+20 Compilation Document* lists as one of its 'concrete' measures recommended: "Support regional and international scientific cooperation in the social and human sciences directed at analysis of climate change impacts and of the policy challenges raised by adaptation to Global Environmental Change" (UNESCO 2012a p14).

Recommendations

16.1 Urge the World Heritage Committee, the Intangible Cultural Heritage Committee and UNESCO to rate climate change as a **top priority** in recognition of its ever increasing threat.

16.2 Further to the recommendation 15.2 that proposes revising the 2007 ***Policy Document on the Impacts of Climate Change on World Heritage Properties*** *(WHC 2007; UNESCO 2008a)* and further to the recommendation 15.10 supporting the development of an ***Intangible Cultural Heritage Climate Change Response and Policy Document,*** it is suggested that consideration be given to:

- Ensuring that 'human impacts' on cultural heritage are included,
- Ensuring that all 'human impacts' policies are **socially inclusive** and
- Ensuring that community **engagement** is included in policy planning and implementation.

This should particularly apply where indigenous communities and minority groups have cultural responsibilities for heritage.

16.3 Consider having the WH and ICH Committees host a **joint meeting** of representatives and relevant experts to address potential *'human impacts'* of climate change on both WH and ICH tangible property and intangible elements in order to contribute to the ***Policy Document on the Impacts of Climate Change on World Heritage Properties,*** the proposed ***Intangible Cultural Heritage Climate Change Response and Policy Document*** and perhaps a joint climate change policy. This meeting could continue the work of the Global Humanitarian Forum in exposing the **'silent crisis,'** in this case, related to heritage.

16.4 Preliminary to the climate meeting consider **surveying** States Parties about the present and possible impacts on their populations of climate change and the consequential **'human impacts' on their cultural heritage**. The World Heritage committee has acknowledged that human impacts will overwhelm the physical and biological impacts of climate change on World Heritage sites but has little knowledge on this.

16.5 Act on the results of the survey to **inform** the joint meeting. Invited experts might include mental health workers, cultural workers, social scientists and scientists and representatives from refugee organisations such as the UNHCR, the UNDP and the major religions. Provide all invited experts with the latest climate change science as background. Aim to predict the climate change impacts on humanity and the 'human impacts' on cultural heritage.

16.6 Consider addressing the following **issues** at the above joint meeting:

- The current wellbeing of the traditional custodians.
- The implications of continuing climate change on cultural heritage 'owners'.
- The role of climate justice in reducing the human impacts of climate change.
- Adaptive interventions to cope with 'human impacts' of climate change on World Heritage sites e.g. capacity building, retention of traditional knowledge, knowledge of climate change science and resilience in World Heritage protection. Are there lessons to be learnt from the Nairobi Work Programme? Are there guidelines to be heeded about policy implementation from 'common pool resource management'? (See Chapter 9).
- Lessons to be learnt about **sustainability** from observations on the survival and collapse of past and present traditional cultures (see Chapter 9). Consider resilient present day traditional societies such as Ladakh, (Trainer 2007; Alexander 2007) (Norberg-Hodge 2009; Rizvi 1993) and Bhutan, the Kwakiutl of the Pacific Northwest, other traditional societies with strategies **for managing local 'commons'** (see chapters 9 and 13) and emerging stories from there and elsewhere of successful and unsuccessful strategies for addressing climate change impacts.
- How can communities, mindful that some World Heritage cultural sites are **community based**, meet the **health** challenges including mental health challenges of an unrelenting climate compromised world? Consider possible resurgence of religiosity with climate change; fatalism; extremism and emergence of iconoclasts and zealots.
- **Migration** and its impacts on World Heritage sites, for instance options for conserving cultural heritage (including cultural memory) prior to or following fragmentation or displacement of communities and even whole States Parties. Consider lessons from exiled cultures such as the Tibetan culture (His Holiness The 14th Dalai Lama of Tibet 2011).
- The question of cultural heritage protection and **loss**. How to decide what to let go? At what cost?
- Consider the contributions of **'holism'**, **Gaia theory** and **the Mother Earth Movement** to 'human impacts' on World Heritage (see Chapter 4).
- Consider innovations in organisational structures relating to possible improvement in implementation of climate change related policy/management plans. Consider **'nested' organisation** see Insert 12.1.

16.7 Continue assistance for regional level **capacity building** and awareness raising workshops including the use of web based communication such as UNESCO's *Climate Frontlines* which has a demonstrable function in non-formal evaluation of climate change schemes such as the REDD scheme.

16.8 Exploit UNESCO's World Heritage status to **attract world attention** to:

- The 'silent crisis' of the climate change impacts on humanity now and the probability of a lack of silence in the future.
- The disconnect of the developing communities at the frontline of climate change from those in industrialised communities who are currently relatively buffered and comfortable.
- The greenwashing practices of many powerful western States Parties.

Cricket Ladakhi-style

Chapter 17

Do the Conventions Respond to Climate Change Induced Displacement and Cultural Loss?

There is really a sense of being alienated from their lands, from their culture, from their livelihoods and just a sense of who they are. In the smaller islands, everybody knows everybody; they identify themselves by their islands or island groupings. They have a sense of spiritual connection with the land, so if they are going to be displaced, they're going to feel like not belonging. They have been disenfranchised from the whole rural community so there's going to be a real social stress on meeting their social obligations and commitments to their people as well as trying to earn a living, because the economic situation in Federated States of Micronesia (FSM) is not very good. Augustina Takashy, a community activist from Chuuk State in FSM (Oxfam 2011 p26).

Do the Conventions address the effects of climate-induced human displacement? How do the Conventions inform and influence climate change induced abandonment of cultural heritage through abandonment of homeland caused by, for example sea level rise, and subsequent potential loss of cultural memory?

The World Heritage Convention

Displacement, an Under- acknowledged Impact of Climate Change

Climate change driven displacement is most obviously happening in the **Small Island Developing States (SIDS)**. Sea level is rising contaminating fresh water sources and gardens. Severe weather events such as storms and cyclones are generating surges which erode coasts and cause flooding. Islanders are moving inland, to mainlands, and overseas. Both forced and voluntary displacement will increase because climate-proofing of islands is beyond the resources of the residents. As sea level rise continues with an anticipated 1.2m rise by 2100, the large proportion of cities which are coastal will be affected, threatened with relocation. More than a billion people live within a 25 metre elevation of sea level, around 100 million people in Asia, mostly Bangladesh, eastern China and Vietnam; 14 million in Europe; and eight million each in Africa and South America. It is not just higher sea levels, but the combination of these with more extreme weather events that risk lives and livelihoods.

A survey of the **World Heritage** cultural sites reveals that many are in serious danger from a 1.2 metre sea level rise. More than 100 sites either whole or in part, are situated on coasts or flood plains, at elevations of less than 25 metres.

The report *Predicting and Managing the Effects of Climate Change on World Heritage* lists migration as an impact on cultural heritage but does not elaborate. No mention of 'migration' as a threat to World Heritage is included in the *Operational Guidelines*. In the UNESCO publication *Case Studies on Climate Change and World Heritage*, migration is recognised as a threat, "...climate change can force populations to migrate (under the pressure of sea level rise, desertification, flooding etc) leading to the break-up of communities and the abandonment of property, with the eventual loss of rituals and cultural memory" (Colette 2007 p65).

Migration is both an adaptation to and consequence of climate change (UNDP 2009). The **World Heritage Convention (WHC)** only offers protection to World Heritage cultural properties if they are abandoned causing them 'serious and specific' danger. The property may then qualify for placement on the List of World Heritage in Danger. If there was a gradual abandonment of a cultural site or an 'occupation' from an outside group as a consequence of displacement caused by climate change, the 'protection' offered by the WHC is unclear. There is no reference to migration or the displacement of people associated with World Heritage sites nor any distinctions made between internally displaced people, migrants and 'climate displaced people'. The latter are persons displaced temporarily or permanently due to environmental causes, (notably by land desertification, sea level rise and weather-related disasters).

In 2009 there were about 26 million climate displaced people (GHF 2009) with the prediction of 200 million by 2050 (Stern 2006). The WHC is not yet equipped to respond to the impacts of climate-induced displacement on world heritage. The WHC centres on protection 'to the utmost'. It does not address loss of cultural heritage although, as mentioned in Chapter 19, there is an acknowledgement that, over time under climate change impact, there will be a decline in the extent of protection possible. Participants of the 2010 ***Conference on European and Pacific Responses to Climate Change in the Pacific*** drew attention to the lack of a special term referring to cross-border displacement due to climate change. This implies "a deficiency in both state protection and international human rights law. The question of the legal status of people displaced by climate change needs to be discussed and the climate change issue be integrated into the Human Rights agenda" (On the Run Final Statement of the participants of the Conference on European and Pacific Responses to Climate Change in the Pacific. 2010). There is a gap in knowledge about the extensive relationship of climate change, displacement and cultural heritage.

Recommendations

17.1 Include all forms of **'displacement' (voluntary, forced, internal, and international) as a risk factor** in the nomination, periodic reporting and monitoring processes. Consider the consequences of displacement on World Heritage sites such as the gradual abandonment of the site by custodians, the severing of spiritual links, the mental health of the absent displaced custodians, the lack of maintenance staff and the potential occupation of the site by non-related destitute people.

17.2 Prioritise **studies** of the potential impacts on World Heritage sites of climate-induced displacement. **Engage** with traditional site custodians and indigenous 'owners'.

17.3 Support **migration policies** that are sympathetic to the World Heritage implications of displacement and are driven by human rights needs rather than by the capacity of affected populations to fit within general migration programmes.

The Intangible Cultural Heritage Convention

Although there is a growing awareness of the perils of climate change, its likely impact on human displacement and mobility has received too little attention. António Guterres, (UNHCR 2009)

The most important impact on humanity of climate change is poverty. This question addresses the next most important impact of climate change - that of 'displacement' - upon the world's holders of intangible cultural heritage (see also Chapter 7). The movement of people may not appear climate induced in the first instance. Crop failure or fresh water shortages may be the initiating motivation, but climate change is often a contributing factor and perhaps the root cause. The Intangible Cultural Heritage Convention has no specified 'migration' or translocation-related measures for addressing climate change impacts and does not discuss climate caused heritage loss.

The issue was raised in 2007 at the *Sub-regional Meeting on the ICHC: Implementation and Inventory-making, 'How can we deal with the issue of migration?'* The answer was, "While the issue of migration will probably require more attention and consideration, the first step might be to foster relationships between host and diaspora communities. In the case of the proclamation of Masterpieces, experts checked multinational files attentively to see if the proposed intangible heritage was shared with other countries or not (UNESCO 2007g)."

Recommendations

17.4 Raise awareness of the momentous impact **climate change-induced migration** will have on ICH. Presently twenty six million are climate displaced people.

17.5 Address **'gaps in knowledge'** by researching the effects of climate change induced displacement on ICH. Incorporate the gained knowledge in the proposed ***Intangible Cultural Heritage Climate Change Response and Policy Document***.

17.6 Make climate change-induced displacement **a priority** consideration in the safeguarding of ICH.

17.7 Support a **migration policy** as recommended for World Heritage. New Zealand has established a settlement visa called the Pacific Access category. While climate change was the impetus behind the programme, eligibility is based on age and language criteria. This limits the effectiveness of that programme to address the HR and ICH needs of those impacted by climate change.

The late Lama Narwang in one of the passageways of the spectacular, precipitously situated, Phuktal Gompa (Monastery) in Zanskar, Ladakh

Chapter 18

How do the Conventions Address Climate Change-Induced Health Impacts on the Custodians and their Communities?

Climate change represents an inevitable, massive threat to global health that will likely eclipse the major known pandemics as the leading cause of death and disease in the 21st century ... The health of the world population must be elevated in this discussion from an afterthought to a central theme around which decision-makers construct rational, well informed action-orientated climate change strategies.
Dr. Dana Hanson, President, World Medical Association

How do the Conventions address climate change-induced health impacts on the custodians and their communities? For example how do the Conventions address the consequences on World Heritage and Intangible Cultural Heritage of custodians who can no longer perform their responsibilities due to climate-induced poverty, displacement or mental stress?

The World Heritage Convention

This question was proposed because of new insights emerging from international, national and local health organisations about the human costs in health presently being borne and of the looming impacts to come from accelerating climate change. The impending threat to our health is extreme and was captured in the opening sentence of *Managing the health effects of climate change* by *The Lancet* and the University College London Institute for the Global Health Commission, "Climate change is the biggest global health threat of the 21st century" (Costello 2009 p1693).

World Heritage properties are vulnerable through losses, due to ill health of cultural custodians, (Colette 2007), traditional owners, stewards and rangers along with their traditional knowledge and skills. The greatest health impacts are from water and food borne diarrhoeal diseases, malnutrition, respiratory ailments, vector-borne diseases such as dengue and malaria, injuries from conflict and stress related disorders. These impacts will be felt most in developing countries because of high levels of poverty, their often tropical situation and the limited capacity of public health systems to respond.

Climate change is a health threat multiplier. It amplifies the vicious cycles of disadvantage making escape an enormous challenge. If a site is threatened by a severe weather event, heat wave, flooding or cyclone then the 'human impacts' related to health can be significant. People may migrate, become traumatised and in some cases succumb to diseases which may emerge as a result of the weather event.

Indigenous and minority groups, especially women and children associated with World Heritage cultural landscapes and properties in developing State Parties will be under the greatest threat (UNDP 2007b). For instance the World Heritage Site of **Tsidilo, Botswana** is a living religious heritage site (in the hostile Kalahari Desert) believed to be frequented by ancestral spirits. Over 4,500 rock paintings are preserved in a 10 square km. area. The archaeological record of the area gives a chronological account of human activities and environmental changes over at least 100,000 years. Local communities visit Tsodilo to worship as do local church congregations and traditional doctors. Tsodilo is a place to pray and meditate (UNESCO 2011j). The potential damage to the Outstanding Universal Value (OUV) of this revered site if the local communities suffered a climate related epidemic or abandonment through climate induced desertification and subsequent starvation, would be catastrophic.

The mental health toll of climate change, including 'solastalgia' (Albrecht 2007), insidious depression, acute 'Broken heart Disease', traumatic stress, and identity loss linked with migration, cultural fragmentation and statelessness, is uncharted.

The background publication, the report *Predicting and Managing the Effects of Climate Change on World Heritage,* does warn of longer term health threats to World Heritage properties. "The annual melting of mountainous glaciers also drives the hydrological cycles of entire regions. But as the ice recedes, there will first be floods, and some time later, water supply will cease to be available, eventually leading to famine and pandemic disease (UNESCO 2006a p26)." However the Convention does not protect World Heritage properties from climate change induced health impacts except as a Human Right.

Recommendations. See ICHC Recommendations below.

The Intangible Cultural Heritage Convention

The **Intangible Cultural Heritage Convention (ICHC)** 'safeguards' by encouraging nomination of elements, implementing practices to ensure their viability, their transmission, their understanding and their tradition including the welfare of those who perform the intangible cultural heritage. The 'safeguarding' does not include any direct health preparedness principles, or plans for preventative measures to address climate change impacts. The Convention does not provision for what will become the ubiquitous climate change-induced health impacts on its custodians, practitioners, performers, transmitters, knowledge holders, customary leaders and craftspeople, except as a human right (ICHC Preamble).

The threat is manifesting in diseases as described. Medical authorities anticipate that as people lose their land, livelihoods, and even nations to climate change, their mental health will be vulnerable (DEA 2009). We have read of the 'psychological pain' expressed by Mali Voi from Papua New Guinea living testament to the psychological damage caused by climate change. The instability may be more acute for indigenous and rural people (Albrecht 2007) and certainly for those displaced by extreme weather events, desertification and sea level rise. Heritage policy makers do need to consider health in general, including mental health, when developing and implementing an intangible cultural heritage climate change policy as proposed in recommendations in chapters 15 and 16. Policy makers should be aware of the possibility of extremism related to poor mental health (Costello 2009).

Recommendations

18.1 Frame 'health' in all World Heritage and Intangible Cultural Heritage policies, documents, guidelines, directives and reporting mechanisms as in the **Universal Declaration of Human Rights (UDHR)**:

- Article 25 "Everyone has the right to a standard of living adequate for the health and well-being of himself and of his family, including food, clothing, housing and medical care and necessary social services, and the right to security in the event of unemployment, sickness, disability, widowhood, old age or other lack of livelihood in circumstances beyond his control" (UN 1948) and

- Article 12(a) of the **International Covenant on Economic, Social and Cultural Rights (ICESCR)** recognises the right of everyone to 'the enjoyment of the highest standard of physical and mental health'

18.2 Encourage **research** into climate change induced health impacts on World Heritage and ICH. Include consideration of the consequences of mental health impacts such as solastalgia.

Chapter 19

How are the Conventions Addressing Climate Change Vulnerability, Risk, Equity and Marginality?

We strongly believe that it is the political and moral responsibility of the world, particularly those who caused the problem, to save small islands and countries like Tuvalu from climate change, and ensure that we continue to live in our home islands with long-term security, cultural identity, and fundamental human dignity. Forcing us to leave our islands due to the inaction of those responsible is immoral and cannot be used as quick-fix solutions to the problem.
Speaking to the UN General Assembly in September 2008, Tuvalu Prime Minister Apisai Ielemi (OXFAM 2009 p7)

How are the World Heritage and Intangible Cultural Heritage Conventions addressing climate change vulnerability and risk status of cultural WH properties and Intangible Cultural Heritage elements? Are developing States Parties marginalised with regard to climate change protection of their cultural heritage?
Do the Conventions address and influence geographic balance and equity in addressing climate change protection and safeguarding of cultural heritage? Is there a bias towards protecting the heritage of States Parties in the northern latitudes? Is the heritage of smaller island states underrepresented? Do the Conventions address and influence 'human development' balance and equity in addressing climate change protection and safeguarding of cultural heritage? For instance, are developing States Parties disadvantaged as regards having their heritage protected?

The World Heritage Convention

Climate Change Vulnerability and Human Development

Vulnerability assessments and risk analyses are critical for informing decision-makers of specific options for alleviating and adapting to the impacts of climate change.

The *Policy Document on the Impacts of Climate Change on World Heritage Properties* emphasises the importance of vulnerability assessments in determining management action. It advises climate change professionals to work with World Heritage managers to address vulnerability issues, especially related to the built environment, using a combination of methods ranging from traditional knowledge through to the latest scientific methods. The *Policy* calls for specific criteria to be developed for identifying those properties most threatened by climate change. These criteria could form the basis for prioritising vulnerability assessment, mitigation and adaptation activities. The WH Committee stated "The need to incorporate these criteria into the *Operational Guidelines* will be considered only after assessing their utility for this purpose". The criteria were amended, see Decision 32COM 7A.32 Chapter 26. The amendments were minimal.

The report *Predicting and Managing the Effects of Climate Change on World Heritage* recognises the lack of research on the vulnerability of cultural heritage under conditions of climate change. It suggests deriving regional vulnerability maps from climatic data, and developing thematic groupings of sites facing similar threats e.g. archaeological, movable, coastal, mountainous or marine. This approach focuses on the biological and physical impacts of climate change and vulnerability. The authors have expanded upon the work in this report (UNESCO 2006a p30) to compile the table ***Tangible Cultural Heritage Risk Assessments,*** see insert 30.3. It includes indicators, risks, physical and biological impacts on tangible cultural heritage and the effects on humans and consequential human impacts on tangible cultural heritage.

Appendix 1 consists of a table containing all base line data used in the charts in this chapter. All nations are listed along with their **World Heritage Convention (WHC)** status; numbers of World Heritage, cultural World Heritage and Intangible Cultural Heritage inscriptions; Small Island Developing States (SIDS) and Least Developed Country (LDC) status, climate change vulnerability; ranked human development; ranked population and ranked area.

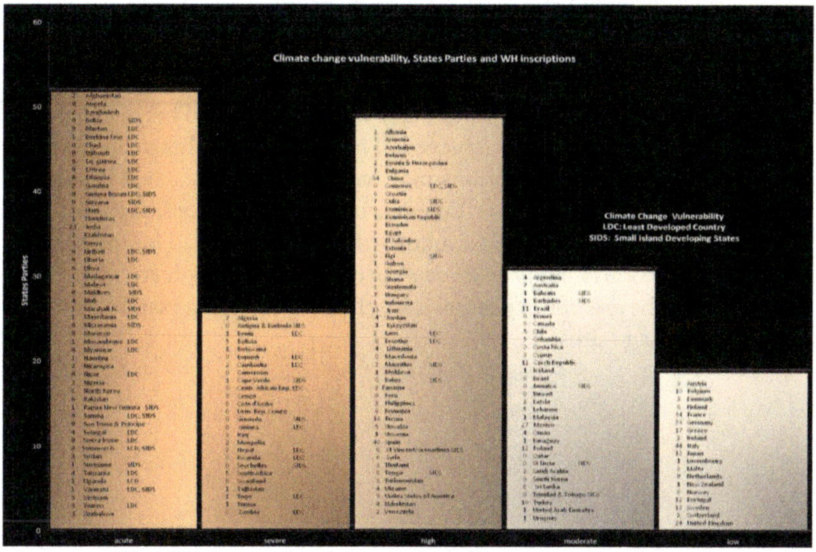

Insert 19.1 World Heritage cultural sites per State Party grouped by climate change vulnerability factor, 2010. Group categories are acute vulnerability, severe vulnerability, high vulnerability, moderate vulnerability and low vulnerability (DARA 2010)

The chart at Insert 19.1 shows the number of World Heritage cultural sites held by each State Party grouped by **climate change vulnerability** based on 2010 data. It may be noted that the States Parties categorised as being of 'acute' or 'severe' climate change vulnerability had the lowest average numbers of inscriptions with a disproportionate share of the least developed and developing nations having no inscriptions. These categories included most of SIDS, large swathes of Asia and Sub-Saharan Africa, but also India which was exceptional with its many inscriptions. By contrast the States Parties of 'low' vulnerability had the highest number of inscriptions and were nearly all European nations which all had inscriptions. The 'moderate and high' vulnerability groups had on average intermediate numbers of inscriptions with the populous nations of China, Iran, Russia and Spain, each with long settled histories, having disproportionately large numbers of inscriptions. The vulnerability factors used above were those derived by Climate Vulnerability Monitor which is "a global assessment of vulnerability to different aspects of climate change including its health impact, weather disasters, human habitat loss and economic stress on key industries and natural resources. It is a new tool aimed at advancing understanding of the impact climate change has on human society and actions needed to address the harm this causes" (DARA 2010).

The United Nations Development Programme's (UNDP) ***Human Development Index*** **(HDI)** is based on three components: a decent standard of living measured by the GDP per capita; "a long healthy life measured by life expectancy at birth; and knowledge measured by the adult literacy rate and a combined primary, secondary and tertiary gross enrolment ratio" (UNDP 2010). Insert 19.2 used the HDI ranking and 'descriptor' categories to plot States Parties according to their 2011 HDI. This measure is tied to the number of cultural World Heritage inscriptions held by each State Party. The relationship revealed is reasonably evident. In general the higher the human development of a State Party the more cultural heritage inscriptions they are likely to have achieved. The relationship between development and numbers of inscriptions may have been strengthened if past human development had also been taken into account. The similarity between the charts is indicative of the common drivers they share. The more 'developed' the State Party, and the more sheltered it is from vulnerability to climate change, the greater in general is its chance of holding inscribed cultural World Heritage sites.

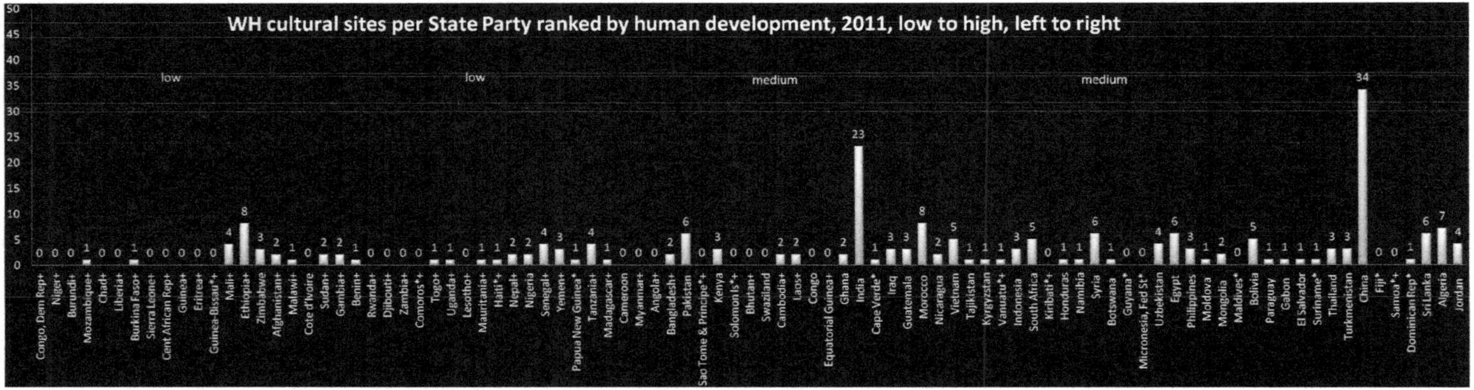

Insert 19.2 World Heritage cultural sites per State Party ranked by human development, 2011. (Cultural sites include mixed sites)

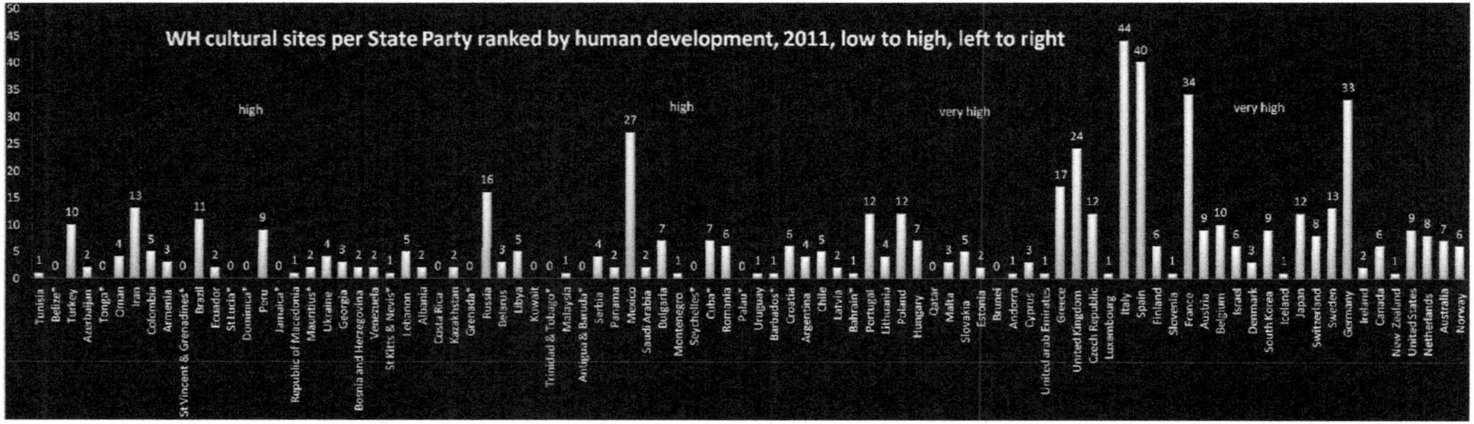

Marginality: Regional, Zonal, Area and Population Distribution

Insert 19.3 and 19.4 show States Parties ranked according to their area and to their population size. The number of cultural World Heritage sites is plotted as a function of the rankings.

As would be expected, as land area and population increase States Parties show a general tendency towards having more inscribed cultural heritage sites. Note the scatter of States Parties with zero inscriptions is more concentrated at the lower area / lower population end of the scales. Unsurprisingly the 'population' chart shows the more easily discernible relationship with number of inscriptions. In general the larger the population of the State Party, the more listed sites they are likely to have achieved.

Both charts clearly show the dominating and anomalous States Parties, those that have 'uncharacteristically' (as far as the trend shows) large numbers of inscriptions. These ambitious States Parties with rich cultural heritages are grouped more according to population than to area. They are all populous States Parties, but not all are of large area. The nine 'stand out' States Parties are China, India, Mexico, Germany, France, United Kingdom, Italy, Spain and Greece, all with long histories of settlement.

Conversely those States Parties with small territories and low populations, mostly 'island' territories, have few or no inscriptions. This is not surprising when one considers that there are only 753 World Heritage cultural sites in a world of 7 billion people. This amounts to on average one site for every 9.3 million of global population. Pleasantly surprising is the high level of representation of small States Parties when population is taken into consideration. When all States Parties are analysed according to their number of inscribed cultural sites per their population size the 13 highest inscription rates are found to be associated with States Parties with populations of less than one million people! At the top of the list is the Holy See with two sites for a population of less than a thousand; probably an unfair inclusion! Others are Malta and Cyprus with 3 inscriptions each and Iceland, Luxembourg, Andorra, Barbados, all with very high HDIs, San Marino (unclassified but wealthy), St Kitts and Nevis with a high HDI, Cape Verde, Suriname and Vanuatu with medium HDIs and the unclassified but relatively poor Marshall Islands make up the remaining ten. Of these Cape Verde was classified as having severe vulnerability, and the last 3 States Parties, acute vulnerability to climate change. Twenty four of the remaining 25 States Parties with populations less than one million have no listed cultural sites. These on average had significantly lower HDI's and higher climate change vulnerabilities than the former group. All in all the 38 States Parties with populations of less than a million had 26 World Heritage sites for a cumulative population of 12 million people, achieving a representation nearly 30 times the global average.

An examination of the World Heritage statistics in Chapter 14 shows two other factors being correlated with inscription status. These factors are 'latitude' and 'regional grouping'. The charts show that in the northern hemisphere there exists the long recognised north-south relationship, with the mostly wealthy industrialised nations of the 'temperate' European and North American regions holding over half the cultural World Heritage sites. Their neighbours in the 'northern tropics' and those further south, especially those in the tropical Asia Pacific and African regions have a disproportionally humble number of collective sites, averaging a little over one site per State Party.

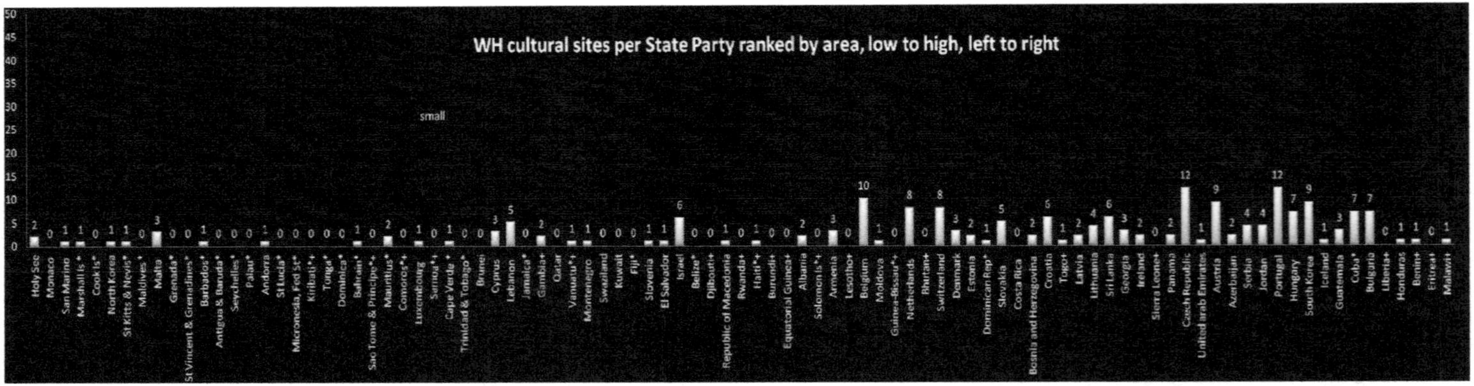

Insert 19.3 World Heritage cultural sites per State Party ranked by area, 2011. (Cultural sites includes mixed sites)

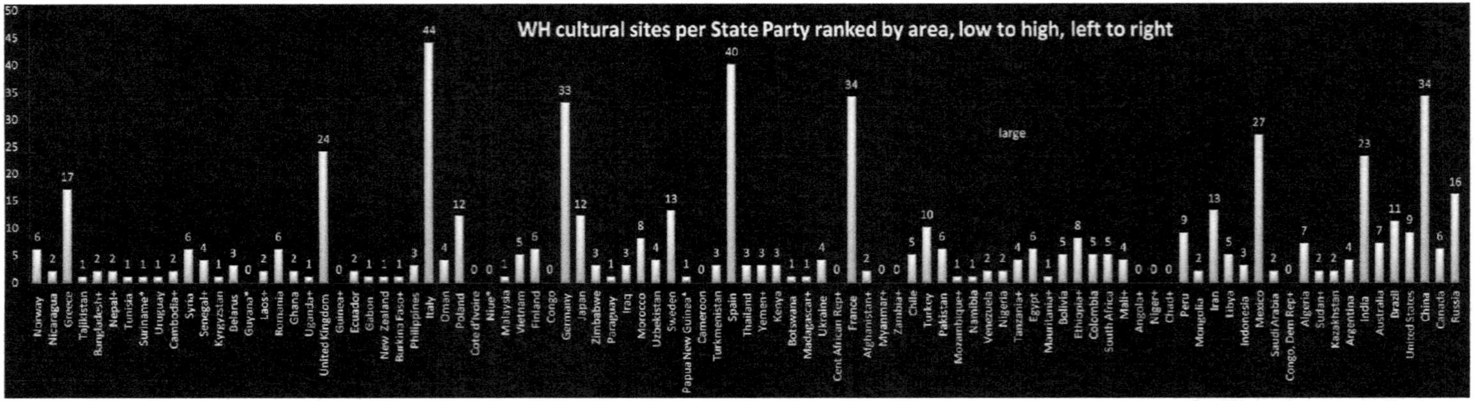

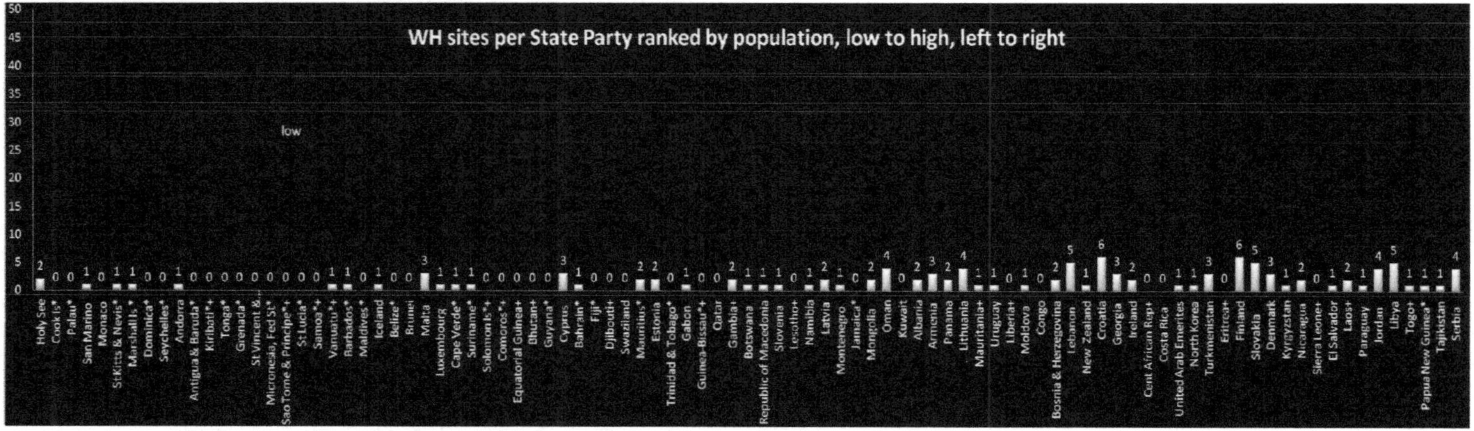

Insert 19.4 World Heritage cultural sites per State Party ranked by population, 2011. (Cultural sites include mixed sites)

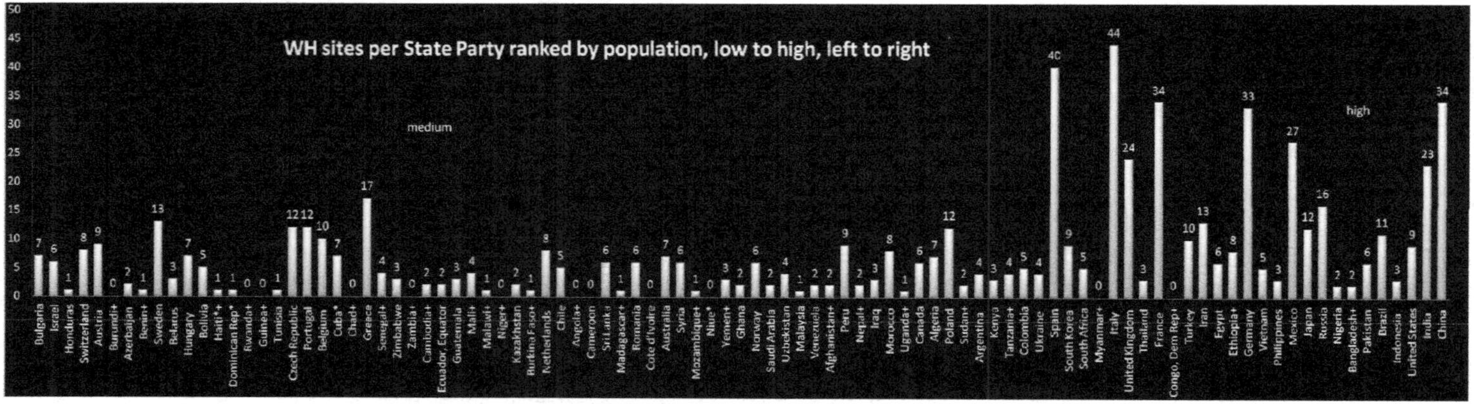

Whilst cultural heritage is not equitably protected by the WHC, if all States Parties irrespective of population size are considered to be equally entitled to have protected sites, it is clear that on a per population basis much has been achieved in addressing this apparent inequity. This **evidence of affirmative action** suggests that the lack of cultural listings for many of the small sized developing island States Parties in the tropics with high climate change vulnerability may be related to factors other than a lack of sufficient help to successfully nominate. After all, as has previously been noted, there is on average only one listed site per 9.3 million of world population. Poorer African States Parties with low human development and 'extreme' climate change vulnerability also lack cultural site inscriptions on the World Heritage Lists. With these States Parties more pressing poverty related priorities are likely to play a major role.

As a generalisation, a large population, a large land area, a northern location with a less than average vulnerability to climate change, a high human development index, a long history of settlement and an ambition to have sites listed are conditions favourable to the holding of cultural World Heritage sites.

These observations mask an underlying positive. The status of ratification alone is a measure of protection for cultural heritage, for under the WHC agreement all nations which have ratified must commit to take legal, scientific, technical, administrative and financial responsibility for the management and protection of their national cultural heritage and must commit to protecting all World Heritage.

Disaster Risk Management

Disaster risk preparedness for the protection of climate change -impacted cultural heritage is recognised in accordance with the WHC (Article 11.4) and its methods of address are evolving in line with new developments, especially of early detection technology. Disaster risk reduction is assisted by the World Heritage Committee Decision 31COM 7.1, (UNESCO 2007a) urging the World Heritage community to 'integrate actions pertaining to climate change risk preparedness into policies and action plans'. Reference is made to the *Policy document on the impacts of climate change on World Heritage properties* and the *Strategy for Reducing Risks from Disasters at World Heritage Sites* for guidance in reducing risks at World Heritage sites so as to protect their Outstanding Universal Value (OUV), authenticity and integrity. This decision followed the 2004 Decision 28COM 10B (UNESCO 2004a) recommending States Parties include risk preparedness as an element in their World Heritage site management plans and training strategies.

- *meteorological:* hurricanes, tornadoes, heat-waves, lightning, fire;
- **climate change: increased storm frequency and severity, glacial lake outburst floods (GLOFs)**
- *biological:*epidemics, pests;
- *hydrological*: floods, flash-floods, tsunamis;
- *geological:* volcanoes, earthquakes, mass movement (falls, slides, slumps);
- *astrophysical*: meteorites;
- *human-induced*: armed conflict, fire, pollution, infrastructure failure or collapse, civil unrest and terrorism.

(UNESCO et al. 2010)

Insert 19.5 Some of the most common hazards that may lead to a disaster

Risk preparedness for an emergency at site-level is included in the nomination process with guidance from the *Operational Guidelines* (118, Annex 2, 4B (iii)). It addresses the immediate threats from natural disasters such as earthquake and hurricane plus human-made hazards such as explosions and fire. It is also addressed in Periodic Reporting (II.5) as a 'factor affecting a property'. Risk preparedness is systematically referred to and cross referenced in all the climate change World Heritage publications

UNESCO has recently upgraded its response to the risk of disasters to World Heritage sites. In a joint undertaking with the International Centre for the Study of the Preservation and Restoration of Cultural Property (ICCROM), the International Council on Monuments and Sites (ICOMOS) and the International Union for Conservation of Nature (IUCN), the World Heritage Centre has published online the first of its World Heritage manuals: *Managing Disaster Risks for World Heritage* (UNESCO et al. 2010). The manual comprehensively covers preparedness processes for common disaster hazards (see Insert 19.5) including those directly related to climate change and does a broad analyses overall of climate change hazards (see in Insert 19.6). The 60 page book also includes the latest web links on early warning systems.

As noted in Chapter 3, the first evidence based studies able to identify the likelihood of human induced climate change contribution to the severity and frequency of extreme weather events have recently been published. In the manual's foreword this is anticipated, "climate change has been associated with the occurrence of more frequent and intense extreme weather events in some parts of the world" (UNESCO et al. 2010; UNESCO 2006a p2). The use of the term natural hazards when describing severe storm damage or fire destruction is becoming increasingly a misnomer as the human-induced component is becoming quantifiable.

Significantly, the manual includes the following 'box' that emphasises the complexity of addressing risk under climate change. The italics are not in the manual, but are added here for further emphasis.

Note It is very important to make a clear distinction between natural hazards and disasters, as this relates to the degree of management intervention that is appropriate for a given World Heritage property. The ability to prevent harm or loss of heritage values as a result of natural processes may be limited. So the response and recovery actions must be carefully studied. ***The general view that static natural or cultural heritage features can be maintained in a changing environment is being replaced with an understanding that some alterations to these values cannot be avoided.*** Therefore assessing disaster risks will become increasingly complex as these properties experience both gradual and sometimes catastrophic affects of ***climate change*** (UNESCO 2010 p24).

Climate change may also increase impacts of disasters on World Heritage cultural properties through its effects on significant underlying risk factors. Any increase in soil moisture, for example, may affect archaeological remains and historic buildings, thereby increasing their vulnerability to natural hazards such as earthquakes and floods.

- sea-level rise
- melting permafrost
- rainfall pattern change
- increased storm severity or frequency
- desertification (UNESCO et al. 2010).

Insert 19.6 Typology of the hazard of climate change

In a warming world, climate change related disastrous events will include health impacts such as epidemics, heat waves and displacement impacts such as abandonment due to cyclone damage. These impacts will affect World Heritage custodians and traditional owners. As well, where World Heritage sites are in daily use and occupied by local communities, whole support regimes may be disrupted by climatic shocks such as storms or local floods. Where cultural sites are concerned, the impact can be even greater. Inscribed buildings and landscapes are often 'living' or 'lived in' heritage. World Heritage examples are the still vibrant ancient trading town of **Hoi An in Vietnam**, the **Sewell working copper mine in Chile**, the bustling crossroads city of **Samarkand** and the pilgrimage site of **Lumbini, Nepal, the Birthplace of the Lord Buddha**. The consequences of climate change related extreme weather events on these sites could lead not only to the degradation of the site but also to the break-up of communities, the fragmentation of cultural practices, psychological distress, displacement and abandonment. The repercussions can be far reaching. As Augustin Colette explains in *Case Studies on Climate Change and World Heritage*, "As far as the conservation of cultural heritage is concerned, this abandonment raises an important concern in contexts where traditional knowledge and skills are essential to ensure a proper maintenance of these properties. In this respect, biological changes (with species shifting ranges) can also have an impact on conservation issues, with the reduction of availability of native species to repair structures and buildings" (Colette 2007 p65).

Recommendations

19.1 Encourage States Parties to consider their national **climate change vulnerability** and to subsequently initiate major site recording programs (for national listing and possibly World Heritage listing) targeting those places at risk of loss or major damage, and for which mitigation cannot be sufficient.

19.2 Acknowledge the **differential climate change impact** on World Heritage related to **geographic location**. Sites located towards the poles and at altitude, (in the Himalayas, sometimes referred to as the third 'pole' (UNESCO, SCOPE, and UNEP 2011) are experiencing temperature rises two or more times greater than the average global change.

19.3 Support policies that favour the nomination of sites from the more **vulnerable** States Parties that lack inscriptions (as advocated at 14.2)

19.4 **Reaffirm the preamble** to the WHC, paragraph 8 "Considering that, in view of the magnitude and gravity of the new dangers threatening them, it is incumbent on the international community as a whole to participate in the protection of the cultural and natural heritage of outstanding universal value, by the granting of collective assistance which, although not taking the place of action by the State concerned, will serve as an efficient complement there to." The purpose of the reaffirmation is to raise awareness of the danger posed by climate change to World Heritage and of **our collective responsibility** to our 'World Heritage commons' to protect it by working through the WHC for world climate change mitigation and a global acceleration to renewables.

19.5 Support the development of "standards, protocols, indicators and databases within the field of cultural heritage and climate change" (WHC 2007 p14) **with particular emphasis on disadvantaged states parties that are especially vulnerable to climate change impacts** on built heritage and indigenous heritage.These developments could be used to prioritise affirmative action when addressing the imbalance of geographic representation, development marginality and equity.

19.6 Address marginality through further **capacity building** and provision of practical assistance, expert advice, and funding as is occurring with the SIDS and *Focus on Africa* priority programs. Consider the resonance that the **Mother Earth movement** is having with the developing world and whether there are opportunities for its strengths to be incorporated into addressing marginality.

19.7 Achieve **a positive inscription status** for all States Parties.

The Intangible Cultural Heritage Convention

Adopt a global contract between industrialized and developing countries to scale up investment in approaches that integrate poverty reduction, climate stabilization, and ecosystem stewardship.
The Stockholm Memorandum. Tipping the Scales towards Sustainability, 2011

Climate Change Vulnerability and Human Development

The chart in Insert 19.7 indicates the numbers of Intangible Cultural Heritage listings for each State Party where the States Parties are ranked according to climate change vulnerability. Ranking is according to *The Climate Monitor's* climate change vulnerability factor 2010 (DARA 2010). The vulnerability groupings are: acute, severe, high, moderate and low. Overall the charts show that there is no obvious relationship between climate change vulnerability of States Parties and their numbers of inscribed elements. Those nations which do not hold inscriptions and those with few inscriptions are spread across the climate change vulnerability spectrum. As previously noted the Least Developed (LDCs) States Parties and the States Parties that are Small Island Developing States (SIDs) are more heavily clustered in the more climate change vulnerable categories.

The charts in Insert 19.8 link States Parties with 'human development'. They show the absence of any clear relationship between the human development indices of States Parties and their numbers of Intangible Cultural Heritage inscriptions. The most notable features are the lower mean number of inscriptions from the States Parties ranked as low on the Human Development Index and the large variability with regard to numbers of inscriptions per State Party within the three higher human development listed index groups. The 40% of States Parties without any inscriptions at all were drawn from all four index groups without any great difference between groups.

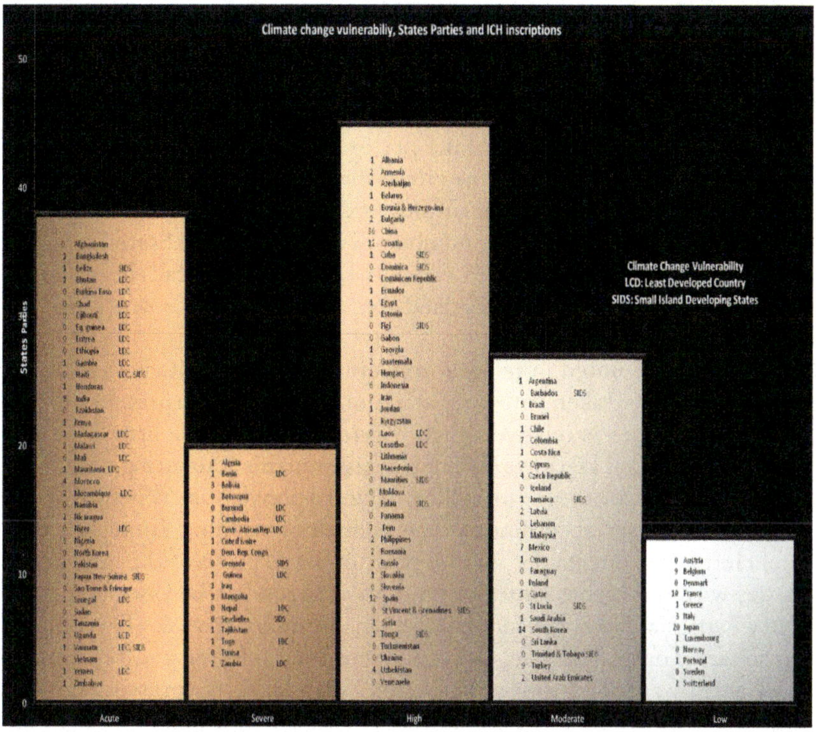

Insert 19.7 Intangible Cultural Heritage inscriptions per State Party grouped by climate change vulnerability factor, 2010. Group categories are acute vulnerability, severe vulnerability, high vulnerability, moderate vulnerability and low vulnerability (DARA 2010)

Keeping in mind that there have only been 3 rounds of nominations to the Intangible Cultural Heritage Lists, these charts indicate a degree of success in representativeness. Generally there are high, medium, low and zero inscription-holding States Parties across the climate change vulnerability and human development spectrum. The States Parties with low or zero inscriptions are not linked clearly with disadvantage in relation to climate change vulnerability or human development. This could indicate that the Intangible Cultural Heritage Committee's advocacy programmes have been well targeted. The charts indicate which States Parties are most vulnerable and disadvantaged and therefore most likely to benefit from the Convention's resources to realize the full benefits of the Convention's safeguarding.

Marginality: Regional, Zonal, Area and Population Distribution

The charts in Inserts 19.9 and 19.10 plot the number of inscribed elements for each State Party. Insert 19.9 *ranks* the States Parties in order of area, low to high, left to right. The graph, insert 19.10 does the same for population. The 'area' chart shows that, in general, the number of inscriptions per State Party is related to the nation's land area, those with smaller areas having fewer inscriptions than those with larger areas. Whilst many of the

States Parties with small land areas have no inscriptions, few with large areas lack inscriptions altogether. Variation in numbers of inscriptions between States Parties appears to increase with increasing land area.

The 'population' chart shows that, in general, with increase in population, the number of inscriptions per State Party also increases. States Parties lacking inscriptions tended to be small in population with few of the States Parties with larger populations lacking inscriptions. Again there is considerable variation between numbers of inscriptions per State Party particularly at the higher population end of the spectrum. Overall the pattern is similar to that observed between land area and numbers of inscriptions per State Party.

As previously noted in relation to cultural World Heritage sites, when the population is taken into consideration the level of representation of small State Parties is relatively high. Currently there are a little over 300 Intangible Cultural Heritage elements registered this amounting to on average one element for every 23.1 million of global population. When all States Parties are analysed according to their number of inscribed Intangible Cultural Heritage elements per their population size, six out of the nine States Parties with the highest inscription rates are found to be associated with States Parties with populations of less than one million people. These States Parties are Tonga, Vanuatu, Belize, Cyprus, Luxembourg and Bhutan. Of these only two are classified as having very high HDIs (Cyprus and Luxembourg). Of the remaining 4 States Parties two are classified as having high HDIs and two medium HDIs. Three of these are also classified as having acute vulnerability to climate change and one as having high vulnerability. The remaining 14 States Parties with populations less than one million have no listed elements. These did not differ significantly from the former group with respect to either HDI or climate change vulnerability. All in all the 20 States Parties with populations of less than a million had 7 Intangible Cultural Heritage elements for a cumulative population of 6.8 million people, achieving a representation nearly 24 times the global average. From this it is apparent that the Intangible Cultural Heritage Committee has initiated **an effective affirmative program**.

Strategists will be interested in recognising that there are States Parties across the spectrum of population and national area with no inscriptions that could benefit from encouragement and follow-up dissemination and implementation sessions. Of particular focus will be the States Parties shown in these charts that cluster at the low population and small national area end of the spectrum, as these States Parties are in general poorer in resources. At an international level the Intangible Cultural Heritage Committee recognises the regional (insert 14.14) and zonal imbalance (insert 14.15) in inscription status as discussed in Chapter 14. The Committee, mindful of its own obedience to equitable geographic member representation and rotation (Article 6), has made strong efforts through 'information-sharing' programmes (see findings to Chapter 23), and direct assistance to inform and influence this inequity and marginality.

The Intangible Cultural Heritage Committee has not developed any criteria for assessing climate change vulnerability or advocating its assessment by States Parties. This is especially problematic. How do practitioners assess the vulnerability of their intangible cultural heritage? There has to date been little consideration of how to marry safeguarding actions with the challenges of climate change. Also how do practitioners manage and secure without stultifying intangible cultural heritage in the face of climate change impacts?

Risk

General assessments of risk in relation to Intangible Cultural Heritage elements at the international level are made by examiners in response to the descriptions on nomination forms on a case by case basis. This happens when the State Party is nominating an element 'under immediate threat' for inscription on the List – *Intangible Cultural Heritage in Need of Urgent Safeguarding*. As a result of the Committee's assessment, the nomination is accepted or rejected, or the decision may be deferred pending more information. To date 27 elements have been inscribed. The ICHC proposes multiple ways of safeguarding intangible heritage including inventory- making, training, educating, awareness raising and capacity building plus 'non-formal means' (Articles 12, 13 14). However apart from keeping "the public informed of the dangers threatening such heritage" (Article 15b) the ICHC gives no direct means of addressing complex mid and long term threats such as climate change, or any clear direction of the need to instigate disaster risk preparedness.

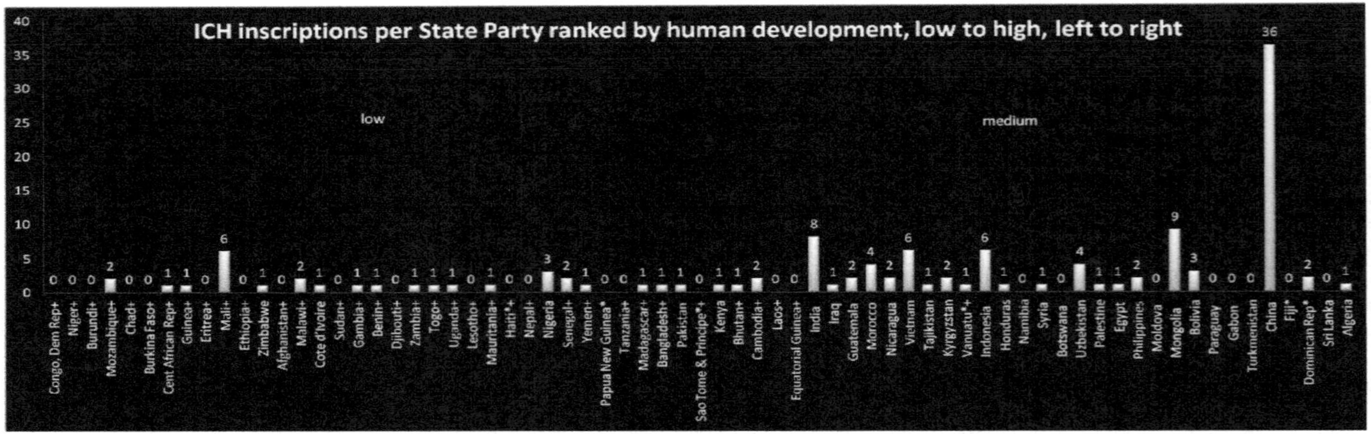

Insert 19.8 Intangible Cultural Heritage inscriptions per State Party ranked by population, 2011

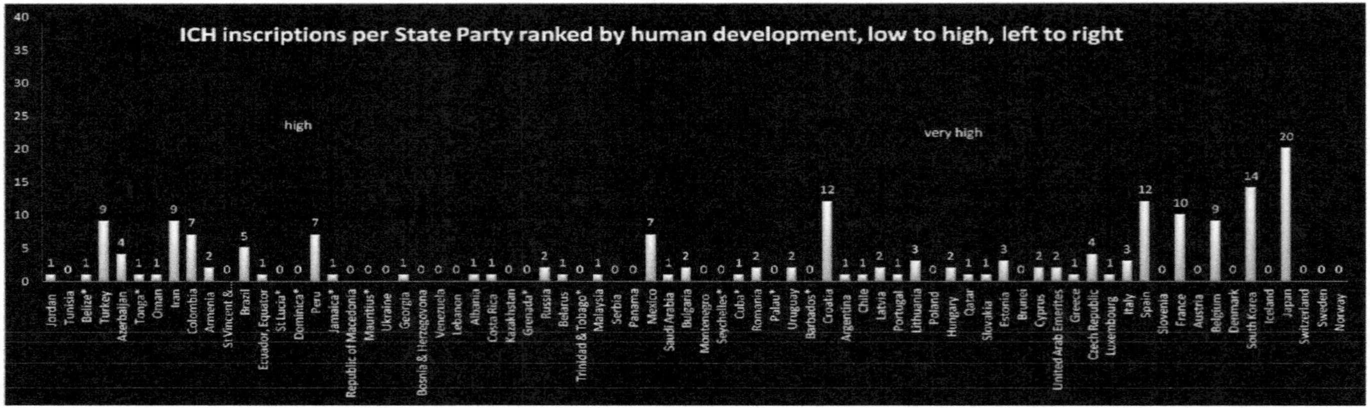

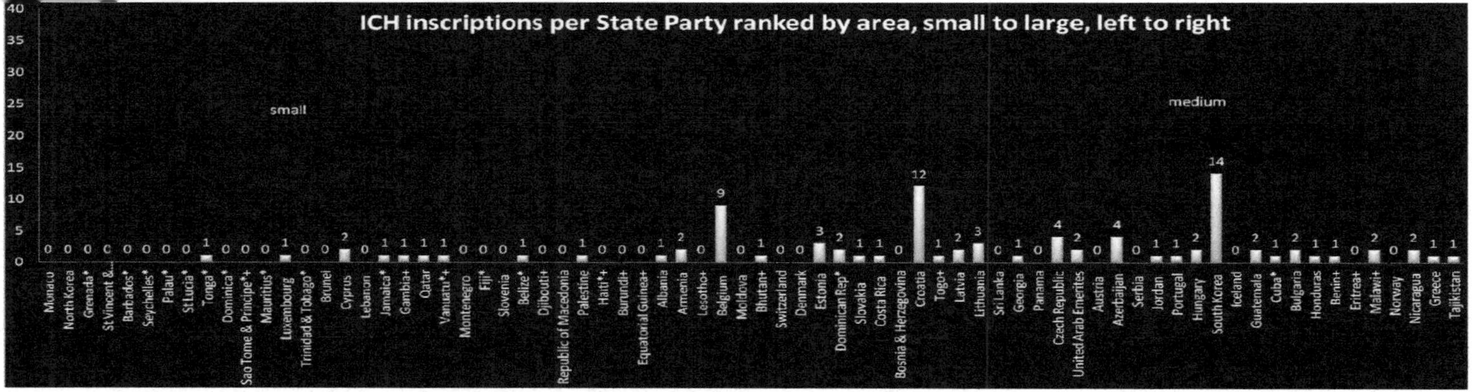

Insert 19.9 Intangible Cultural Heritage inscriptions per State Party ranked by area, small to large

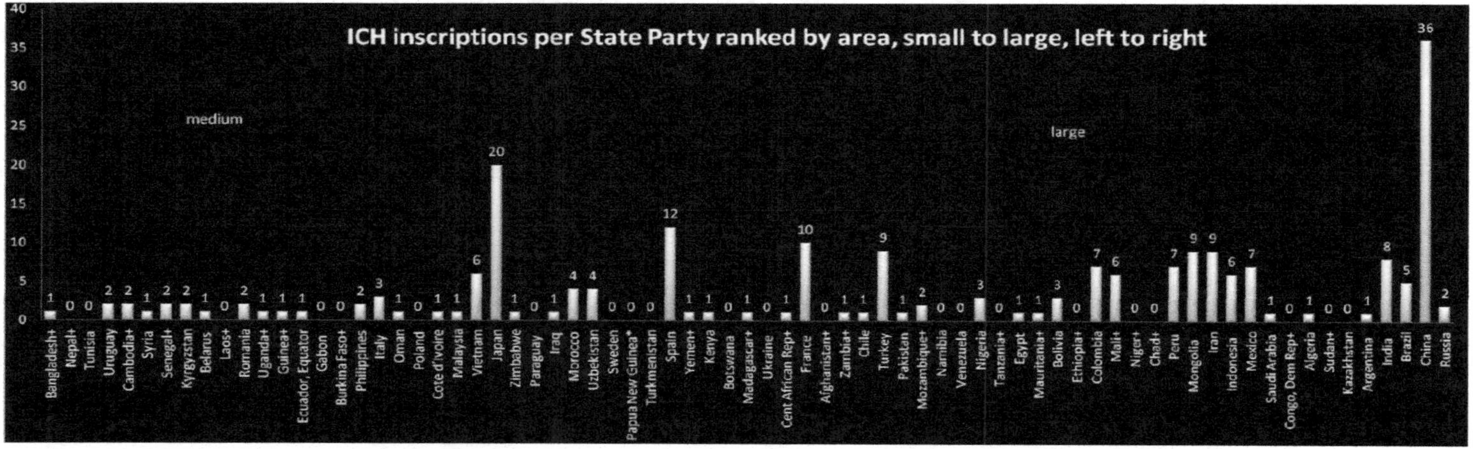

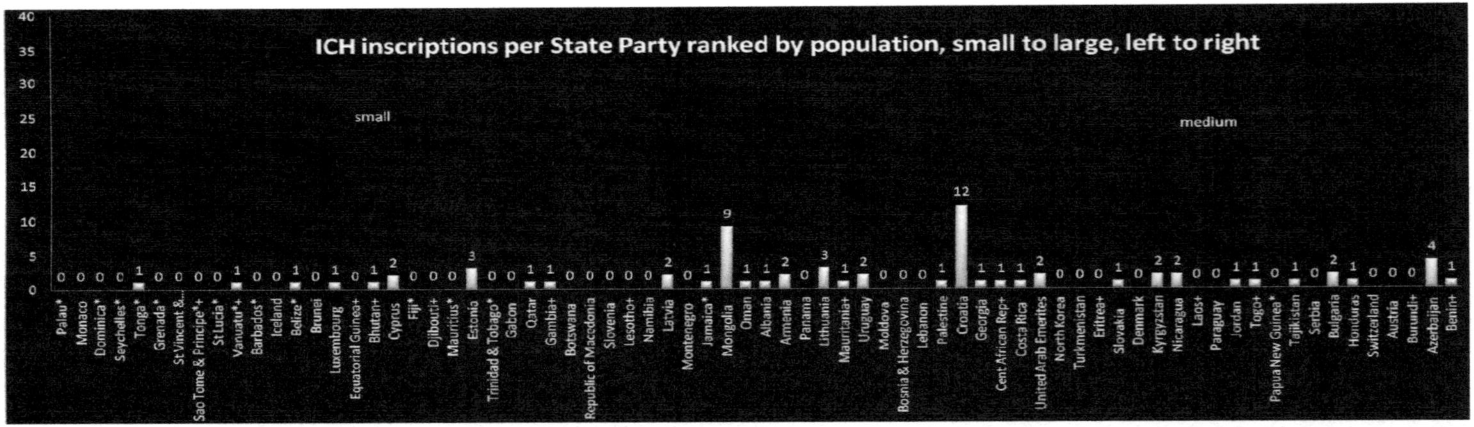

Insert 19. 10 Intangible Cultural Heritage inscriptions per State Party ranked by population, 2011

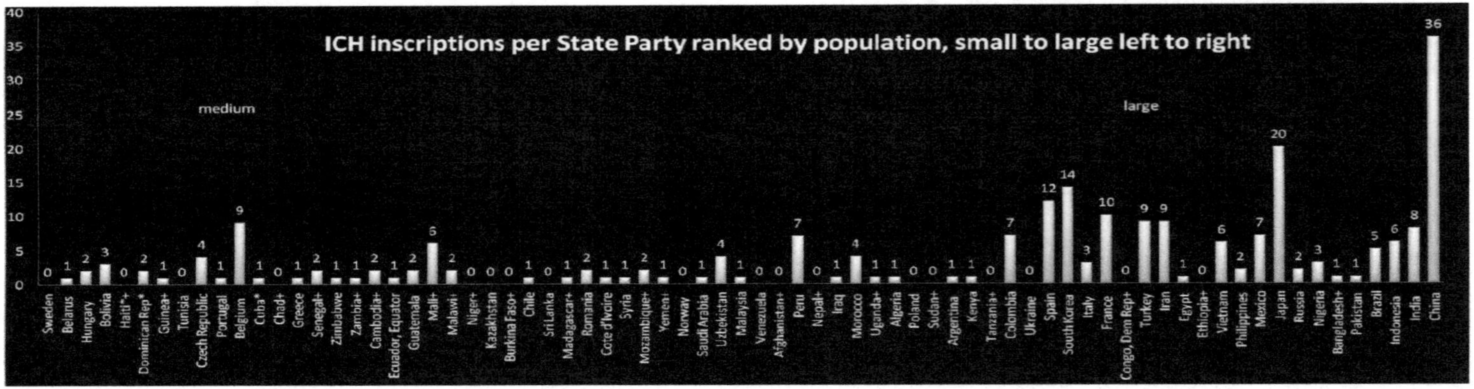

Many analysts are drawing attention to the need for local, national and international vulnerability studies as the first step in facing anticipated climate change problems. The ICHC and *Operational Directives* do not address vulnerability in their processes. Although intangible cultural heritage such as **Chinese calligraphy** can be portable and moveable (UNESCO 2009f), and therefore may appear more immune from climate change impacts than immoveable heritage, this is not so. Intangible cultural heritage is not mobile by definition. What often does define it is **the *context/environment*** on which it is based. It is this environment, this context that must be safeguarded as part of safeguarding the element's practice. As noted, in the nomination documents of the Listed Mexican element, **Places of memory and living traditions of the people of Tolimán: the Peña de Bernal, guardian of a sacred territory** (UNESCO 2009l) specific reference is made to the *context* which is being impacted by climate change.

The particular challenge for intangible heritage inventory recorders, policy makers and managers is to have vulnerability assessment tools which can document and assess the element in its entirety including this *context,* now, and assess its vulnerability to a climate changed future.

In response, the authors have developed the table ***Assessing Intangible Cultural Heritage Vulnerability to Climate Change*** with reference to Schroter's 8 step approach, see chapter 30, insert 30.1 (Schröter 2005). It tabulates qualifiers under the headings 'Climate change induced culture indicator', 'Culture risk' and 'Impact on intangible cultural heritage'. They have also compiled a table similar to that described above relating to climate change and risks to tangible heritage. It also appears in chapter 30 at insert 30.3 and is titled ***Tangible Cultural Heritage Risk Assessments.***

Recommendations

19.8 Acknowledge achievements to date in attaining **inscriptions for smaller States Parties.**

19.9 Continue to preferentially encourage the submission of, and favour the acceptance of, intangible heritage proposals from States Parties classified as being at acute, severe or of high **vulnerability from the effects of climate change.**

19.10 Continue to preferentially assist those States Parties with small populations and low areas which appear to be **disadvantaged** as far as the number of inscribed elements held.

19.11 Formulate intangible heritage climate change vulnerability assessment **tools**, for use at community, regional and national levels for inclusion in the proposed *Intangible Cultural Heritage Climate Change Response Document*. Vulnerability assessment research should invite full community **participation.** As this is a relatively new **area of inquiry** participants should be aware that current knowledge is limited and will benefit from interdisplinary contributions. A challenge will be, as with the ICHC nomination processes, to prioritise community engagement and make the vulnerability assessment tools accessible at village level.

19.12 Consider designing an online **joint World Heritage and Intangible Cultural Heritage climate change toolkit** that might incorporate the '*Framework for addressing climate change impact*' (Chapter 28) and the risk and vulnerability assessment tables in chapter 30, insert 30.1 *Assessing Intangible Cultural Heritage Vulnerability to Climate Change*; insert 30.2 *Intangible Cultural Heritage Risk Assessments* and*Tangible Cultural Heritage Risk Assessments* at insert 30.3. Consider referring to the 'toolkit' at the nomination and reporting stages, in the *WHC Operational Guidelines* and the *ICHC Operational Directives.* Propose including examples of Vulnerability Assessment Case studies of cultural heritage, see chapter 7. Such **case studies could be likened to List 3** of the ICHC List 3 *Practices and Process of Safeguarding Intangible Cultural Heritage*.

19.14 Be aware of the **complexity** of vulnerability processes. Reviews of vulnerability 'tools' may be useful in determining their merits and demerits.

19.13 Support the new manual: *Managing Disaster Risks for World Heritage*. Consider making additions to future editions to include Intangible Cultural Heritage.

Chapter 20

How do the Conventions Address the Need for Climate Change Mitigation?

If we're honest, we can have some hope of change. If we're not honest, we can do nothing.
Kevin Anderson, Tyndall Centre, University of Manchester July 2011 (Anderson 2011).

Do the World Heritage Convention and the Intangible Cultural Heritage Convention have the capacity for a multilevel mitigation responsibility and mitigation mechanisms (including monitoring and reporting) for addressing climate change impacts on cultural heritage?

The World Heritage Convention

Site-based Actions

The WHC has the capacity through Articles 11.7, 5 and 29 to protect World Heritage sites from climate change impacts. To date the World Heritage Committee has done this by adopting site-specific actions such as the preventative measures of monitoring key factors as for example, increase in pest and biological infestations in the Australian World Heritage site **Royal Exhibition Building and Carlton Gardens** (Heath 2008 p20). WHC reporting mechanisms e.g., Periodic Reporting, provides further preventative opportunities. As the required studies and research of the site (Article 11.7) expose the present and potential impact of human–induced climate change as a threat, then the State Party is committed to "do all it can" (Article 4) to protect the site . This would allow on-site mitigation measures to be adopted such as installing solar panels to generate energy. The preventative measures discussed in the report *Predicting and Managing the Effects of Climate Change on World Heritage, A Strategy to Assist States Parties to Implement Appropriate Management Responses* and *Policy Document on the Impacts of Climate Change on World Heritage Properties* focus on the physical and biological, with little mention of the climate change induced human impacts on cultural sites, other than to state that little is known of their effects.

A review of over 70 World Heritage Decisions and draft Decisions since 2006 that specifically included climate change in their wording (tabled in Chapter 26) showed that only one specified climate change mitigation. This is not to ignore the mitigation programmes World Heritage States Parties are independently adopting such as the REDD scheme, but it signifies that despite climate change being increasingly a topic of decision-making for the World Heritage Committee, climate change mitigation programmes are not high on the agenda.

The *Operational Guidelines* (IV.A.V Annex 7) state that monitoring, reporting and mitigation are integral to all levels of World Heritage management with the use of any effective method encouraged. These *Operational Guidelines* are backgrounded in the four World Heritage documents:*Policy document on the impacts of climate change on World Heritage properties, A Strategy to Assist States Parties to Implement Appropriate Management Responses, Strategy for Reducing Risks from Disasters at World Heritage Sites and* the report *Predicting and Managing the Effects of Climate Change on World Heritage.* When nominating a property for inscription, evidence of monitoring the properties condition is required. Findings can be reviewed regularly so as to give an indication of trends over time. Reactive Monitoring, Reinforced Monitoring, and Periodic World Heritage Reporting have already been explained. However it is necessary to become familiar with the level of monitoring required in Periodic World Heritage Reporting to fully appreciate the comprehensiveness of the protection sought for World Heritage cultural properties.

States Parties have the responsibility to provide up-to-date ***state of conservation*** information on the physical condition of any property including any threats to the OUV of the property and the conservation measures being implemented at the property (OG Paragraph 132). These become key indicators, for example, in a historic town the protective action, 'monuments need to be repaired' is reported, as well as the scale and duration of any recent or forthcoming major repair projects. This anticipated action becomes a key indicator for monitoring to measure and assess the state of conservation of the property, the factors affecting it, conservation measures at the property, the periodicity of their examination, and the identity of the responsible authorities. At least six key indicators for each site are requested. Up-to-date information should be provided in respect to each of the key indicators. Care should be taken to ensure that this information is as accurate and reliable as possible, for example by making observations in a consistent manner, using similar equipment and methods, and making measurements at the same time of the year and day.

Understanding the breadth of details of periodic World Heritage Reporting and the depth of detail in reactive monitoring and reinforced monitoring is of direct interest when considering how to measure climate change induced 'human impacts' on World Heritage cultural properties.

According to the publication *A Strategy to Assist States Parties to Implement Appropriate Management Responses,* site specific mitigation measures are *limited,* to reducing emissions at site level, advertising these as best practice, exploiting carbon sequestration opportunities, and using improved technology. Information might also be provided to the IPCC and the UNFCCC. The WHC's position on mitigation at the national and international level is determined by the World Heritage *Policy*. The *Policy* was prepared in 2007 by a specially convened 'working group'. Of interest are the documents drawn upon by the developers of the *Policy*. The documents record how forthrightly framed advice was given by a representative from Australia (which has one of the highest GHG emissions per capita). This advice was echoed less dogmatically by other industrialised States Parties, telling the working group that the WHC should confine itself to mitigation action at site level only. Submissions from developing States Parties (which have miniscule emissions per capita by comparison) appealed to the meeting for the World Heritage Committee to adopt widespread advocacy at all levels for reduction in GHG emission levels. Advice from an **International Union for the Conservation of Nature (IUCN)** legal submission for obligatory GHG emission reduction directives to be developed, and from another State Party for a dedicated 'climate change endangered' List to be considered, were sidelined (UNESCO 2007e). The evolution of this mitigation position taken by the World Heritage Committee subsequent to the working group recommendations is addressed in Chapter 26 comments column. As a result the *Policy document on the impacts of climate change on World Heritage properties,* which was endorsed in 2007, takes an inadequate stance on the 'big picture' of climate change mitigation deferring mitigation issues at the national and international level to the UNFCCC.

The *Policy* echoes the World Heritage Committee's "characterization of the UNFCCC and the IPCC as the primary international institutions to address climate change," indicating "that the World Heritage Convention should focus on its 'comparative advantage' of management of outstanding cultural and natural properties" (Burns 2009 p155). It has since been equally energetically argued that the UNFCCC and the IPCC are not necessarily the 'primary' institutions for World Heritage purposes (Burns 2009 p161), that having more than one agency dealing with the same issue is not necessarily a disadvantage and that overlap in monitoring climate change indicators already exists between agencies. Environmental lawyer W.C.G. Burns argues that the UNFCCC might be characterized as a quintessentially failed regime, with burgeoning global greenhouse gas emissions and critical thresholds for severe impacts looming for both human institutions and ecosystems. Thus, it would be reasonable for the WHC to step into the vacuum created by the UNFCCC to fashion remedies to protect the sites within its trust. This, of course, would not preclude the WHC from reverting to a more secondary role on this issue should the parties to the UNFCCC commit themselves to the kind of deep cuts in greenhouse gas emissions that will be critical to protect the world's natural and cultural heritage (Burns 2009 p160).

The *Policy Document on the Impacts of Climate Change on World Heritage Properties* does countenance that some properties may be able to be involved in sequestration and carbon offset activities at a national level and it does take the initiative to advise the World Heritage community to reduce air travel, recycle and purchase Gold Standard offsets.

The Asia and Pacific region is particularly exposed to disasters and the short and long term impacts of climate change. However, managers of most World Heritage properties in the region, especially of cultural and mixed heritage sites, are not aware of the specific risks that may affect their sites, nor have they put in place adequate plans to reduce these risks. The *Policy Document on the Impacts of Climate Change on World Heritage Properties* in recognising this, states: "Priority in all climate change related actions under the Convention will be given to properties in Africa and SIDS" (UNESCO 2008a p4). Although the World Heritage Committee explains that some mitigating measures are seen as "beyond the scope of the World Heritage Convention" (Colette 2007 p15) it is beholden to all in the World Heritage community to advocate for global mitigation.

UNESCO has advised in its book *Case Studies on World Heritage and Climate Change* that the World Heritage network is a useful tool to share and promote lessons learnt and best practices, as well as to raise awareness regarding climate change impacts (on World Heritage properties) using their iconic value (Colette 2007). As has been mentioned this approach has been taken up by the *UNESCO Climate Change Initiative* in its *Global Climate Change Field Observatory of UNESCO Sites programme* (Insert 15.3 and 15.4). This 2009 *Field Observatory* programme uses UNESCO World Heritage Sites as priority reference sites for understanding the impacts of climate change on human societies and cultural diversity, biodiversity and ecosystems services, the world's natural and cultural heritage and the possible adaptation and mitigation strategies.

It is appropriate that this *Climate Change Initiative* links World Heritage with climate change mitigation. Too long has the World Heritage Committee remained publically uninvolved with reducing climate change in other than, according to some, the weak and ineffectual ways described in the 2006 *Publication, A Strategy to Assist States Parties to Implement Appropriate Management Responses,* (Huggins 2007). The report *Predicting and Managing the Effects of Climate Change on World Heritage* called for "the implementation of regional and/or trans-boundary mitigation and adaptation strategies that reduce the vulnerability... of World Heritage sites" (UNESCO 2006a p53), but this advice was never followed up with guidelines or a framework for action.

Broad Based Mitigation, an Inconvenient Truth

As is unequivocal now, according to environmental lawyer Erica Thorson, World Heritage sites will never be preserved for transmission to future generations unless the States Parties, led by the World Heritage Committee, act more proactively than merely supporting site specific mitigation (Thorson 2007). As Anna Huggins has argued in her article *Protecting world heritage sites from the adverse impacts of climate change: obligations for States Parties to the World Heritage Convention*, through "its mandate to protect and preserve places of OUV the WHC provides an unlikely yet effective tool in global efforts to mitigate climate change" (Huggins 2007 p121). She explains that the iconic nature and high profile of many World Heritage sites makes them ideally suited to build public and political support for greater action to ameliorate climate impacts. This is reinforced by the idea that heritage sites are 'places in the heart', that is 'places and objects [that] contribute to a sensory and emotional perception of belonging, of home and community' (Lyster et al. 2007). As a result, current and future climate change impacts on these sites are likely to resonate more emotionally and immediately with the public than the science of climate change such as the impacts of the acidification of the oceans. There is a potential flow on effect. Efforts to protect World Heritage sites from climate change will also protect the surrounding areas (Huggins 2007 p122). As States Parties are obliged to protect the World Heritage sites in their territory, so they are obliged to make meaningful reductions to GHG emissions and to support a fair, ambitious and binding international agreement through the UNFCCC. States Parties, of course, could choose to make more meaningful reductions outside the UNFCCC agreement. How much more outraged will the reaction be when people realise that their beloved World Heritage sites are knowingly being mismanaged because of the reluctance of their State Party to commit to deep cuts to CO_2 pollution and of the reluctance of the World Heritage Committee to commit to encouraging deep cuts in CO_2 pollution. Huggins concludes "In order to ensure the continuing success of the WHC, States Parties must engage in extensive mitigation strategies without delay" (Huggins 2007 p127).

Terrill argues that the WHC's role in climate mitigation should be confined to registering an interest in national and international action (see Chapter 15). His assertion that the WHC has no role in addressing climate change beyond the site level because the World Heritage Centre and Committee lack expertise and the WHC lacks effective leverage to influence States Parties appears defeatist. His wish that the WHC "could establish regular contact with the UNFCCC at secretariat level" (Terrill 2008 p7) has not happened. Terrill believes that "Other challenges are currently more pressing (than climate change) although over time the balance will shift. Planning for adaptation to climate change impacts must commence in the short term, although an ongoing challenge will be to keep the level of effort commensurate with the threat." Three years on, climate scientists are telling us that such ideas are dangerously out of date.

James Hansen warns "If ice sheets begin to disintegrate, there will not be a new stable sea level on any foreseeable time scale. Instead, we will have created a situation with continual change, with intermittent calamities at thousands of cities around the world. Because the ocean and ice sheets each have response times of at least centuries, change will continue for as many generations as we care to think about. Change will not be smooth and uniform. Instead, local catastrophes will occur in association with regional storms. Given the enormous infrastructure and historical treasures in our coastal cities, it borders on insanity to suggest that humans should work to 'adapt' to climate change, as opposed to taking actions needed to stabilize climate' (Hansen 2009 p85).

Wil Burns concludes "The World's Heritage Convention's failure to take the opportunity to address the potential effects of climate change in a meaningful fashion is lamentable given the virtual abdication of responsibility by the parties to the UNFCCC to confront one of the most pressing issues of this generation, and many more to come. If our world heritage is truly 'a gift from the past to the future', then every effort must be made to address the climatic threats to both natural and cultural sites. Should the parties to the Convention continue to ignore these threats, essentially indulging the fiction that other regimes will act effectively, it will have abdicated its responsibility to both this generation and those that follow" (Burns 2009 p163).

Recommendations

20.1 Recognising that the World Heritage Committee in its Decisions (30COM 7Ba, 31COM 7.1 chapter 26) and climate change publications has repeatedly called for synergy with the UNFCCC and

- Taking into consideration the difficulties facilitating this, namely that the UNFCCC and the WHC share no single authority that can make decisions about their work, and have no cabinet through which to coordinate and
- Acknowledging that **cooperative action is not occurring**

It is proposed that the World Heritage Committee **promote mitigation action** at the State Party and global levels by adopting **Resolution 16 GA 10 A** (15.1) and associated recommendations 15.1, 15.2, 15.10, 19.4, 20.3, 20.4, 24.4, 25.2, 26.1, 26.2, 26.3, 26.4.

20.2 Support and publicise collaborative World Heritage-related climate change mitigation programmes **between developed and developing** States Parties.

The Intangible Cultural Heritage Convention

The ICHC does not make specific provision for mitigation measures for addressing climate change impacts on intangible cultural heritage. The ICHC has the opportunities through Articles 7a,13,29 and *Operational Directives* (Paragraphs 1-7, 79-99, 100-123, 151-164) to introduce measures and to access the expert support needed (as outlined in chapters 15 and 19) to formulate 'preventative' safeguarding policies, strategies and reporting directives by which climate change impacts may be addressed through mitigation, monitoring and reporting. Through the ICHC, traditional knowledge about the history of the element in the light of contemporary climate analyses could be used to help communities make 'preventative' decisions to safeguard their intangible heritage.

Recommendations

20.3 Propose that the ICH Committee join with the WH Committee to consider developing a **joint World Heritage and Intangible Cultural Heritage climate change policy** which would include a strong stance on mitigation at the international and national levels, see 15.2, 15.21, 20.1, 20.5, 21.6, 24.8, 25.2, 26.1.

20.4 Integrate **traditional knowledge** into an emerging *ICH body of knowledge* on climate change and its mitigation.

20.5 Monitor and report climate change impacts to safeguard ICH. Presently reporting opportunities exist in the Reporting Forms under 'assessment of the element's viability and current risks'. Climate change impacts could be recorded here but **specific climate change information** seeking *Operational Directives* are preferred as the latter enable better standardised data gathering.

20.6 Be prepared, when monitoring and reporting, **to address questions about how to record the 'changing' nature of intangible heritage.**

20.7 Be prepared also to suggest **new ways of documenting** a 'practice,' 'performance' or 'custom' to climate-protect it for survival.

20.8 Consider developing a self monitoring periodic **reporting** form for ICH (similar to that for World Heritage) that factors in potential climate change impacts with input from community **custodians.**

A mosque in lowland Ladakh, 150 km downstream of Leh

Chapter 21

How do the Conventions Address the Need for Climate Change Adaptation?

There are two main building blocks for a productive response to the adaptation challenge. The first is to make sure we have a strong, flexible economy, with smoothly functioning markets. The second is to make sure we have sound information about possible impacts of climate change on various regions and activities and that information is disseminated in easily useable forms.
These are the most valuable things that we could bequeath those who come after us as they do their best in a world of climate change. Adaptation policy is first of all about doing these things well (Garnaut 2011 Ch.8)

Do the World Heritage Convention and the Intangible Cultural Heritage Convention have the capacity for a multilevel adaption responsibility and mechanisms, such as maintenance, repair expertise, and financial resources to address climate change impacts on cultural heritage?

The World Heritage Convention

The World Heritage Committee has made concerted efforts to promote adaptive measures at a global, regional and site level. Perusal of the climate change related Decisions in Chapter 26 is testimony to this. The publication, *A Strategy to Assist States Parties to Implement Appropriate Management Responses,* and the report *Predicting and Managing the Effects of Climate Change on World Heritage* give theoretical guidelines proposing that climate change adaptation be integrated with climate change threat identification, climate change vulnerability analysis, climate change risk assessment (including Disaster Risk Management (DRM)) and management plans. The Periodic Reporting regime provides feedback on which maintenance, repair and preparedness programmes can be built. A diverse and only partly comprehensive set of 9 cultural examples are presented in the book *Case Studies of Climate Change and World Heritage* (Colette 2007). They clearly show how adaptive measures require cross disciplinary expertise, assiduous selection of assessment tools and financial support, sometimes of immense proportion. As has been referred to, the UNESCO Climate Change Initiative is drawing on World Heritage sites for information on which to build adaptive practices (UNESCO 2009j).

The separation in the report *Predicting and Managing the Effects of Climate Change on World Heritage* of adaptive actions for cultural and for natural properties may prove limiting as the best adaptive measures are holistic taking in cultural and natural aspects, regional and site factors. As the report well recognises, the competition for water needs between people and natural sites will become increasingly acute as climate change more fully impacts and undermines initial 'adaptive' measures. Indeed, the competition for land for settlement and land for food production will increase. The report anticipates the loss of cultural properties both from physical damage and social abandonment leading to the eventual loss of cultural memory (UNESCO 2006a p53), but does not elaborate adaptive measures to adjust to loss, or offer any guidance on how communities may have to decide what to support and what to let go.

UNESCO CLIMATE CHANGE INITIATIVE:

Core Program:

Climate Change, Culture and Biological Diversity and Cultural Heritage

.... Many small islands, rural and indigenous communities are already facing the impacts of climate change. Their high vulnerability relates to their reliance upon resource-based livelihoods and the locations and configurations of their lands and territories. The wealth of their knowledge and experiences to observe, cope with and adapt to change is increasingly being recognized as an important building block in the development of both climate change observation and adaptation strategies.... (UNESCO 2009j)

Insert 21.1

There are significant barriers to implementing adaptation. These include both the inability of natural systems to adapt to the rate and magnitude of climate change, as well as technological, financial, cognitive and behavioural, and social and cultural constraints. There are also significant knowledge gaps about adaptation as well as impediments to flows of knowledge and information relevant for adaptation decisions as has been raised by the IPCC in Chapter 10 of their 2007 Assessment Report (IPCC 2007). For example despite the well documented resilience of island communities to climate change, (Insert 21.1) and the growing recognition that this resilience is partly the result of strong cultural solidarity, unsettling changes have been observed. The cultural change and the increased 'individualism' associated with economic growth in Small Island Developing States (SIDS) have eroded the sharing of risk within extended families, thereby reducing the contribution of this social factor to adaptive capacity (Pelling 2001).

In summary, the (WHC) supports climate change adaptation measures, but with the monumental and unrelenting nature of climate change impacts, their implementation will, under Common but Differentiated Responsibilities (CBDR), require support from wealthy States Parties.

Recommendations

21.1 Propose that climate change adaptation in World Heritage management be **prioritised.** Include climate change adaptation as a major consideration in WHC implementation.

21.2 Disseminate and exchange up-to-date **climate change knowledge** with stakeholders, including indigenous communities and minority groups, prior to developing adaptive management policy and projects.

21.3 Ensure the dissemination process is **culturally appropriate**. For instance, climate change science can be meaningfully interpreted through a Buddhist perspective (His Holiness the 14th Dalai Lama of Tibet 2011) (Wangchuk 2007).

21.4 In consideration of recommendation 15.2 and as a consequence, propose revisions of the report ***Predicting and Managing the Effects of Climate Change on World Heritage*** and the publication, ***A Strategy to Assist States Parties to Implement Appropriate Management Responses.*** Add to the revisions a more detailed evaluation of climate change assessment options, (including assessment of physical, biological and 'human impacts'), advice on technical assistance available for assessments and illustrated examples of cultural World Heritage site assessments.

21.5 Adopt the approach of the **Yamato Declaration on Integrated Approaches for Safeguarding Tangible and Intangible Cultural Heritage.** Extend these to climate change impacts on World Heritage properties and ICH. **Engage** traditional custodians and their communities in these processes.

The Intangible Cultural Heritage Convention

The (ICHC) does not make reference to adaptative safeguarding measures for addressing climate change impacts on intangible cultural heritage. The ICHC has the capacity through Article 13 to 'foster scientific studies' and 'to keep the public informed of the dangers threatening such heritage' to partly address climate change. It also has opportunities to report on the dangers (Article 29, and *Operational Directives*) and to access expert support, 'international cooperation and assistance' (Article 1). The ICHC safeguards intangible heritage by multiple adaptive and management means (Article 2.3) but not, as yet, by overarching climate change specific safeguarding policies, strategies and reporting directives.

Recommendations

21.6 Suggest that the proposed *Intangible Cultural Heritage Climate Change Response and Policy Document*, (15.15) **includes the safeguarding actions of adaptation management, monitoring and reporting** of climate change impacts and responses.

21.7 Be aware that the vulnerability assessment knowledge needed to make adaptation-related decisions will itself raise **new areas of inquiry** for community leaders, heritage owners and practitioners.

21.8 Be prepared to consider **questions** such as how do communities make judgements about their changing craft or performance in the light of a confronting and climate compromised future? How will communities predict the changing environment in which their ritual or ceremony will be performed? How can communities safeguard their intangible heritage when that heritage is vulnerable to change? Are there lessons to be learnt from the evolution or demise of past customary practices? Which approaches offer better means to assess ICH risk and vulnerability to climate change?

A Chorten near Bagso, Ladakh being built with a combination of modern and traditional materials and methods by local workers, including women

Chapter 22

How do the Conventions Address Climate Change Threats to the Sustainable Development of Cultural Heritage?

How do the World Heritage Convention and the Intangible Cultural Heritage Convention address climate change-induced threats to the sustainable development of World Heritage and Intangible Cultural Heritage? For example how do the Conventions address breakdown in site/element maintenance caused by eco-tourism income loss that is itself caused by climatic extremes affecting both the associated tourist facilities and the traditional community of heritage custodians and/or practitioners and tourism staff?

- *Clearly we need to break with* ***'business as usual'****. Green economies are an important means to achieve what sustainable development ultimately aims at: the wellbeing of people while respecting the environment. UNESCO, Irina Bokova.*
- *Clearly, in order to close the implementation gaps,* ***we need to break with 'business as usual'****. This calls for a holistic response which addresses in an integrated and comprehensive manner the social, economic and environmental issues... Rio+20. (UNESCO 2012a)*

- *The challenges of the 21st century – resource constraints, financial instability, inequalities, environmental degradation are a clear signal that **'business-as-usual' cannot continue**. The planetary risks we are facing are so large that **business-as-usual is not an option** We have had a good run, but **business-as-usual cannot continue** ..., we are beginning to live off the Earth's capital, rather than the interest. (3rd Nobel Laureate Symposium on Global Sustainability 2011)*

The World Heritage Convention

The WHC makes no direct reference to **sustainable development (SD)** but the World Heritage Committee has interpreted Articles 4 and 5 of the WHC to introduce the concept into the *Operational Guidelines*. The 2011 *Operational Guideline* states in paragraph 6 "Since the adoption of the *Convention* in 1972, the international community has embraced the concept of sustainable development. The protection and conservation of the natural and cultural heritage are a significant contribution to sustainable development" (UNESCO 2011a). The 2008 *Operational Guidelines* included in paragraph 119 "... The State Party and partners must ensure that such sustainable use (of a World Heritage site) does not adversely impact the OUV, integrity and/or authenticity of the property. Furthermore, any uses should be ecologically and culturally sustainable". This guideline has now been revised as recorded later in this chapter.

Climate change is a direct threat to sustainable development as it undermines the three accepted pillars – environmental, social and economic, and the one widely advocated and proposed pillar – culture, on which sustainable development is built. Since climate change will only intensify, 'resilience to climate impacts' in the face of climate change is the proposed and justifiable 5th pillar of support (see Chapter 11). Sustainable development is an objective not a reality. It must now plan to be 'climate-proof' as integral to being 'environmentally responsible,' 'economically healthy,' 'culturally vital' and 'socially just,' see Insert 11.1 titled *Culturally Sustainable Development in a climate-changing world.*

According to World Heritage Committee's recent expert advice on the relationship between the WHC, conservation and sustainable development "The possible conflict between conservation and development should be therefore addressed through a balanced analysis (of the pillars) taking into account legitimate interests while reconciling global and local values" (Rossler 2010 p5). The emphasis has shifted from the fundamentality of the sustainable development UN definition of 1987 with its emphasis on 'without compromising' (recorded in Chapter 11) to one that is open to compromise and dependent on 'balanced analysis' variously also qualified by the World Heritage Committee as 'careful consideration' or 'appropriate and equitable consideration', or 'balanced compromise'. As climate change bites more viciously eroding economies and dislocating societies it can only be hoped that people demand that this 'balance' be based on long term survival rather than short term expediency.

As has been addressed in Chapter 11, the empowerment of women and family planning are additional and important factors in sustainability and sustainable development. There are no references in the WHC documents relating to these issues in regard to cultural World Heritage. However UNESCO's *Preliminary Proposals by the Director-General for priority fields of action for document 36 C/5 2012-2013* (UNESCO 2010n) include supporting:

- Women's participation in sustainable development, including climate change.
- science education for girls
- empowerment of women as agents of change and decision-makers working on climate change adaptation, mitigation and sustainable development

Much energy, innovation and ground breaking initiative has been invested in sustainable developments associated with and in World Heritage properties (Galla 2002). There has been a groundswell of enthusiasm and success originating in developing States Parties to embrace 'sustainable heritage development' resulting in many museums and eco-tourist programmes. For instance, the emergence of the concept of the eco-museum, and its successful 'model' – the ***Ha Long Ecomuseum*** in the World Heritage site of spectacular Ha Long Bay, Vietnam has been an inspiration for other communities and has invigorated sustainable development in the museum sector in Vietnam. The Museum was developed by the local fishing community in 2003 and has remained sustained by community management since. The IPCC 4th Assessment (2007) in its deliberations on climate change and sustainable development has drawn attention to integrating the two way relationship of sustainable development and climate change; that is recognising that the pace and character of development influences adaptative capacity and that adaptative capacity influences the pace and character of development.

The impacts of climate change are being felt particularly acutely on a number of small Pacific Islands. Mindful of the 'two way relationship' UNESCO has directed much energy and resources into promoting sustainable development and education to this region through its **Intersectoral Platform on Small Island Developing States** (UNESCO 2009v) and its contributions to the **Mauritius Strategy for the implementation of the programme of action for the sustainable development of small island developing states.** The projects involved, build upon and reinforce the resilience and strength that characterizes island societies. Recognition is paid to the importance of the cultural identity of people in advancing sustainable development. *Local and indigenous knowledge* is recognized and reinforced in SIDS education and environmental management as a response to climate change. Integrated heritage policies have been developed especially for SIDS. The number of successful nominations from SIDS on the World Heritage List has improved. Capacities for sustainable conservation have been enhanced and intangible cultural heritage safeguarding plans developed (UNESCO 2010p). In 2011 the World Heritage Committee reported that a project proposal on ***Capacity Building to Support the Conservation of World Heritage Sites and Enhance Sustainable Development of Local Communities in Small Island Developing States (SIDS)*** had been developed to build regional capacity building programmes for Pacific and African SIDS, and to further strengthen the existing capacity building programme for Caribbean SIDS (CCBP). However, despite this tenacious planned programming there has been little development (except in the recognition of the role of traditional knowledge) in formally recognising that World Heritage site management should protect sustainable development programmes from climate change.

A review of World Heritage programmes addressing management and sustainable development goes some way to explain this lack of 'protective' consideration. The review demonstrates how the WHC has been used, to pursue sustainable development at the expense of properly addressing its global climate change threat.

Management Strategies 1992-2011

The 1992 **Strategic Objectives**, the 2002 **revised Strategic Objectives**, the 2002 **Partnerships for Conservation Initiative (PACT)** and the 2006 **World Heritage Centre's Natural Heritage Strategy** mark stages in the focussed effort to improve the management of World Heritage properties' sustainable development. Mechtild Rössler from the World Heritage Centre explains that it is a challenge for a State Party to make the "paradigmatic shift" from working through the World Heritage listing process (when nominating a site for inscription) to achieving "best practice management and standards in the conservation" of that site. It has been made even more so because the management of cultural heritage under the WHC "is only defined in very broad terms as each site and type of site differs: cultural landscapes or living cities, single monuments or large scale archaeological sites have completely different resources. They are managed through processes by which the OUV of the property is protected and cultural heritage resources are given consideration in both the local and global contexts, including issues such as local population pressures, increasing international tourism and climate or global change"(Rossler 2010 p4). The difficulty of achieving good management has been a dilemma that has demanded much energy and resources from the World Heritage Committee. The consideration of the 'global contexts of climate change' has been, as documented in Chapter 26, either directed elsewhere (to the UNFCCC) or been overlooked in the determination to embed sustainable development as a necessity for World Heritage conservation.

The Strategic Orientations (1992) adopted by the 16th session of the WH Committee were supported by the WHC Articles 4 and 5. They comprised **5 Strategic goals** and their attendant objectives (UNESCO 2002b p21). The goals are 1. Promote completion of the identification of the world heritage. 2. Ensure the continued representativity and credibility of the World Heritage List. 3. Promote the adequate protection and management of the WH sites. 4. Pursue more systematic monitoring of WH sites. 5. Increase public awareness, involvement and support. These were developed as a direct influence of the **1992 United Nations Conference on Environment and Development (Rio Summit)** and marked a change in the implementation of the WHC. Management was recognised as a tool in best practice conservation (Rossler 2010 p3). In 2002 the World Heritage Committee, acknowledging the continuing difficulty of achieving good management, called for nations and other partners to join together and co-operate in the protection of heritage to ensure the WHC *"remained an instrument for the sustainable development of all societies through dialogue and mutual understanding"* (UNESCO 2002c). The outcome was the 2002 **Budapest Declaration**. The Declaration stresses the need to "ensure an appropriate and equitable balance between conservation, sustainability and development, so that World Heritage properties can be protected through appropriate activities contributing to the social and economic development and the quality of life of our communities" (UNESCO 2011d). It invited all partners to support World Heritage conservation through **four revised Strategic Objectives:**

- Strengthening the **Credibility** of the World Heritage List;
- Ensuring the effective **Conservation** of World Heritage properties;
- Promoting the development of effective **Capacity building** in SPs and
- Increasing public awareness and involvement and support for World Heritage through **Communication**. Known as the **4 'Cs'**, a **5th** was added in 2007:
- To enhance the role of **Communities** in the implementation of the WHC (supported by *Operational Guideline* Paragraph 12).

The **World Heritage Partnerships for Conservation Initiative (PACT)** was launched in 2002 (UNESCO 2007) to raise awareness about World Heritage and to implement sustainable partnerships based on the Strategic Objectives. This is provided for by Article 7 of the WHC. The World Heritage Centre manages PACT by building 'international co-operation and assistance' through relationships with the private sector, including intellectual and technically based partnerships, and by creating networks of exchange and technical assistance for World Heritage conservation. It raises funds in support of these activities by soliciting and/or assessing proposals and expressions of interest from a wide range of non-governmental, civil society and private sector institutions including for profit organizations interested in long term World Heritage conservation. The World Heritage Centre also works with SPs, to involve local communities in setting up partnerships at World Heritage sites. PACT operates under a review mechanism and a regulatory framework where partnerships are based on principles that ensure the integrity of actions taken and the neutrality of positions adopted (UNESCO 2002e).

Launched in 2006 the **World Heritage Centre's Natural Heritage Strategy** works to further advance the influence of the '5Cs' and PACT. Its mission is "To promote the fullest and broadest application of the WHC by all relevant stakeholders, from site level individuals to global organizations, in the pursuit of long-term conservation of biodiversity and sustainable development" (UNESCO 2006b). Entrepreneurial by nature, this strategy solicits, initiates, and pursues support through a range of partnerships promoting the WHC's 'comparative advantage' and its conservation leveraging potential. In this pursuit to continually improve World Heritage site management capacities and to counter threats to World Heritage in Danger sites, the World Heritage Centre has and is developing new cooperation between NGOs and the private sector. Specifically stated is the intention to help State Parties deal with new and emerging management challenges at World Heritage sites, such as those resulting from the impacts of climate change, by mobilising technical and financial resources for developing and implementing adaptation measures. To this end the World Heritage Centre and the **Division of Ecological and Earth Sciences** intend cooperating on climate change and its impacts on World Heritage sites, and to explore opportunities of raising finance through the emerging market in emissions trading (UNESCO 2006b p10). Additionally under this heritage strategy partnerships are integrating an ecosystem approach to sustainable development and conservation at World Heritage sites and are supporting the growth of sustainable livelihoods of local communities, which directly or indirectly assist site conservation. The diverse partnerships that have been achieved range from establishing a "Friends of World Heritage" group to partnering with Shell and Earthwatch to implement the five-year 'Business Skills for World Heritage Programme'. The latter aims to train managers of World Heritage sites to develop and implement business strategies to promote the sustainable development and management of these sites. This strategy does not cover cultural sites but its incorporation of the climate change threat sets an example for future World Heritage sustainable development at cultural sites.

Sustainable Tourism

The World Heritage Centre identified 6 priority areas in sustainable development which were given 'thematic programme' status. One was the **Sustainable Tourism Programme** aimed at increasing World Heritage site capacity to manage tourism, promoting alternative livelihoods for local communities and engaging the tourism industry to affect increased conservation benefits. In 2002 it published a user friendly manual ***Managing Tourism at World Heritage Sites: a Practical Manual for World Heritage***

Site Managers. Climate change was not identified though the process described for identifying threats to World Heritage sites would accommodate its appraisal. The 6 principles of the **ICOMOS International Cultural Tourism Charter** were included (see Insert 22.1) (Pedersen 2002 p94).

In 2009 100 experts and representatives from 21 countries met in China for an international workshop ***Advancing Sustainable Tourism at Cultural and Natural Heritage Sites*** (APEC 2010). Their report to the World Heritage Committee (UNESCO 2010h) stated that "Participants were keen to see the establishment of a process that brings the same level of awareness of the impacts of unsustainable tourism as there is for climate change (APEC 2010 p8; UNESCO 2010h p6). They felt that "tourism is mainstream business for World Heritage and should be integrated into mainstream processes". They felt that the World Heritage Committee "should see tourism as an opportunity (successful sites control their relationship with tourism, rather than vice versa)." They agreed that WH listing makes many sites more attractive to visitors and that World Heritage should be recognised as "an integral part of the tourist/visitor sector (not part of an 'us and them' relationship) (UNESCO 2010h p1)".

The Workshop recognised the problem of popularity; "while conferring internationally recognised designation to protect and conserve very special places and making them accessible as intended by the WHC, tourism also brought many new challenges for site management and protection (UNESCO 2010h p1). The 45 page document goes to great lengths to emphasise the centrality of protecting the OUV of a site. However when threats posed by the fast-growing tourism industry were considered, climate change was not recorded.

Principle 1 Since domestic and international tourism is among the foremost vehicles for cultural exchange, conservation should provide responsible and well managed opportunities for members of the host community and visitors to experience and understand that community's heritage and culture at first hand. **Principle 2** The relationship between heritage places and tourism is dynamic and	**Principle 1 Contribution to WH objectives.** Tourism development and visitor activities associated with WH Properties must contribute to and must not damage the protection, conservation, presentation and transmission of their heritage values. Tourism should also generate sustainable socio-economic development and equitably contribute tangible as well as intangible benefits to local and regional communities in ways that are consistent with the conservation of the properties. **Principle 2 Cooperative partnerships.** WH Properties should be places where all stakeholders cooperate through effective partnerships to maximise conservation and presentation outcomes, whilst minimising threats and adverse impacts from tourism. **Principle 3 Public awareness and support.** The Promotion, Presentation and Interpretation of WH Properties should be effective, honest, comprehensive

may involve conflicting values. It should be managed in a sustainable way for present and future generations. **Principle 3** Conservation and Tourism Planning for Heritage Places should ensure that the Visitor Experience would be worthwhile, satisfying and enjoyable. **Principle 4** Host communities and indigenous peoples should be involved in planning for conservation and tourism. **Principle 5** Tourism and conservation activities should benefit the **host community.** **Principle 6** Tourism promotion programmes should protect and enhance Natural and Cultural Heritage characteristics.	and engaging. It should mobilise local and international awareness, understanding and support for their protection, conservation and sustainable use. **Principle 4 Proactive tourism management.** The contribution of tourism development and visitor activities associated with WH Properties to their protection, conservation and presentation requires continuing and proactive planning and monitoring by Site Management, which must respect the capacity of the individual property to accept visitation without degrading or threatening heritage values. Site Management should have regard to relevant tourism supply chain and broader tourism destination issues, including congestion management and the quality of life for local people. Tourism planning and management, including cooperative partnerships, should be an integral aspect of the site management system. **Principle 5 Stakeholder empowerment.** Planning for tourism development and visitor activity associated with WH Properties should be undertaken in an inclusive and participatory manner, respecting and empowering the local community including property owners, traditional or indigenous custodians, while taking account of their capacity and willingness to participate in visitor activity. **Principle 6 Tourism infrastructure & visitor facilities.** Tourism infrastructure and visitor facilities associated with WH Properties should be carefully planned, sited, designed, constructed and periodically upgraded as required to maximise the quality of visitor appreciation and experiences while ensuring there is no significant adverse impacts on heritage values and the surrounding environmental, social and cultural context.

Insert 22.1 ICOMOS International Cultural Tourism Charter

Insert 22.4 Principles for Sustainable Tourism at World Heritage Properties

The Workshop report acknowledges that not all World Heritage properties can and want to embrace visitors. Indeed some sites were not suitable for visitation. However it is quick to generalise that "Almost all individual World Heritage properties are significant tourism destinations" (APEC 2010h p16). It was concluded that "properties should base their planning mechanisms on the solid foundations of understanding the heritage values of the place and the needs of visitors". Strengthening the sustainable development tourism agenda further, the Workshop reported that "Planning needs to acknowledge that at most sites, visits by the public and the implementation of interpretation of all the significant values are integral elements of conservation" (APEC 2010h p11).

The Workshop report's glossary defines **Sustainable Tourism** as referring "to a level of tourism development and activity that does not compromise or degrade the heritage values of a place, including World Heritage properties, over the long term. It can be maintained because it results in a net benefit for the social, economic, cultural and natural environments of the area in which it takes place". **Cultural Tourism** is essentially that form of tourism that focuses on the culture, and cultural environments including landscapes of the destination, the values and lifestyles, heritage, visual and performing arts, industries, traditions and leisure pursuits of the local population or host community. It can include attendance at cultural events, visits to museums and heritage places and mixing with local people. It should not be regarded as a definable niche within the broad range of tourism activities, but encompasses all experiences absorbed by the visitor to a place that is beyond their own living environment" (UNESCO 2010h p17).

III. Process for the inscription of properties on the WH List
132. 5. Protection and... ***Management: Sustainable development principles should be integrated into the management system.***
Annex 5 Format for the nomination of properties for inscription on the World Heritage List.
5. e Property management plan or other management system.
Sustainable development principles should be integrated into the management system.
Paragraph 155: The Statement of OUV should include a summary of the Committee's determination that the property has OUV, identifying the criteria under which the property was inscribed, including the assessments of the conditions of integrity or authenticity, and of the protection and management in force and ***the requirements for protection and management.*** The Statement of OUV shall be the basis for the future protection & management of the property. (UNESCO 2011c)

Insert 22.2 Revision of Operational Guidelines relating to sustainable development, 2011. (Amendments in bold)

As the Workshop failed to identify climate change as a serious threat it is not surprising that the report did not address climate change impacts on the heritage custodians and/or element practitioners and their communities and the tourist facility staff on whose presence sustainable development depends.

The Workshop's outcomes paved the way for a hefty revision of some *Operational Guidelines* (UNESCO 2011c) to provide guidance on managing tourism at World Heritage sites. Insert 22.2 specifies that sustainable development should be integrated into the World Heritage management system. Insert 22.3 details how visitation management is to be described in the nomination process. ***Six Principles for Sustainable Tourism at World Heritage Properties*** (see Insert 22.4) were developed drawing to an extent on past documents. It was reported that "The Principles as agreed provide a best practice framework that all stakeholders should apply in the achievement of their specific objectives so that heritage resources will be protected and conserved and the many positive benefits of tourism realised" (UNESCO 2010h p4). Comparing the two sets of principles and considering the depth of detail now mandatory for nomination, it is clear that the World Heritage Committee has endorsed a well resourced move to substantially revise World Heritage management to embrace sustainable development particularly in the form of sustainable tourism.

The Workshop also proposed management "best practice policy guidance", "strategies and methods" and means for sharing best practices at cultural and natural heritage sites from around the globe" (APEC 2010). In the 2010 **Decision: 34 COM 5F.2**, the World Heritage Committee adopted the revision of the *Operational Guidelines* and the *policy orientation document* which defined "the relationship between World Heritage and tourism"; discussed "management responses" and highlighted "the responsibilities of different actors in relation to World Heritage and tourism" (APEC 2010 p23). The policy orientations are quite detailed and include commercial strategies such as to "*reward best practice examples of World Heritage properties and businesses within the tourist/visitor sector* (APEC 2010 p17)". The Decision resulted in the closure of the World Heritage Tourism Programme, the instigation of the proposed **World Heritage and Sustainable Tourism Programme** (which, if adopted by the World Heritage Committee at the 36th session will be launched during the 40th anniversary celebration of the WHC (UNESCO 2012j)) and the opportunity for the World Heritage Centre to adopt the Principles for Sustainable Tourism at World Heritage Properties.

Annex 5, Point 4.b: Factors affecting the property ***Responsible visitation at World Heritage sites.*** ***Provide the status of visitation to the property (notably available baseline data; patterns of use, including concentrations of activity in parts of the property; and activities planned in the future). Describe projected levels of visitation due to inscription or other factors. Define the carrying-capacity of the property and how its management could be enhanced to meet the current or expected visitor numbers and related development pressure without adverse effects. Consider possible forms of deterioration of the property due to visitor pressure and behaviour including those affecting its intangible attributes.***
Annex 5, Point 5: Protection & Management of the Property 5.h Visitor facilities and infrastructure ***The*** section ***should*** describe the ***inclusive*** facilities available on site for visitors, and ***demonstrate that they are appropriate in relation to the protection and management requirements of the property. It should set out how the facilities and services will provide effective and inclusive presentation of the property to meet the needs of visitors, including in*** relation to the provision of safe and appropriate access to the property. The section should consider visitor facilities that may include interpretation/explanation, (***signage***, trails,

guides, notices or publications, ***guides)***; property museum/***exhibition devoted to the property***, visitor or interpretation centre***; and/or potential use of digital technologies and services***(overnight accommodation; restaurant; car parking; lavatories; search and rescue ***etc).***
5. j Staffing levels and *expertise* (professional, technical, maintenance) Indicate the skills and ***qualifications*** available ***needed for the good management of the property including in relation to visitation and future training needs. (UNESCO 2011c)***

Insert 22.3 Revision of Operational Guidelines. Amendments to the format for the nomination of properties for inscription on the World Heritage List (in bold)

The 34th World Heritage session decided to introduce policies and procedures that, together with maintaining the OUV of properties, would make the contribution to sustainable development ***an explicit and intentional objective*** of World Heritage conservation. The World Heritage Committee had observed that despite the promotion of the '5Cs' sustainable development was not being translated into actual policies and procedures within the WHC (Decision: 34 COM 5D). "The World Heritage Convention focused on maintaining the OUV of World Heritage sites without considering the possible implications in respect of their wider social, economic and environmental context, except when these implications engender a risk for the heritage. A certain degree of ambiguity, therefore, appears to exist at present as regards the functional relationship within the Convention, the practice of conservation promoted by it and the goal of sustainable development (UNESCO 2010e p3)". Consequentially the World Heritage Committee defined sustainable development in the World Heritage context, see Insert 22.7. A *careful balance of dimensions* is needed to meet the needs of current and future generations.

Sustainable Development is defined as "the *careful balance* of environmental, social and economic dimensions, in order to meet the needs of current and future generations". "Environmental sustainability requires that... the extraction of renewable resources not exceeds the rate at which they are renewed; social sustainability involves a fair and equitable society able to work towards common goals, where basic individual needs, such as those for health and well-being, nutrition, shelter, and education, are met. Economic sustainability... occurs when development, which moves towards social and environmental sustainability, is financially feasible" Cultural sustainability "enables continuities in cultural values, expressions, identities, and knowledge systems of particular groups associated with heritage sites (UNESCO 2010g)p3.

Insert 22.7 Sustainable Development

Next the World Heritage Committee moved to bring a **global recognition** of the contribution of World Heritage to sustainable development. World Heritage contributes to sustainable development as it protects heritage and in so doing "plays a fundamental role in fostering strong communities, supporting the physical and spiritual well-being of its individuals and promoting mutual understanding and peace" A 'well protected' World Heritage property via a "variety of goods and services and as a storehouse of knowledge" very often contributes directly to livelihoods and sustainable development (where sustainable development is) intended as a development where each of the three pillars, the environmental, the economic and the social – including intra and intergenerational equity - is given *adequate consideration* (UNESCO 2010g p4). Sustainable development is claimed as ***an essential condition for successful conservation*** as unsustainable development was thought to be "perhaps the most significant threat to heritage conservation" in all countries. Thus a **sustainable conservation** of heritage will take into account and integrate a concern for the social, economic and environmental dimension of development (UNESCO 2010g p4).

To anchor sustainable development in the WHC context the Committee decided to **mainstream** sustainable development. World Heritage concern for sustainable development should be integrated at an **early stage** in planning for heritage conservation. Sustainable development relies on an established system of governance plus staff, resources, community engagement and a capacity and expertise to plan and anticipate. These should be incorporated at an early stage too. The following **methodologies** for successful integration of sustainable development at World Heritage properties were identified:

- First carry out an analysis of the socio-economic context of World Heritage properties and of local and national stakeholders' aspirations.
- Engage all stakeholders.
- Fully explore the development potential of heritage sites, including alternatives to tourism, which may benefit local communities;
- Empower local communities (UNESCO 2010g p5).

A stand alone 16 point methodology was reproduced in this World Heritage report as an example methodology. Point 14 (out of 16) reads: "Promote mitigation and adaptation strategies to the impacts of climate change through planning and design (UNESCO 2010g p11)".

To implement these methodologies the World Heritage Committee recommended the following **tools:**

- **Cultural mapping** is a tool by which the relationship between various elements, their flows, interactions, and processes can be visualized spatially thus identifying less visible aspects of places. Cultural information may be based on detailed observations, community mapping, interviews, and other ethnographic techniques and may include practices of resource extraction, use, and management of ecologies;
- The **development of indicators** to help reconcile sustainable development and heritage conservation goals;
- The synthesis of a) and b) to provide **diagnostic elements** for better decision making and

- **Promotional tools** acting on the economic dimension including incentives systems, fundraising mechanisms and the possible establishment of a World Heritage Tax (UNESCO 2010g p6).

Management systems: Paragraph 110 ***Impact assessments for proposed interventions are essential for all World Heritage properties.***

Paragraph 111 In ***the monitoring and assessment of the impacts of trends, changes, and of proposed interventions;***

Paragraph 112. Effective management involves a cycle of ***short, medium*** and long-term actions to protect, conserve and present the nominated property. ***An integrated approach to planning and management is essential to guide the evolution of properties over time and to ensure maintenance of all aspects of their OUV. This approach goes beyond the property to include any buffer zone(s), as well as the broader setting.***

Sustainable use: Paragraph 119 World Heritage properties may support a variety of ongoing and proposed uses that are ecologically and culturally sustainable and ***which may contribute to the quality of life of communities concerned.*** The States Parties and its partners must ensure that such sustainable use or any other change does not impact ***adversely*** on the OUV ***of the property.*** For some properties, human use would not be appropriate. ***Legislations, policies and strategies affecting World Heritage properties should ensure the protection of the OUV, support the wider conservation of natural and cultural heritage, and promote and encourage the active participation of the communities and stakeholders concerned with the property as necessary conditions to its sustainable protection, conservation, management and presentation.*** *(UNESCO 2011c)*

Insert 22.5 Revision of Operational Guidelines relating to management, sustainability, 2011. (amendments in bold)

The World Heritage Committee agreed that successful integration of sustainable development relied on the quality of the **initial identification, assessment and listing** of a property. If this was not well managed it could be a threat to a site's OUV or impact adversely on the traditional ways of life of those who live nearby. The inscription process was thought crucial to establishing a **management framework** to safeguard the OUV and integrity of the property, for the benefit of current and future generations, taking into consideration the likely impacts arising from the inscription. To ensure this enhancement of, not damage to, a site's values, (APEC 2010 p4) the Committee **amended the *Operational Guidelines*** to more rigorously describe and manage the OUV of a site. The revisions have been included in Insert 22.5 and 22.6 to demonstrate their scope and detail. The Committee closed the session by welcoming the proposed Action Plan for 2012 to strengthen linkages between the WHC and other multilateral environmental agreements, but the UNFCCC was not named.

The 2011 publication ***Sustainable tourism & natural World Heritage*** identifies a range of factors that support and hinder sustainable tourism. It sums up with advice from the Deputy Director of IUCN's Global Business

& Biodiversity Programme Giulia Carbone: "Careful planning is a the heart of ensuring that World Heritage sites benefit from the high-profile that comes with their global status, through collaboration between the private sector, local communities and site managers" (Borges et al. 2011). Climate change is not mentioned.

Sustainable Development and Culture

Although climate change does not appear to have come in from the cold as far as being recognised as a significant threat to cultural heritage, the contribution of culture in development has. In December 2010, the UN General Assembly adopted a **resolution** entitled **"Culture and Development"** that reaffirms the role of culture in development and calls for its integration in global development policies. There is a need to underscore how World Heritage properties, through management that involves a wide range of stakeholders, can be seen as contributing to the **Millennium Development Goals (MDGs)** and to sustainable development The World Heritage 2011 **Draft Decision: 35 COM 5E** acknowledged the important progress made in building links between biological and cultural diversity to ensure environmental, economic, social and cultural sustainability and human well-being. Sustainable development can be seen as a mechanism for drawing together more closely the cultural and natural sides of the WHC.

> **3. Justification for inscription.** ***The justification should be set out under the following sections.*** This section must make clear why the property is considered to be of "OUV". The whole of this section of the nomination should be written with careful reference to the ***requirements*** of the *Operational Guidelines*. It should not include detailed descriptive material about the property or its management, which are addressed in other sections, but should convey **the key aspects** that are ***relevant to the definition of the OUV of the property.***
>
> 3.***1. a Brief synthesis. The brief synthesis should comprise (i) a summary of factual information and (ii) a summary of qualities. The summary of factual information sets out the geographical and historical context and the main features. The summary of qualities should present to decision makers and the general public the potential OUV that needs to be sustained, and should also include a summary of the attributes that convey its potential OUV, and need to be protected, managed and monitored. The summary should relate to all stated criteria in order to justify the nomination. The brief synthesis thus encapsulates the whole rationale for the nomination and proposed inscription.***
>
> **3.1 b Criteria under which inscription is proposed (and justification for inscription under these criteria).** See Paragraph 77 of the *Operational Guidelines*. Provide a separate justification for each criterion cited. State briefly how the property meets those criteria under which it has been nominated (where necessary, make reference to the "description" and "comparative analysis" sections ***of the nomination,*** but do not duplicate the text of these sections) ***and describe for each criterion the relevant attributes.***

3.1 c Statement of integrity. The statement of integrity should demonstrate that the property fulfils the conditions of integrity set out in Section II.D of the Operational Guidelines, which describe these conditions in greater detail. The Operational Guidelines set out the need to assess the extent to which the property: includes all elements necessary to express its OUV; is of adequate size to ensure the complete representation of the features and processes which convey the property's significance; suffers from adverse effects of development and/or neglect (Paragraph 88). The Operational Guidelines provide specific guidance in relation to the various WH criteria, which is important to understand (Paragraphs 89–95).

3.1 d Statement of authenticity (for nominations made under criteria (i) to (vi)The statement of authenticity should demonstrate that the property fulfils the conditions of authenticity set out in Section II.D of the Operational Guidelines, which describe these conditions in greater detail. This section should summarise information that may be included in more detail in section 4 of the nomination (and possibly in other sections), and should not reproduce the level of detail included in those sections. Authenticity only applies to cultural properties and to the cultural aspects of 'mixed' properties. The Operational Guidelines state that 'properties may be understood to meet the conditions of authenticity if their cultural values (as recognized in the nomination criteria proposed) are truthfully and credibly expressed through a variety of attributes' (Paragraph 82). The Operational Guidelines suggest that the following types of attributes might be considered as conveying or expressing OUV: form and design; materials and substance; use and function; traditions, techniques and management systems; location and setting; language and other forms of intangible heritage; spirit and feeling; and. other internal/external factors.

3.1 e Protection and management requirements This section should set out how the requirements for protection and management will be met, in order to ensure that the OUV of the property is maintained over time. It should include both details of an overall framework for protection and management, and the identification of specific long term expectations for the protection of the property. This section should summarise information that may be included in more detail in section 5 of the nomination document (and also potentially in sections 4 and 6), and should not reproduce the level of detail included in those sections. The text in this section should first outline the framework for protection and management. This should include the necessary protection mechanisms, management systems and/or management plans (whether currently in place or in need of establishment) that will protect and conserve the attributes that carry OUV, and address the threats to and vulnerabilities of the property. These could include the presence of strong and effective legal protection, a clearly documented management system, including relationships with key stakeholders or user groups, adequate staff and financial resources, key requirements for presentation (where relevant), and effective and responsive monitoring.

Secondly this section needs to acknowledge any long-term challenges for the protection and management of the property and state how addressing these will be a long-term strategy. It will be relevant to refer to the most significant threats to the property, and to vulnerabilities and negative changes in authenticity and/or integrity that have been highlighted, and to set out how protection and management will address these vulnerabilities and threats and mitigate any adverse changes. As an official statement, recognised by the WH Committee, this section of the Statement of OUV should convey the most important commitments that the SP is making for the long-term protection and management of the property.

3.2 **Comparative analysis (including SOC of similar properties).** The property should be compared to similar properties, whether on the WH List or not. The comparison should outline the similarities the nominated property has with other properties and the reasons that make the nominated property stand out. The comparative analysis should aim to explain the importance of the nominated property both in its national and international context (see Paragraph 132). ***The purpose of the comparative analysis is to show that there is room on the List using existing thematic studies and, in the case of serial properties, the justification for the selection of the component parts.***

3.3 Proposed statement of OUV. ***A Statement of OUV is the official statement adopted by the WH Committee at the time of inscription of a property on the WH List. When the WH Committee agrees to inscribe a property on the WH List, it also agrees on a Statement of OUV that encapsulates why the property is considered to be of OUV, how it satisfies the relevant criteria, the conditions of integrity and (for cultural properties) authenticity, and how it meets the requirements for protection and management in order to sustain OUV in the long-term. Statements of OUV should be concise and are set out in a standard format. They should help to raise awareness regarding the value of the property, guide the assessment of its state of conservation and inform protection and management. Once adopted by the Committee, the Statement of OUV is displayed at the property and on the UNESCO WH Centre's website. The main sections of a Statement of OUV are the following: a. Brief synthesis; b. Justification for criteria; c. Statement of integrity (for all properties) d. Statement of authenticity (for properties nominated under criteria to vi) e. Requirements for protection and management. (UNESCO 2011c)***

Insert 22.6 Revision of Operational Guidelines Annex 5 Format for the nomination of properties for inscription on the World Heritage List. Explanatory notes (amendments in bold)

Sustainable Development and Climate Change

How does this history of sustainable development within the WHC help answer the original question: How does the WHC address climate change-induced threats to the sustainable development of World Heritage? World Heritage sustainable development policy 'orientations' do not identify climate change as a significant threat. Climate change adaptation has been the subject of various sustainable development programmes set up by the **World Heritage Natural Strategy** but these interventions have not spurred any policy development. Considering the acknowledged 'renown' of climate change by the sustainable tourism workshop participants, its failure to gain formal integration into World Heritage sustainable tourism management policy is baffling.

Brokpa men, as well as women from the Dha-Hanu region of Ladakh (near the border with Pakistan) traditionally wear flowers on their hats

It is evident that the World Heritage Committee understands that underpinning every major *Operational Guideline* change, sustainable tourism principle and sustainable development policy orientation, is the short, medium, and long need for protection of the OUV of a property. As scientists are reporting and as the history of site management attests in Chapter 26, climate change is threatening the OUVs of many World Heritage sites today. No matter whether the development in World Heritage sites is judged sustainable by 'balanced analysis' or 'appropriate and equitable consideration' or 'balanced compromise', it will be threatened in the next 20-30 years by serious climate change impacts caused by GHG emissions generated by the powerful and populous SPs which have much to gain from tourism at World Heritage sites. If sustainable development is to be integral, an essential condition of management and a necessary condition of conservation it must be climate change ready. As environmentalists have pointed out 'Mother Nature doesn't do bailouts'. It does seem problematic that the 'new' sustainable development management of World Heritage (and it is expected that this will apply in almost every site) needs to consider sustainable tourism, the carbon footprint of which, can only contribute to further undermining the long term OUV. Unless **visits and developments are carbon neutral,** more sustainable development will overall necessitate more 'conservation' to combat the added pollution and the damage it will cause. Emphasis on intense supportive local visitation and online tourism would seem to be far more in line with long term protection of the OUV of World Heritage properties.

The sustainable tourism industry is big business. It has initiated an academic green eco-discipline generating standards, strategies and a plethora of indicators (Bloyer, Gustke, and Leung 2004). Recent studies give insight into the energy going into tourism and climate change policy and practice, especially into the use of offset schemes and efficiency drives by national and state tourism instrumentalities. These actions raise awareness and go some way to de-accelerating the rising carbon dioxide emissions. However in the global perspective, both time scale and pollution scale, these actions are like preparing for a tsunami with water wings. For instance a 2011 study ***Green Tourism Futures: Climate Change Responses by Australian Government Tourism Agencies*** (Zeppel and Beaumont 2011) identified the state that was most enthusiastic about implementing sustainable tourism schemes. This same state is the largest coal exporting state in Australia which is the largest coal exporting country in the world, and the state that for economic reasons is endangering the OUV of a natural World Heritage property without even notifying the World Heritage Centre.

Many developing States Parties, especially the smaller island States Parties are heavily dependent on tourism. Tourism is a leading source of foreign exchange for virtually all non-oil exporting least-developed countries (LDCs) (Honeck 2008 p4). Tourism is important for many more developing countries than any other export. It would appear that the new requirements in the Nomination process, property identification and assessment, management structure and those relating to visitor management will further challenge developing States Parties in achieving inscriptions. Extra requirements in the reporting processes relating to sustainable tourism will also add to the difficult demands on developing States Parties which are already being strained more than developed nations by the impacts of climate change. This spiral of challenges is made the more poignant by the realisation that poor developing State Parties, especially LDCs have the most to benefit from the tourism associated with World Heritage inscriptions.

Mani stones carrying the Buddhist mantra 'om mani padme hum' have been built up over centuries to form Mani walls. The mantra is very powerful as it encapsulates the entirety of Buddhist teaching. Stone carving is a winter activity done when agriculture demands are low due to the intense cold and in some places, deep snow. Mani walls often lead to monasteries

The wealthy SPs will also benefit from a sustainable tourism income derived from World Heritage properties, especially Australia. Australia is steward to 19 World Heritage sites and one of the chief backers of the moves to reorient the WHC to embrace sustainable tourism. It is understandable that with tourism making AUD $5 billion or 8% of total export earnings (2010-11), (2.5% of the GDP) (ABS 2011) and employing over half a million residents it has a lot to gain from the opening up of hitherto relatively undeveloped World Heritage properties. World Heritage visitor research in

Australia has shown that "Visitor use is a key pressure or threat...", that "Assessing the ecological impacts of visitors is still limited in all (Australian) World Heritage Areas" and that "Improved processes for collecting data, reviewing data and reporting and applying results are needed as currently sustainable visitor use of World Heritage Areas is a work in progress" (Hill and Pickering 2008 p58). An earlier study found that "...visitor numbers at Australian World Heritage Areas are commonly up to an order of magnitude higher than at comparable control sites." However the researcher warned that "...this cannot necessarily be ascribed specifically to World Heritage branding, but may be associated more with political controversy over listing". The Australian study concluded "it does appear that World Heritage designation yields significant increases in proportions of international visitors to individual sites" (Buckley 2004 p70). A recent working paper concluded that "For psychological, institutional and political reasons, World Heritage listing of natural or cultural sites plays a positive role in their conservation" (Tisdell 2010 p30) and that "As illustrated by the Australian case, World Heritage listing can also provide political leverage".

When viewed from a State Party perspective, it could appear that this important reorientation to more fully embrace sustainable development, particularly in the form of sustainable tourism, is a way of making World Heritage properties self funding. Certainly an income stream could help in climate change related 'adaptations' and 'maintenance' and 'monitoring' and site based mitigation e.g. using solar generation, but it is a precarious 'protection' in the medium and long term....... akin to 'future eating', and in this case 'future OUV eating'. It will be inevitable that as national budgets are eroded by climate change related problems e.g. rehabilitation after extreme weather event damage, State Party funding to World Heritage properties will be reduced.

Appreciated from another angle, this World Heritage history of sustainable development integration has demonstrated the capacity of the WHC to be reinterpreted. It has been used to disentangle the status of sustainable development and to support its mainstreaming. This capacity strongly supports the possibility of the WHC backed by committed supporters being able to make **addressing climate change at all levels** ***as explicit and intentional an objective*** as sustainable development has become.

UNESCO, World Heritage, Rio+20 and Sustainable Development

From Green Economies to Green Societies UNESCO's Commitment to Sustainable Development, 2011 and **UNESCO's Input to the Rio+20 Compilation Document** are UNESCO's contribution to the **UN Conference on Sustainable Development (Rio+20)**, Brazil, June 2012. UNESCO sees Rio+20 as offering a unique chance to place the world on a sustainable development path and to commit to a long term agenda on sustainable development. The conference has three main objectives: to secure renewed political commitment to sustainable development, to assess progress and gaps in implementation of agreed commitments, and to address new and emerging challenges. The two themes of the Conference are a *green economy within the context of sustainable development and poverty eradication*, and *the institutional framework for sustainable development*.

The two documents present a powerfully worded, integrated and focussed view of sustainable development. Their rationales are expressed with a touch of revolutionary fervour and disclose the appalling facts relating to the state of the world's underprivileged. Both documents are calls to action (UNESCO 2011e), (UNESCO 2012a). The 'Green' document is an attractive, generously illustrated 80 page online report complete with case studies presented for the public. The second is a 23 page document of thematically listed and reasoned recommendations following a vision statement. Both declare that society must break with BAU in development. This necessitates holistic thinking and action with ***"new indicators to guide us"***, which will address the social, economic and environmental issues facing the world today. Both call for urgent and long term commitment and support for a new 'green' approach to development that will transform our lives. "The time when we could put off difficult choices is over. There are no more shortcuts" says Irina Bokova, Director General of UNESCO (UNESCO 2011e p5).

The new approach should embrace "the principles of fairness, social inclusion and equity, solidarity, mutual respect, gender-equality, human rights and peaceful coexistence (which are fundamental ingredients for poverty reduction and sustainable development) within the limits and thresholds of the natural system" (UNESCO 2012a). An institutional framework for sustainable development beyond 2015 'must be established'. The principles it elucidates link with the Education for all (EDA) goals, Millennium Development Goals (MDG) and the Internationally Agreed Development Goals (IADGs).

Climate change pervades both documents, the terminology appearing once or twice a page in the *Rio+20 document*. Climate change is discussed specifically in relation to climate science, migration, adaptation and mitigation, access to climate knowledge and to partnerships to address climate related risk reduction and adaptation efforts. There are 9 recommendations in the Rio+20 document addressing peace, education, science, technology and innovation (STI), ocean and freshwater resources, biodiversity conservation, the 'leveraging' potential of culture and the role of the media. All have a bearing on the sustainable management and development of World Heritage properties, especially now that the World Heritage Committee has made sustainable development a condition of conservation and strongly supports sustainable tourism development at World Heritage sites where this is feasible. World Heritage is a major resource for the tourism sector because almost all individual World Heritage sites are potentially significant tourist destinations. The World Heritage brand is a visitor magnet. It can have more impact upon tourism to lesser known properties than to iconic properties (APEC 2010 p15).

The 'Rio+20' document brings together climate related programmes already discussed such as the Climate Change Initiative, the World Heritage Field Observatory and the World Heritage Centre's Natural Heritage Strategy.

Seabuckthorn collecting. An indigenous vitamin-C rich fruit used to make jam and soft drink. These are sold at the Leh health food shop as a seasonal speciality

The two UNESCO publications unambiguously advocate the relationship between science, climate change, sustainable development, culture and biological and cultural diversity. Although this compilation is just that – an input tackling big ideas but without detailed backup - it promises a breakthrough in redefining our lives - cutting our coats according to our cloth. Culture has been brought into the equation – it is no longer going to be the last cab off the climate rank. For instance attention is drawn to "local and indigenous knowledge systems including women's specific knowledge and management practices, and environmental management practices provide valuable insight and tools for tackling ecological challenges, preventing biodiversity loss, reducing land degradation, and mitigating the effects of climate change" (UNESCO 2012a p17).

The Rio+20 document explains that the aim of sustainable development "is to improve human well-being, reduce inequalities, and alleviate environmental risks as well as ecological scarcity". As such sustainable development has a cultural dimension encompassing cultural heritage, creative and cultural industries, cultural tourism and cultural infrastructure. Culture is described as a driver for effective sustainable development. To protect cultural heritage is to reinforce sustainable development. Culture is also a vehicle for sustainable, pro-poor, green development, particularly for developing countries. It is a powerful global economic engine ...". (UNESCO 2011e p67). Cultural World Heritage sites contribute towards the green economy by generating income and jobs from tourism. But 'greenness' or lack thereof depends on the way the tourists travel, what they eat and where they stay. As sustainability is tied to GHG pollution sustainable tourism will require carbon auditing to measure its protection of the OUVs of WH sites.

The World Heritage cultural references include the physical impacts of climate change on sites and the emerging recognition of climate change impacts on heritage custodians. Two site based case studies are described and passing mention is made of DRM and sustainable tourism. The Climate Change Initiative's position is reported: that natural World Heritage sites are favourable places for "developing innovative climate change adaptation and mitigation responses, including through increased resilience and through the conservation of their vast carbon stocks (in natural heritage) which actions position them as places of learning and demonstration of appropriate sustainable development" (UNESCO 2012a p14). The Rio+20 Conference may realise a more radical rethinking that has the potential to use the WHC even more deeply to contribute to the greening of society.

The documents do have drawbacks suggesting that diplomacy has intervened when describing climate change causes and referring to climate science. The 'Rio+20' document states that "There is *strong* evidence to suggest that humanity is responsible for climate change". Actually, the evidence is so strong that it is *unequivocal*. Both documents reposition science as a cornerstone for sustainable development policy. In the 'Green' report there is advocacy for 'proper information' (p11) and 'resolute science', (p29) and in the 'Rio+20' document 'quality science' (p8) and open access to scientific information which "is a prerequisite for generating knowledge for sustainable development". However a following statement relates "Scientific evidence and ethical principles should inform behaviours, policy action and governance decisions to strengthen sustainable development agendas. Impartial reference to the climate science over the past decade would reword this statement such that 'strengthen' was replaced by 'determine'. As the 'Green' report says "Science must drive the green transition to more equitable and sustainable development" and UNESCO must be seen as abiding by its own advice and being honest in facing the science. Good science is apolitical and speaks the universal language of reason. Uncensored and heeded fearlessly by decision makers it can offer solutions to the most difficult of global problems.

Many experts agree that in order to avoid catastrophic effects of climate change, global temperature averages must not rise above 2 degrees Celsius. Indeed, already today, at a temperature rise of less than 0.8 degrees Celsius, the impacts of climate change are substantial. This implies not only contraction (decreasing the amount of emissions in the atmosphere) but, on a medium time-frame, also convergence (people everywhere pollute to a similar, and for some very much lower, amount). Within globally sustainable levels of GHG concentration, emerging economies and developing countries must aim at increasing development without reproducing the per capita emission levels of industrialized countries today. Likewise, the emission levels of industrialized nations will have to achieve aggressive reductions. (GHF 2009e p8).

Insert 22.7 Convergence and Contraction

The climate-proofing of sustainable development related to World Heritage has potentially taken a considerable conceptual leap forward through the rethinking of development and the green orientation of these documents. The 'Green' document reports on the outcomes of meetings with UNESCO, the **International Council for Science (ICSU),** natural scientists, social scientists and engineers. Their recommendations are far reaching. They include promoting a development paradigm change which focuses on achieving **sustainable development objectives** ***through science,*** allowing scientists to disseminate the results of their research even when that ***may call official positions into question.*** "(Doing so must not jeopardize their livelihood or their access to research support)" and promoting an ***ethical view of the principles and vision that guide sustainable development,*** and which must also guide the science and technology community. Ethical principles help to build trust among various sectors of society and among nations (UNESCO 2011e p33).

There are reservations. UNESCO received 677 submissions from political groups, member states, regional meetings, the UN and major groups contributing to the Rio+20 Input Compilation (UNESCO 2012a). The submissions put forward priorities for sustainable development and recommendations on the Conference objectives and themes. While 'Climate change' registered 2,681 times in 450 submissions, 'crisis' registered 1,062 in 268 submissions and GHG were referred to 366 times in 145 submissions. The equity principle associated with addressing GHG emissions 'common but differentiated responsibility (CBDR)' was mentioned 161 times in 77 submissions. *Neither GHG emissions nor CBDR were referred to in either of the documents.* It is uncharacteristic for any climate change discussion not to reference these terms as they are core to addressing climate change. UNESCO promotes consideration of science and technology in an ethical framework by initiating and supporting the process of democratic norm building (standard setting) but does not link this with the pivotal need to generate norm building of the climate change crisis by open inclusion of all the profoundly significant contributing factors – the main one being fossil fuel emissions.

UNESCO's ethical approach is founded upon its ideal of "true dialogue, based upon respect for commonly shared values and the dignity of each civilization and culture". The 13th concrete measure recommended in the Rio+20 document in relation to ***Strengthening disaster preparedness and climate change adaptation and mitigation*** is: Strengthen the capabilities of civil society, particularly social groups at risk and at disadvantage, to effectively communicate their rights and increase their participation in the formulation and implementation of national and regional adaptation policies to **Global Environmental Change (GEC).** It is unclear whether UNESCO is referring here to CBDR.

While calling for 'holistic' responses to break with BAU and a *holistic approach to successfully addressing climate change* and for promoting sustainable development education, and access to and sharing of knowledge for a sustainable future, and for the mobilizing of science (UNESCO 2011e p9), the report sidelines addressing the need to reduce GHG emissions. As many of the Rio+20 submissions agreed, there is no other action (except our dithering), including geoengineering interventions, that will have a more profound effect on heritage protection and sustainable development and the chance to enjoy both, than deep cuts in GHG emissions. Mitigating GHG is the only medium and long term response that will stabilise the climate and give heritage a better chance of surviving. Reducing GHG and moving to 100% renewables lies largely within the political/industrial sphere, as it deals with preventive measures on a geopolitical level (Cassar 2005 p14). However in a document of this import, climate change should be addressed squarely. At the very least, and in the highest interest of heritage protection and sustainable development, UNESCO should make a statement acknowledging that the most important action to stabilise climate is to reduce GHG to 350 parts per million (ppm) of CO_2 which is what many scientists, climate experts, and over 112 national governments say is the safe upper limit for CO_2 in our atmosphere.

The ancient Alchi Monastery is renowned for its magnificent 11th and 12th century murals. Alchi is 70 km downstream of Leh on the Indus River

A responsible stance on sustainable development necessitates sustainable development being understood in the perspective of a GHG polluted world, the pollution of which can only be addressed through rapid deep cuts to GHG emissions under an internationally agreed contraction and convergence strategy (Insert 22.7) (GHF 2009) accompanied by a move to 100% renewable energy.

UNESCO's proposed institutional framework for sustainable development holds promise for medium term commitments and opportunities for Rio+20 to declare targets and a hierarchy of goals. The Director General of UNESCO has declared that we need a change of culture to address climate change (UNESCO 2011e p4). To give this change a better chance of happening peacefully "all sectors of society must come together to restore confidence within the international community with ambitious action to address implementation gaps and take further action on the core principles." To this may be added the need for leaders and international organisations that will do what is needed and explain why it is needed – not what is the best that can be done at present.

The document has missed an opportunity to make a public statement affirming the work of the UNFCCC and to make a 'concrete measure recommended' that the high emitting nations get on board the 'green' path by aiming to cut their GHG by an ambitious percentage by 2020.

To return to the question *How do the Conventions address climate change threats to the sustainable development of cultural heritage?* the WHC does have the capacity to protect those who are involved with sustainable developments that are climate change impacted and in WH properties. But this has not been acted upon. There do not appear to be any think tanks putting together a case to present to the WH Committee. It can be argued that climate change threats, including to those who are working in sustainable development, can be addressed through 4 yearly or 6 yearly Periodic Reporting and consequent monitoring. However it would be more proactive, especially under the more rigorous amended *Operational Guidelines*, to have climate change a specific priority in sustainable development management plans from the outset. With a medium term perspective in mind, understanding that climate change impacts are accelerating, it would seem that a more 'protective-of- OUV' position would be to prepare State Parties (not just for DRM) and their World Heritage site managers with a sustainable development and sustainable tourism management strategy with a heavy emphasis on climate change preparedness. As is becoming the sustainable mantra –'we must break with BAU', sustainable development management will need to become holistically conceived, scientifically based, solution (not fixes) driven with protection the primary consideration.

A World Heritage site under climate change threat.
'The Ruins' were inscribed on the WH List in 1981 as a cultural property because they "provide exceptional architectural, archaeological and documentary evidence for the growth of Swahili culture and commerce along the East African coast from the 9th to the 19th centuries, offering important insights regarding economic, social and political dynamics in this region. The Great Mosque of Kilwa Kisiwani is the oldest standing mosque on the East African coast and, with its sixteen domed and vaulted bays, has a unique plan" (UNESCO 2012h).

In 2004 the site was inscribed on the List of WH in Danger because of deterioration and decay leading to the collapse of the historical and archaeological structures for which the property was inscribed. The deterioration was brought about by a number of factors the chief being invasion by vegetation and inundation by the sea. In 2008 the World Heritage Committee noted with concern the challenges faced by the property from climate change, leading to among others beach erosion (UNSCO 2008 p23). In 2009 it was reported that beach erosion was critical in 4 locations and the on-the-beach structure of the Makutani Palace where very urgent shoring was needed. The World Heritage mission advised low-cost and straightforward solutions, such as rows of wooden poles in the water to dampen wave action and currents, with gabion walls at the beach and terraced cliff sides behind, as opposed to complex and full-blown engineering works. While each of these low-cost solutions requires research, including measuring impacts on the bio-physical environment and on the OUV of the property, such an approach would allow for the use of local labour, require limited investment and phased implementation capable of monitoring, for any necessary in-process adjustment (UNESCO 2009b p51). In 2010 the State Party reported that a sea wall, built of coral blocks embedded in traditional mortar, at a depth of 1.5 metres and a length of 150 metres had been constructed in front of the Malindi sea front which is considered to be effective. There are plans to construct a second wall at the southwest corner of Makutani Palace in the near future (UNESCO 2010c).

The site's OUV includes: *Key management issues including climate change impact due to increased wave action and beach erosion. The ability of the sites to retain their authenticity depends on implementation of an ongoing conservation programme that addresses all the corrective measures necessary to achieve removal of the property from the List of World Heritage in Danger. Long term major threats to the site will be addressed and mechanisms for involvement of the community and other stakeholders will be employed to ensure the sustainable conservation and continuity of the site, (UNESCO 2011b p33)*. The site has a threat intensity coefficient of 75 (100 is the maximum). This is based o the frequency at which the World Heritage Committee has deliberated over this property over the past 15 years. With over a metre rise in sea level expected by 2100, this site will require an enormous investment to protect.

UNESCO has provided US$41,370 worth of assistance to develop 2 management plans and a tentative list.

Insert 22.8 The Ruins of Kilwa Kisiwani and Ruins of Songo Mnara (United Republic of Tanzania)

Insert 22.8 gives an overview of just one - sea level rise - of a number of threats to a Tanzananian World Heritage site on the 'in Danger' List. It demonstrates the difficulty of addressing global climate change impacts and indicates the daunting task of ensuring protection from the anticipated sea level rise, increased wave action and erosion. It is challenging to envisage how sustainable tourism could be managed at this site without exacerbating the site's deterioration.

Recommendations

22.1 Propose making 'culture' the 4th pillar of Sustainable Development (SD). Support the integration of *culturally* sustainable development into both the revised *Policy Document on the Impacts of Climate Change on World Heritage Properties* and the *Operational Guidelines.* As UNESCO's DG Irina Bokova says 'We need a change of culture to tackle climate change" (UNESCO 2011e p4).

22.2 Ensure that climate change is fully integrated into SD to prioritise scientifically established needs of mitigation and maintenance over economic goals by adopting **'climate change resilience' as the 5th pillar of sustainable development.** This will necessitate a strong definition of sustainability, one that may be **operationalised** to plan World Heritage development including 'sustainable tourism'. This should focus support on developing and using globally adopted 'boundary conditions for redesign' for solutions (not fixes) to attain WH site sustainability and the adoption of scientific management strategies such as the *Framework for Strategic Sustainable Development* (FSSD) (Blue Planet Laureates et al. 2012).

22.3 Strongly support the recommendations in *UNESCO's Input to the Rio+20 Compilation Document* that include World Heritage and its role in building 'green' societies. Its determined approach plus its advocacy for breaking with BAU in development, hold promise for holistic thinking and action leading to truly sustainable management of heritage.

22.4 Strongly support the concept for the proposed UNESCO Rio+20 **'new indicators' for development** (UNESCO 2012a), including 'development' in WH properties, especially sustainable tourism.

22.5 Propose that these 'new indicators' have a strong environmental advocacy base and include a requirement for measuring the **carbon footprint** of sustainable tourism enterprises and sustainable developments in world heritage areas.

22.6 Support the showcasing of World Heritage sites that demonstrate appropriate SD 'best practices' and innovation in conservation (UNESCO 2012a p14) as driven by *UNESCO's Climate Change Initiative* and endorsed by *UNESCO's Rio+20* vision.

22.7 Exploit World Heritage's 'green' credentials' to promote the protection of the 936 WH sites through the **heritage site vulnerability campaign** recommended at 15.4.

22.8 Emphasis **'local' tourism** (involving only short journeys) and **online tourism** as the preferable and responsible forms of 'sustainable tourism'.

The Intangible Cultural Heritage Convention

The ICHC Article 2 states, that "For the purposes of this Convention, consideration will be given solely to such intangible cultural heritage as is compatible with existing international human rights instruments, as well as with the requirements of mutual respect among communities, groups and individuals, and of sustainable development."

In relation to *Donations* in support of sustainable development *Operational Directives* paragraph 73 says: "No contributions may be accepted from entities whose activities are not compatible with the aims and principles of the Convention, with existing international human rights instruments, with the requirements of sustainable development ..." and in paragraph 111 under *Communications and the media*: "The media are encouraged to contribute to raising awareness about the importance of the Intangible Cultural Heritage as a means to foster social cohesion, sustainable development and prevention of conflict,..."

Mauritius Strategy

In 2005, the **Mauritius Strategy** identified culture as an integral element in the promotion of **sustainable development (SD).** On that occasion, the SIDS countries committed themselves to "developing measures to protect the natural, tangible and intangible cultural heritage and increase resources for the development and strengthening of national and regional cultural initiatives" (UNESCO 2010p). UNESCO's activities for SIDS in the field of intangible cultural heritage are guided by this Strategy. Through the ICHC, UNESCO is assisting SIDS in developing integrated heritage policies, improving safeguarding measures for intangible cultural heritage and enhancing capacities for sustainable management of intangible heritage while strengthening international cooperation. For example in Samoa, June

2010, UNESCO gave a workshop on the ICHC and its *Operational Guidelines*. As a group exercise, participants prepared a strategy and action plan for intangible cultural heritage safeguarding in Samoa. National experts made presentations on different aspects of intangibe heritage ranging from language, tattooing, traditional boat building, handicrafts, traditional medicine, performing arts, environmental conservation practices, intellectual property protection and intangible heritage education at school.

In 2008 UNESCO developed demonstration tools through more systematic collection of cultural statistics, inventories, and national and regional "mapping" of cultural resources. As part of their Medium Term Strategy, 2008 – 2013, UNESCO recognised "There is a need to develop a coordinated approach to the heritage in all its various forms and its triple role – as a foundation of identity and a vector for development and as a tool for reconciliation. In response UNESCO adopted **a Strategic Programme to strengthen the contribution of culture to sustainable development**, adding, "UNESCO will endeavour to promote participatory and inclusive policies and measures that concomitantly address the requirements of conservation and development and foster social cohesion, innovation and peace by raising awareness of a shared heritage and a common past (UNESCO 2008c p26)."

In relation to intangible heritage UNESCO explains that "The social and economic value of this transmission of knowledge (intangible cultural heritage) is relevant for minority groups and for mainstream social groups within a State Party, and is as important for developing States Parties as for developed ones" (Khaznadar 2009). UNESCO's Director-General has recently repeated that awareness-raising on the importance of intangible cultural heritage and its specific contribution to sustainable development is a proposed priority field of action for 2012-2013 (UNESCO 2010n). The Director General's preliminary proposals for this same period noted that culture is a key element for sustainable development and that the links between culture and development were insufficiently recognized. The proposals highlighted the close relationship between culture and economic development and recommended promoting cultural industries by the promotion and marketing of creativity, traditional and local handicrafts, and folk music and by microfinancing policies for cultural activities.

Robes and masks that are worn by monks during the annual festival of the oracles, hanging in the museum at Matho Monastery, Ladakh

In 2009, aid-funded researchers evaluating the **Festival of Pacific Arts** which celebrates the intangible cultural heritage of the Pacific found that many islanders benefitted economically from selling their products which arose through using their intangible cultural skills and craftsmanship at the festival. Their report noted discontinuation and lack of sustainability of activities was usually due to lack of ongoing resources and/or technical know-how. One of the researchers found that "Being on the front line of climate change and sea-level rise also makes Kiribati a significant venue to not only promote the rich cultures of the Pacific Island region but also to draw the world's attention to cultures of Kiribati and other low-lying islands that are threatened. Kiribati is already losing land and 72 people each year are migrating to New Zealand under the climate change relocation agreement." Islanders reported that the festival helped to raise awareness of the importance of Kiribati's climate change threatened culture. "To the I-Kiribati, it is important to continue to promote their sense of identity and their crafts and dances" (Pennington, Leahy, and Yeap 2009).

In 2011 the General Assembly of the UN approved the following resolution presented by Bolivia. Titled "Harmony with Nature," it requests approval to convene an interactive dialogue on **International Mother Earth Day** on April 22nd, 2011. Topics included methods for promoting a holistic approach to harmony with nature, and an exchange of national experiences regarding criteria and indicators to measure sustainable development in harmony with nature. This resolution (UN 2010) recognizes that "human beings are an inseparable part of nature, and that they cannot damage it without severely damaging themselves".

Despite the recognition, as outlined above, that sustainable development and intangible cultural heritage are of immense benefit to developing States Parties, the Convention does not 'safeguard' the enterprises against climate change impacts.

Recommendations

22.9 Include the current and anticipated impacts of climate change on ICH **practices and practitioners associated with SDs** in the proposed *Intangible Cultural Heritage Climate Change Response and Policy Document* and in the response requirements of the *Operational Directives*.

22.10 Support **research** into the safeguarding of ICH and its practitioners associated with sustainable tourism from the impacts of climate change.

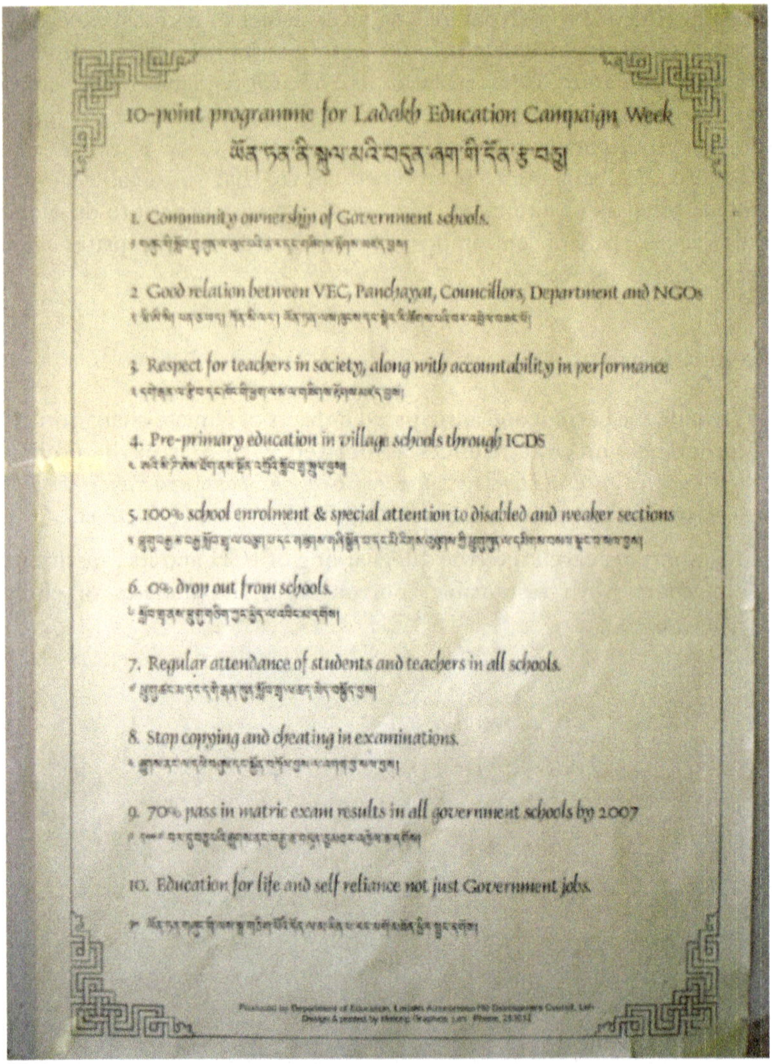

This 10 point programme for a Ladakh Education Campaign Week indicates Ladakhi priorities for community engagement, inclusiveness and straight talking. VEC is the abbreviation for Village Education Committee. The village Panchayat is the traditional local assembly of elected decision makers. ICDS is the abbreviation for the Integrated Child Development Scheme. The notice was on display in a Leh guesthouse

Rio+20 postscript

In many respects the UN's largest conference - *Rio+20 United Nations Conference on Sustainable Development* **(Rio+20)** (United Nations 2012) was disappointing and disheartening. Yet, it offers clear indications for future directions in the care of world heritage and the roles of the WHC and the ICHC.

Rio+20 was held in Rio de Janeiro on 20-22 June, 2012. It was accompanied by the parallel event the Peoples' Summit. As discussed earlier in this chapter, UNESCO's preparations for Rio+20 boldly pointed out the 'high stakes' that were at risk in not achieving a sustainable development approach that broke with BAU to build a green economy with new guiding indicators and the need for a 'new culture to tackle climate change'. In our view these moves heralded positive outcomes for better long term protection of world cultural tangible and intangible heritage. Unfortunately the 49 page, 283 paragraph *Rio+20 outcome document The Future We Want* fell well short of expectations and was in fact a great disappointment. A consequence of its failure may be to alienate many in the development movement and to drive future activism, policy making and innovation away from international multilateral treaty negotiation attempts.

Rio+20 was damaged by its obvious and deliberate corruption by governments that shared an overriding interest in BAU growth economies. For the first time the term "sustained growth" (sometimes linked with 'inclusive' and sometimes with 'equitable') was used: used 16 times interchangeably for 'sustainable development' and 'sustainability'. Patently, to all thinking people, '***sustained growth' is the antithesis of sustainable development***. As *The Guardian* journalist George Monbiot commented "Sustained growth on a finite planet is the essence of unsustainability (Monbiot 2012c)." Renowned economist Professor Tim Jackson went further: "The most staggering linguistic turnabout for me is the one that equates green economy with 'sustained economic growth'. This is hidebound recidivism at its very best. We're no longer even using the terminology of green growth or sustainable growth. Instead of accepting the responsibility of the richest to develop a new economic model, this language has set back by a decade any attempt to question the model that led us to the brink of financial disaster, perpetuates huge consumption inequalities and is driving us towards ecological collapse."..."Disappointment doesn't quite cover it. It's a staggering failure of responsibility." "Promoting sustained, inclusive and equitable economic growth" and increasing "sustainable consumption"... "disregards and intensifies the degradation of the planet's environmental 'base'"(Confino 2012d).

This terminology starkly exposes the negotiators' denial or lack of ability to 'get it' as far as the imperative imposed by climate change and their ability to pay lip service to 'science' without being influenced by its seriousness. If global leaders do 'get it' then this is a disingenuous treaty that reveals the developed world's contempt for the lives of those in the developing world.

The final draft was signed by 188 nations in the presence of 130 heads of state, representatives from 190 countries and at least 50,000 people, guarded by 15,000 soldiers and police (Vidal 2012a). The *Rio+20 outcome document* is a declaration which in essence acknowledges the world's dire environmental and social problems, without indicating how to deal with them, without making provision for adequate funding and without posing a meaningful time frame (Watts and Ford 2012). Aspirational goals are expressed in terms of encouragement (53 times), recognition (160 times), and concern (12 times) causing growing fear that the political will for global change is diminishing, paralysed by the spectre of a deepening global financial crisis and the fear of economic collapse and social turmoil (Confino 2012d).

The laboured platitudinous text, gutted of any strong language of action or accountability, was watered down to the extent that the 're' words - reaffirmed, reiterated, renewed, recalled, reinvigorated and recommitted appeared 89 times. There is little commitment to specific actions with almost no timetables, definitions or ways to monitor proposed sustainable development goals which are disconnected from scientific need and environmental boundaries. The *Rio+20 outcome document* will not, as UNESCO had advocated, "strongly commit nations to move to a green economy that integrates environmental and social costs into decision-making"(Watts and Ford 2012) (despite *strengthen* appearing 68 times). Weakened negotiations were informally blamed on vested interests including big business, the Vatican and national parochialism, with delegates making calls back to their capitals every time there was any suggestion of a change in text (Confino 2012a); "squabbling governments fought over the leadership that the rich industrialised nations should provide and the money the poor should receive to relieve them of destitution" (Vidal 2012a). This was despite the need never being more apparent. States were divided by the influence of corporate lobby groups and suspicion born of political economy, history and culture: factors that will not be overcome by agreements at any single point in time (Rowley 2012b). More than 1,000 businesses at the more progressive end of the spectrum descended on Rio to push for new regulatory frameworks to incentivise moves to a green economy while at the same time calling for the end of perverse subsidies. What they found when they got to Rio was that the fossil fuel lobby had got there first (Confino 2012c). Because of the fossil fuel lobby's influence the urgency of ending fossil fuel subsidies was absent from the final text. Professor Jackson asserted that "These tensions, and this complexity, means we have yet to establish a mature and consistent environmental politics, and with it the policies and incentives that these global problems require" (Rowley 2012b).

The director of the UN Environment Programme Achim Steiner said in mid-conference that negotiations revealed "a world at a loss what to do" (Pearce 2012a). On the final day, he concluded: "We can't legislate sustainable development in the current state of international relations" (Pearce 2012b).

Why the Rio+20 Conference was disappointing was that this perversion of sustainable development and the pervasion of BAU undermined serious and genuine attempts at addressing the aim of the conference. The conference did agree to start talks on setting ***sustainable development goals*** to augment the existing Millennium Development Goals in 2015, but could not agree on what themes they might cover, now to be left to an ***"open working group"***. An ***intergovernmental committee***, comprising 30 experts nominated by regional groups, with equitable geographical representation, will implement this process concluding its work by 2014. It also agreed to encourage corporate sustainability reporting, develop *broader measures of progress to complement GDP,* bolster science in policy making and strengthen The *UN Environment Programme (UNEP)* with powers to initiate scientific research and coordinate global environment strategies. The Conference also established a ***"high-level" forum*** to coordinate global sustainable development, though its format is still to be defined (Watts 2012) and its aims are couched in 'coulds' not 'wills'. An ***inter-agency technical support team*** and ***expert panels*** will propose options on an effective sustainable development financing strategy.

Paragraph 86 of the *Rio+20 outcome* document declares "We decide to launch an intergovernmental and open, transparent and inclusive negotiation process under the General Assembly to define the high level forum's format and organizational aspects with the aim of convening the first high level forum at the beginning of the 68th session of the General Assembly. We will also consider the need for promoting intergenerational solidarity for the achievement of sustainable development, taking into account the needs of future generations, including by inviting the Secretary General to present a report on this issue." This made disturbing reading as we would have thought that *intergenerational solidarity* rather than being a *consideration* would be at the heart of any document entitled "The Future we want".

The *Rio+20 outcome document* presents positively when endorsing initiatives taken by other bodies such as the launching of the UNFCCC's ***Green Climate Fund*** (paragraph 191) and the establishment of the IUCN's ***Intergovernmental Science-Policy Platform on Biodiversity and Ecosystem Services*** (paragraph 204). The pledge in relation to Overseas Development Aid, paragraph 259, "to make development more effective and predictable by providing developing countries with regular and timely indicative information on planned support in the medium term" was also a positive move.

No finite commitments were made to support mandatory environmental and sustainability reporting by large corporations; to bolster human rights especially in situations where they are being sacrificed in the name of development, nor to recommit to protecting women's sexual and reproductive rights (Egeland and Evans 2012). No commitment was made to rescue the high seas that are outside national jurisdictions because of objections from the US, Nicaragua, Canada and Russia. The IUCN called the decision a "deep disappointment" (Watts and Ford 2012b). Paragraph 224 does say that "fundamental changes in the way societies consume and produce are indispensable for achieving global sustainable development", but it does not say what those fundamental changes should be.

The Future

Unsurprisingly the *Rio+20 outcome document* resulted in a spectrum of frustration, an alienation from the failed international approach and a turning to other ways to action change. The perceived weak leadership prompted many in civil society to rethink their strategies, some backing civil disobedience and moves to a war footing (Confino 2012a). Many delegates, civil society activists, business leaders and many in the massed ranks of media declared multilateralism had failed (Monbiot 2012d). George Monbiot commented that "Without monster social movements, without the kind of confrontation required to revitalise democracy, everything of value is deleted from the political text"... "But we do not mobilise, perhaps because we are endlessly seduced by hope" (Monbiot 2012b). Sharan Burrow, General Secretary of the International Trade Union Confederation, said a "red/green alliance was the only way forward". If the current development model doesn't change, "we are going to see economic dislocation greater than we're facing now," she said. "There will be more wars around water and energy, so we need labour and environment walking hand in hand" (Watts and Ford 2012b).

Side Shows

The Rio conference was in another way an acknowledged success in that it created opportunities for the coming together of the like minded through the side events organized by Governments, the UN system and other international organizations including NGOs. Progressive, innovative and informed ideas were considered and adopted at these events. For example at the side event on ***Educating for a sustainable future*** Ms Bokova, the Director-General of UNESCO, explained the need for revising curricula, integrating key sustainable development challenges like *climate change* into teaching and learning, and adopting "new methods to motivate and empower learners to change behaviours and become actors for sustainable development" (UNESCO 2012f). Similarly at the ***Enhancing science-policy links for Rio+20: the Future Earth initiative*** address, Ms Bokova underscored UNESCO's support for the UN initiative ***Future Earth*** clarifying that "the conviction that *interdisciplinary knowledge* is essential to solve complex issues is now a consensus" (UNESCO 2012c). This was in keeping with the beliefs of the eminent economist Jeffrey Sachs who for many years has been advocating "As we learn intellectually how to harness these linkages (between the social and physical sciences, humanities and arts, with cultural and religious traditions) we will be more effective at facing our most fundamental challenge which is the wellbeing and sustainability of the planet" (Confino 2012b).

People's Summit

The 10-day "Summit" was a colourful and extraordinary show of 3,000 political, social, technological and commercial events. Countering the official proceedings of the *Rio+20 Conference* were the hundreds of 'green' decisions and clinched green deals in the form of genuine partnerships, alliances and real projects made outside the conference between an estimated 1,500 corporate leaders, civil society organizations, entrepreneurs, diplomats and NGOs. Agreements were made on investing in public transport, adopting green accounting by corporations and projects planned by cities and judicial bodies on reducing environmental impacts (Watts and Ford 2012b). Campaigns that 'caught on' related to reducing plastics in the ocean, creating a new sanctuary in the Arctic and the abolition of fossil fuel subsidies. Evident was a new generation of business and political leaders who want to connect company success with social and environmental issues that were previously the concern only of NGOs. Many more south-based social and justice movements, technological innovators and social media joined the Rio throng (Vidal 2012b). At meetings involving what some describe as the counter-establishment - groups made up of civil society, environmentalists, social activists, youth, the scientific community and NGOs, the dispossessed, the indigenous communities, and human rights, ecological and other social justice advocates created a public opinion sending a message to leaders that science must be understood and fully responded to. Andrew Deutz, director of international government relations at *The Nature Conservancy*, worked with Indonesia, Australia and Colombia to ensure they all made strong commitments to protecting oceans within their national waters. The *Conservancy* got them to act "not by exhorting them to "save the planet" but instead by pragmatically getting them to see the value of ocean protection measures as a way to ensure the future food security of their citizens" (Platt 2012). Progressive businesses were drawn to interacting with innovative local, national and regional alliances which are going to be more effective in creating the foundation for a new economic paradigm – in other words a bottom-up approach. One journalist observed that "This was expressed in the most often used phrases in the many meetings I attended; the need to create 'coalitions of the willing' and a recognition that 'all issues are inter-connected' and cannot be viewed in silos" (Confino 2012c).

Rio+20's Secretary-General, Sha Zukang asked governments, development banks, the corporate sector and civil society groups to register voluntary commitments and initiatives as a way of bypassing the challenges of multilateral agreements (Irwin 2012). At the Rio+20 Conference closing the UN Secretary-General Mr. Ban Ki-moon emphasized the importance of the more than 700 commitments and advances made by individual countries, companies and other organisations, in total worth about $500 billion. The Conference demonstrated "the growing capacity of grass-roots organizations and corporations to mould effective environmental action without the blessing of governments" (Platt 2012) pointing to a way forward that may lie outside multilateral agreements and political determination.

The Bottom up approach

International lawyer, climate change negotiator and advisor to government Neil Rowley wrote during the Rio conference "The lessons from Copenhagen are many, but key is the need for action rather than analysis, words, agreements or even the science to drive progress. It is by demonstration that renewed momentum can achieve international agreement, not the other way around: new policies, incentives, technologies and low carbon infrastructure established, built and implemented" (Rowley 2012a).

As has been discussed in Chapter 9, Elinor Ostrom has championed the sustainable management of the commons for over 30 years. She wrote her last treatise (Ostrom 2012) for the Rio+20 summit on 12th June 2012 the day she died. Neil Rowley explained her response to climate change and sustainability "She argues that the progress to agreement must be incremental, tangible and measureable. No single approach adopted at a global scale can generate sufficient trust between governments. It is only decisions and deeds at the multiple levels of firms, investors together with national, state and local governments that will now drive the response to the climate problem" (Rowley 2012b; Ostrom 1990) and that it is "through these decisions that the world can reduce the risks of climate change and place our common relationship with the Earth's natural systems into the balance argued for 20 years ago" (Rowley 2012b). In the WH context, Ostrom's rationale for this evolving climate change policy making could be relevant. Expanding on one of her examples - many WH sites and ICH holders are/ and live in cities which are on coasts, on river banks or spread across deltas and are acutely vulnerable to rising water levels and flooding to which adaptation is necessary but mitigation is better. Thus the ground is laid for inter-city, joint city or city clusters to develop policies to reduce GHG emissions. However as time is of the essence here and Ostrom's incremental climate change policy enlightenment may take more time than we may have, the Rio+20 Conference had, in Ostrom's view, a huge role to play in rousing the planet to adopt sustainability as its lifeblood and to hold to global sustainability development goals (Rowley 2012b). The Rio+20 Conference, ineffectual as its outcome document largely is, may yet have achieved its 'rousing' role through the non-multilateral events surrounding it.

Heritage and culture

The Rio+20 outcome document makes no reference to the WHC and the ICHC despite their inclusion in the lead up documents and despite the numerous other conventions and international agreements incorporated into it. Neither is there mention of a "new culture to tackle climate change". ***Cultural heritage*** only rates three references these being in relation to firstly "green economy policies in the context of sustainable development and poverty eradication" because of its centrality for 'poor' people, indigenous communities and secondly "for conservation, as appropriate" ... "of human settlements". The third reference follows: Paragraph 41 "We acknowledge the natural and cultural diversity of the world and recognize that all cultures and civilizations can contribute to sustainable development."

UNESCO brought to Rio+20 a strong emphasis on culture. The Director General Ms Bokova explained the "need to build on the transformative power of education, culture and sciences to empower people for green societies" and that "access to culture for all can unleash considerable forces of creativity and innovation for sustainable development." She was careful to correct the error of making narrowly defined universal sustainable development policies and goals. "This is why it is critical to integrate the cultural dimension of societies into development strategies, in order to better mobilize populations and design people-centred sustainable development strategies" (UNESCO 2012f). Ms Bokova in a meeting with the President of the Economic and Social Council of the United Nations, later noted that unfortunately culture and development had "not found a full-fledged reflection in the *Rio+20 Outcome Document*, although the culture sector was one of the fastest growing and more dynamic segments of economies, especially in the developing countries" (UNESCO 2012d).

Climate change

A tragedy of the *Rio+20 outcome document* was that although explicitly "affirming that climate change is one of the greatest challenges of our time, and" ...in expressing "profound alarm that emissions of greenhouse gases continue to rise globally" and noting "with grave concern the significant gap between the aggregate effect of mitigation pledges by parties in terms of global annual emissions of greenhouse gases by 2020 and aggregate emission pathways consistent with having a likely chance of holding the increase in global average temperature below 2°C, or 1.5°C above pre-industrial levels" **it failed to set any new commitments to fight climate change.** The disconnection between politicians and climate change continues to widen. Neil Rowley believes "climate change is highly confronting for our ethical, religious and personal values" ... "even among those who 'accept' the scientific evidence".... "Very few, if any, of the signatories to the text coming out of Rio this week accept that their state's performance on emissions reduction and clean growth should be, or will become the fundamental criteria whereby countries and economies are assessed over the coming century. International diplomacy is no longer the means to place climate and sustainability at the core of international geo-politics. It is new affiliations between investors, entrepreneurs, city governments and regional alliances between

States, which will demonstrate the environmental, economic and human benefits of more efficient, clean, low emissions activities and so defeat political and policy complacency and create the momentum for meaningful, enduring international agreement" (Rowley 2012b).

Despite the 18 references made to climate change in the text the only proactive action was to welcome the launching of the aforementioned ***Green Climate Fund*** and "call for its prompt operationalisation so as to have an early and adequate replenishment process" .... "to support nationally appropriate mitigation actions, adaptation measures, technology development and transfer and capacity-building in developing countries".

Thermal inertia, tipping points, planetary boundaries, the 'human-induced' component of present extreme weather events and the 'human impacts' of climate change plus the imperative to act were largely ignored prompting one commentator to accuse those involved of perpetrating ecocide (Dyer 2012). According to the OECD, global greenhouse gas emissions could rise by 50% by 2050, if fossil fuels continue to dominate the energy mix. The global economy in 2050 will be four times larger than it is today and the world will use around 80% more energy. This, the OECD said, could drive up the average global temperature by 3 to 6 degrees Celsius by 2100 (OECD 2012).

Science

To the dismay of scientists the *Rio+20 outcome document* did not give science the same prominence (as a critical component of sustainable development solutions) as it did at the 1992 summit. There is no 'Science' section, no mention of new findings on the value of multidisciplinary enterprises being more suited to tackling the complex problems inherent in achieving a green economy and no reference to the well recognised need for new knowledge for the green economy (UNESCO 2012e). Environmental concerns are overshadowed by economic and social concerns. Scientists had hoped that the text would express urgency over the accumulated evidence that many of the planet's systems are now under dangerous stress but the outcome is "sustainable development as usual", rather than a call for action on the scale that the scientific evidence now demands, said Gisbert Glaser, senior advisor at the International Council for Science (ICSU) (Irwin 2012). The *Rio+20* text is characterised by a plethora of support for programmes very few of which reflect the climate change time imperative let alone setting timetables to embrace the emergency. The science is sidelined.

Stronger initiatives came from outside the Conference. Ban Ki-moon requested UNESCO to take the lead in creating a ***Scientific Advisory Board*** and the IUCN was the major negotiator in the establishment of the ***Intergovernmental Science-Policy Platform on Biodiversity and Ecosystem Services*** (UNESCO 2012g).

Sustainable tourism

The *Rio+20 outcome document* contains two pledges under the heading 'Sustainable tourism'. Although World Heritage sites are not specifically mentioned, these pledges echo the support for sustainable tourism advocated by the World Heritage Committee. Again the economic imperative is unrestrained by mention of a specified requirement for scientifically based mandatory environmental accounting mechanisms. It is to be hoped that *sustainable tourism'* will not degenerate to *'sustained tourism'* as has *'sustainable development'* to *'sustained development'*. It is also hoped that it will not follow the World Heritage's example of becoming conditional. The World Heritage Committee now claims sustainable tourism as a condition of conservation of World Heritage sites. Following an extensive stakeholder consultation process the WH Committee at its thirty sixth session, 24 June – 6 July 2012, drew up Draft Decision 36 COM 5E to adopt the new and inclusive ***World Heritage and Sustainable Tourism Programme*** (UNESCO 2012b). It "will take a holistic and strategic approach to WH properties and destinations that will include bottom-up as well as top-down measures to ensure sustainability that reflects not only high-level goals but also local needs and the ability to attain these goals". Of concern is that *high-level goals* and *local needs* do not include any specified scientific input or assessment of vulnerability to the overarching challenge of climate change. It is again to be hoped that the *Programme's* planned support for 'relevant data generation and quality research' and the 'work with relevant international agencies and organisations' will address this fundamental threat to genuinely sustainable development.

New indicators

Despite 'new indicators' being mentioned in the pre-Rio+20 documents and nations raising the difficult issue of measuring sustainable development within a 'green economy' at Rio+20, these 'broader measures of progress to complement GDP' (mentioned above and discussed earlier in this chapter and its recommendations) were not realised in the *Rio+20 outcome document* except in aspirational terms. *Sustainable Development Goals:* paragraph 250 reads, "We recognize that progress towards the achievement of the goals needs to be assessed and accompanied by targets and indicators, while taking into account different national circumstances, capacities and levels of development". This limited response was a disappointment.

The emergence of the call for non-economic indicators such as natural capital measures and social capital measures is of relevance to both World Heritage and Intangible Cultural Heritage. For example the UNEP and the UN University have introduced the ***Inclusive Wealth Index*** while the UNDP has launched its "conceptual framework" for a ***human sustainability***

index (Ford and Watts 2012) that would recognise rates of human development while also weighing up the cost of progress to future generations. In addition the World Bank has introduced ***natural capital accounting*** making environmental protection an investment, rather than a cost. Other alternative measures incorporate consumption and production rates, well-being and happiness. These initiatives point to the future adoption of *cultural measures* to compliment GDP, measures which should include WH sites and ICH elements.

Post-Rio+20 possibilities for the WHC and the ICHC

Reflections on the limitations of Rio+20 suggest that achieving the objectives of the WHC and ICHC may be assisted through a consideration of additional means to those currently employed. Four suggestions are:

1. The WHC and ICHC should remain firmly committed to their aims resisting the threats posed to heritage by governments that are unduly influenced by big business and are desperate for economic growth.
2. Successful pro-active **top down** change may be made by Convention leaders taking more initiative to establish panels and support groups to provide advice and assist with important issues.
3. As a consequence of the demonstrable difficulties in obtaining satisfactory outcomes from multilateral processes at Rio+20 the World Heritage and the Intangible Cultural Heritage Committees should expand their capacities to engage directly with local communities and respond to petitions from concerned citizens and beneficent groups to more effectively pursue their objectives.
4. The Rio+20 experience has made it clear that further bottom up strategies to protect heritage could be adopted. These include:
    1. Obtaining greater **multidisciplinary** help to address WH or ICH problems
    2. Initiating **local actions** to protect and safeguard heritage (e.g. WH site and ICH care groups be initiated to build *local* identity and reduce carbon footprints assisted by local WH/ICH presence in regions with WH sites and ICH practices,) and
    3. Encouraging and facilitating **independent relationships** between and among outside groups (e.g. councils, governments, private enterprise, educational institutions and community groups) and local WHC and ICHC groups supported by online 'headquarter' staff.

Bibliography

Confino, Jo. 2012a. *Furious Greenpeace moves to 'war footing' at Rio+20*. The Guardian 2012 [cited 22.6.12 2012]. Available from http://www.guardian.co.uk/environment/2012/jun/19/greenpeace-rio-20-civil-sobedience.

———. 2012b. *Rio+20: Jeffrey Sachs on how business destroyed democracy and virtuous life*. The Guardian 2012 [cited 22.06.12 2012]. Available from http://www.guardian.co.uk/sustainable-business/rio-20-jeffrey-sachs-business-democracy.

———. 2012c. *Rio+20: Reflections on the way forward for sustainable business*. 2012 [cited 27.6.12 2012]. Available from http://www.guardian.co.uk/sustainable-business/rio-20-reflections-way-forward-sustainable-business.

———. 2012d. *Rio+20: Tim Jackson on how fear led world leaders to betray green economy.* . The Guardian 2012 [cited 25.6. 2012 2012]. Available from http://www.guardian.co.uk/sustainable-business/rio-20-tim-jackson-leaders-green-economy.

Dyer, Gwynne. 2012. *When they make ecocide illegal it will be too late*. 2012 [cited 26.06.12 2012]. Available from http://www.nzherald.co.nz/environment/news/article.cfm?c_id=39&objectid=10815434.

Egeland, Jan, and Jessica Evans. 2012. *Rio+20 missed an opportunity to bolster human rights. BST guardian.co.uk).* . The Guardian 2012 [cited 27.6.12 2012]. Available from http://www.guardian.co.uk/global-development/poverty-matters/2012/jun/27/rio20-missed-opportunity-human-rights.

Ford, Liz, and Jonathan Watts. 2012. *UNDP revealed a template for human sustainability index at Rio+20* The Guardian 2012 [cited 21.6.12 2012]. Available from http://www.guardian.co.uk/global-development/2012/jun/21/undp-human-sustainability-index-rio20.

Irwin, Aisling. *Scientists criticise lack of urgency in Rio+20 accord*. Science and Development Network 2012 [cited 25.06.12. Available from http://www.scidev.net/en/science-and-innovation-policy/science-at-rio-20/news/scientists-criticise-lack-of-urgency-in-rio-20-accord.html.

Monbiot, George. 2012d. *End of an Era* George Monbiot 2012 [cited 25.6.12 2012]. Available from www.monbiot.com/2012/06/25/end-of-an-era/.

———. 2012c. *How "Sustainability" Became "Sustained Growth"*. The Guardian 2012 [cited 22.6.12 2012]. Available from http://www.monbiot.com/2012/06/22/how-%e2%80%9csustainability%e2%80%9d-became-%e2%80%9csustained-growth%e2%80%9d/.

———. 2012b. *Rio 2012: it's a make-or-break summit. Just like they told us at Rio 1992*. The Guardian 2012 [cited 18.6.12 2012]. Available from http://www.guardian.co.uk/commentisfree/2012/jun/18/rio-2012-earth-summit-protect-elites.

OECD. 2012. *OECD says emissions set to surge 50% by 2050*. EurActiv 2012 [cited 16.03.12 2012]. Available from http://www.euractiv.com/climate-environment/oecd-emissions-set-surge-50-2050-news--511532.

Ostrom, Elinor. 1990. *Governing the Commons*: Cambridge University Press.

———. *Green from the Grassroots*. Project Syndicate 2012 [cited. Available from http://www.project-syndicate.org/commentary/green-from-the-grassroots.

Pearce, Fred. 2012a. *Beyond Rio, green economics can give us hope*. The Guardian 2012 [cited 28.6.12 2012]. Available from http://www.guardian.co.uk/environment/2012/jun/28/rio-green-economics-hope.

———. 2012b. *Corporate money men fill the political void at Rio+20*. New Scientist Environment 2012 [cited 25.6.12 2012]. Available from http://www.newscientist.com/article/dn21971-corporate-money-men-fill-the-political-void-at-rio20.html.

Platt, Roger. 2012. *Reflections on Rio+20: Who Are These New Environmentalists Undaunted by Political Gridlock?* 2012 [cited 28.06.12 2012]. Available from http://www.huffingtonpost.com/roger-platt/reflections-on-rio-20-who_b_1632138.html.

Rowley, Nick. 2012a *Rio+20: Multi-lateralism staggers; how to make it run?* (20 June 2012). The Conversation 2012 [cited. Available from http://theconversation.edu.au/rio-20-multi-lateralism-staggers-how-to-make-it-run-7647.

———. 2012b *Rio+20: Small steps could get us out of the climate quicksand* The Conversation 2012 [cited. Available from http://theconversation.edu.au/rio-20-small-steps-could-get-us-out-of-the-climate-quicksand-7755.

UNESCO. 2012d. *At Rio +20, Irina Bokova highlights the critical role of education in reaching sustainable development*. 2012 [cited 22.06.12 2012]. Available from http://www.unesco.org/new/en/media-services/single-view/news/at_rio_20_irina_bokova_highlights_the_critical_role_of_education_in_reaching_sustainable_development/.

———. 2012. *At Rio+20, the Director-General meets the President of ECOSOC*. 2012 [cited 21.06.12 2012]. Available from http://www.unesco.org/new/en/unesco/about-us/who-we-are/director-general/singleview-dg/news/at_rio_20_the_director_general_meets_the_president_of_ecosoc/.

———. 2012g. *Ban Ki-moon calls on UNESCO to lead implementation of recommendations on science for sustainable development.* 2012 [cited 22.6.12 2012]. Available from http://www.unesco.org/new/en/natural-sciences/special-themes/biodiversity-initiative/news-single-view/news/ban_ki_moon_calls_on_unesco_to_lead_implementation_of_recommendations_on_science_for_sustainability/.

———. 2012e. *Rio +20: Irina Bokova addresses the first High Level Roundtable.* 2012 [cited 21.06.12 2012]. Available from http://www.unesco.org/new/en/rio-20/single-view/news/rio_20_irina_bokova_addresses_the_first_high_level_roundtable/.

———. 2012c. *Rio+20: Irina Bokova underscores UNESCO's support for the Future Earth Initiative.* UNESCO 2012 [cited 22.06.12 2012]. Available from http://www.unesco.org/new/en/rio-20/single-view/news/rio_20_irina_bokova_underscores_unescos_support_for_the_future_earth_initiative/.

———. 2012b. World Heritage Committee. 36th Session. Saint Petersburg, Russian Federation, 24 June - 6 July, 2012. 5E: World Heritage Tourism Programme.

United Nations. 2012. Rio+20. United Nations Conference on Sustainable Development. Outcome document. The Future We Want.

Vidal, John. 2012a *Rio+20: Earth summit dawns with stormier clouds than in 1992.* The Guardian 2012 [cited. Available from http://www.guardian.co.uk/environment/2012/jun/19/rio-20-earth-summit-1992-2012.

———. 2012. *Rio+20: reasons to be cheerful.* The Guardian 2012 [cited 27.06.12 2012]. Available from http://www.guardian.co.uk/global-development/poverty-matters/2012/jun/27/rio20-reasons-cheerful.

Watts, Jonathan, and Liz Ford. 2012b. *Rio+20 Earth Summit: campaigners decry final document.* 2012 [cited 23.6.12 2012]. Available from http://www.guardian.co.uk/environment/2012/jun/23/rio-20-earth-summit-document.

———. 2012a. *Rio+20: anger and dismay at weakened draft agreement.* The Guardian 2012 [cited 19.6.2012 2012]. Available from http://www.guardian.co.uk/environment/2012/jun/19/rio-20-weakened-draft-agreement.

Chapter 23

Do the Conventions Have Means of Sharing Knowledge to Address Climate Change Impacts on Cultural Heritage?

Without climate information a farmer is like a mouse in a bottle.
70 year old farmer, Bamako, Mali (Konate 2004)

Do the World Heritage Convention and the Intangible Cultural Heritage Convention have knowledge sharing measures (capacity building, training, awareness raising, education, cooperation and communication programmes, research dissemination, and networking) for addressing climate change impacts on cultural heritage? How are States Parties informed of the latest science in relation to climate change impacts?

The World Heritage Convention

The World Heritage Convention (WHC) robustly encourages knowledge generation, gathering and sharing. Articles 4, 5, 6, 7, 8 and 23 provide multiple opportunities for climate related threats to cultural heritage to be researched, preferably cooperatively, and the gained information disseminated. Dissemination may by through teaching, training, conferencing and capacity building. Finally the knowledge should be incorporated into climate change management plans and the plans funded and acted upon.

The *Operational Guidelines*, the *Policy document on the impacts of climate change on World Heritage properties,* the report *Predicting and Managing the Effects of Climate Change on World Heritage* and the publication, *A Strategy to Assist States Parties to Implement Appropriate Management Responses,* emphasise knowledge sharing from site level to the global level. With climate change complexity becoming better understood the WH Committee is advocating collaboration with other branches of UNESCO and with other relevant international organisations. Three years after the World Heritage climate change publications first emphasised the need for knowledge sharing, UNESCO launched its interdisciplinary climate change strategy - the *Climate Change Initiative,* which incorporates World Heritage knowledge sharing efforts, see Insert 15.3 and 15.4. **The World Heritage Information Network (WHIN)** is another means through which lessons learnt and best practices may be shared and promoted to raise awareness of climate change impacts on World Heritage sites. WHIN is a partnership between the UNESCO **World Heritage Centre,** the Advisory Bodies, States Parties, and those managing World Heritage sites. It was created in 1995 in order to foster the exchange of information between partner networks and World Heritage sites (UNESCO 1995). Although it does not maintain a web site, news is circulated through WHNEWS, the e-mail newsletter, the printed World Heritage Newsletter and the partner websites. To facilitate World Heritage information documentation and communication UNESCO has established the applied programme, the **Information Technology and Heritage Initiative** which aims to increase the use of information technology including computers, digital field instruments, satellites, and the internet.

Insert 23.1

UNESCO CLIMATE CHANGE INITIATIVE: Core Program: Climate Change: Education for the 'Overall Context of Education for Sustainable Development'

Through its Climate Change Education for Sustainable Development programme, UNESCO aims to make climate change education a central and visible part of the international response to climate change. The programme aims to help people understand the impact of global warming today and ***increase "climate literacy" among young people.*** It does this by strengthening the capacity of its Member States to provide quality climate change education; encouraging innovative teaching approaches to integrate climate change education in school and by **raising awareness** about climate change as well as enhancing non-formal education programmes through media, networking and partnerships (UNESCO 2009g).

The directive from the World Heritage Committee that new knowledge about the threats of climate change be shared with contributing frontline communities and the plea from the Global Humanitarian Forum that the latest climate change science should be disseminated as a matter of urgency (GHF 2009c) should both be observed. The preliminary proposal from the UNESCO Director-General to develop online open educational resources for isolated communities using vernacular languages and indigenous knowledge is to be encouraged (UNESCO 2010n).

Capacity development is reported as the potential key to the success of UNESCO's response to climate change (UNESCO 2009o). It is also central to addressing the ethical dilemma of climate change that is discussed in Chapter 24 (UNESCO 2009aa) (COMEST 2010). The Intergovernmental Panel on Climate Change (IPCC) and other involved groups such as the Australian Climate Justice Program point to the potential of capacity development to disseminate knowledge of climate change science (ACJP, CANA, FOE Australia 2008). UNESCO, being an educational exemplar, has championed capacity development for both its potential in climate change education and its contribution to the whole of society, see Insert 23.1. Its **Intersectoral Platform on Climate Change** supports the implementation of the **UNESCO Strategy for Action on Climate Change and its Enhanced Plan of Action** that aims to develop an effective response to climate change and calls for action in all of the Organization's fields of competence. In these major educational drives the consideration of culture and the cultural conventions remains stalled: cultural heritage is still unrecognised for its role in facilitating climate change knowledge sharing and for its imperilled position under the impacts of climate change.

The World Heritage Committee keenly follows up on its advocacy of professional development. It initiates programmes and publicises regional courses, workshops and meetings that draw on expert knowledge. Online perusal shows the extraordinary diversity and reach of their more recent knowledge sharing initiatives. For instance within the span of the months of writing of this book there will be held an *International Training Course on Disaster Risk management of Cultural Heritage* in Japan (at which participants develop their own action plans for a 'Strategy for Rebuilding Risks from Disasters at World Heritage Properties'), a world congress on *World Heritage Cities and Climate Change* in Portugal and a *"I know where I'm going", Remote Access to World Heritage Sites from St Kilda to Uluru Conference* in the UK. UNESCO encourages imaginative programs such as the *Ifugao Province, Philippines and Italy's Cinque Terre Park sign Twinning Program* which recognises how outstanding and evolving organic cultural landscapes could benefit from mutual cooperation for the sites' conservation and sustainable development. UNESCO's partnership with the *Japan Aerospace Exploration Agency* which will assist in bringing the benefits of space technology to the monitoring of World Heritage sites is another example.

The aforementioned timely online initiative 'Climate Frontlines' (UNESCO 2010) is evidence of grass roots climate change information sharing using information technology. This initiative, adapted to local accessible media e.g. radio in Africa, (GHF 2009) has potential to meet the present need for sharing climate adaptation experiences. The UNEP co-ordinates a programme called 'Many Strong Voices' which brings together the peoples of the Arctic and SIDS to share knowledge about the adaptations they are making and to tell their stories online about how they are coping with the serious impacts of climate change (UNEP 2012). 'The Nairobi Work Programme' has an online 'coping strategies database' which allows inquiries at multiple levels and provides examples of adaptation programmes that match inquiry criteria.

The World Heritage Centre has recently published two well illustrated online manuals: *Managing Disaster Risks for World Heritage* and *Preparing World Heritage Nominations*. They were developed in response to feedback through Periodic Reporting and from expert meetings. Analysis of the Periodic Reports showed that more focused training and capacity development was needed in specific areas where World Heritage site managers required greater support (UNESCO et al. 2010).

It is extremely disappointing to trace the history of the proposed World Heritage Manual **Vulnerability Assessment of World Natural and Cultural Heritage to Climate Change**. In 2008 the World Heritage Committee listed it as 8th on its priority list, rising to 3rd at the next session (Decision 32COM 18). However in 2009 the General Assembly only approved the preparation of two titles, the manual no longer being included.

While acknowledging the reputation of the Nobel Prize winning IPPC 4 Assessment Report for its science and its achievement as an internationally signed off 'document', it is now dangerously out of date in critical areas of assessment. The UNFCCC authorities are well placed to have an online interface that rectifies this problem. Unfortunately there is no referral from the IPPC 4 web page to other credentialed internet sites from which the latest and most reputable and vital science can be accessed. It has been reported that the World Heritage Committee is not well briefed on climate science (Terrill 2008) and relevant climate predictive research is not systematically made available to World Heritage field staff.

Recommendations

23.1 Propose that the WH Committee request UNESCO to urge the **IPCC** to maintain an **inter-assessment updating service** that is active between issues of the seven yearly IPCC assessment reports. This might prevent the **inappropriate** referencing of IPCC *Assessment Reports* when they become outdated.

23.2 Include in the proposed ***WH:ICH:CC Database & Network*** a subsidiary portal to access studies, activities and events relevant to the impacts of **climate change on culture**.

23.3 Advocate **museums** take on the vital role of becoming places to learn about climate change generally and specifically through graphically illustrated local and regional case studies that the museums' natural and cultural history collections can tell.

23.4 Support the World Heritage initiative to raise the 'overall profile of science' (UNESCO 2010) and the introduction of a new category of World Heritage inscription - **Heritage linked to Science and Technology.**

23.5 Provide authenticated and usable direct **feedback** to the source communities (especially indigenous communities) from which the climate data and information used in planning adaptive interventions is derived.

23.6 Appreciate that learning about climate change should be a **two way process involving specialists and traditional knowledge holders.** On this basis, combine the experience and observations of the traditional holders with the latest scientific information to arrive at the best possible shared knowledge base on which all can act.

The Intangible Cultural Heritage Convention

Investment in girls' education and health is the best possible investment for managing climate change.
Michael Keating, Director, Africa Progress Panel

Through opportunities in the Preamble, Article 1, 14, and 19, the ICHC has many knowledge sharing mechanisms from village level to international diaspora stage to address climate change impacts on heritage. What is not addressed by the intangible cultural heritage community is the need to recognise and address climate change as an ever growing, ever more destructive threat to intangible heritage. While the ICHC could be seen as attracting a select range of listed elements, the diminution of intangible heritage due to, for instance, climate-related encroaching desertification or loss of fish stocks, passes unrecognised. States Parties making their national Intangible Cultural Heritage inventories will be safeguarding threatened practices by documenting and encouraging such, whilst the fundamental threat goes largely unrecorded and the intangible cultural heritage grows ever more vulnerable. The opportunity to have practices and ceremonies highlighted as listed elements through the Convention's excellent information sharing mechanisms, may thus be lost. Rationally the 'List of Intangible Cultural Heritage in Need of Urgent Safeguarding' should be growing in a race against time to safeguard vulnerable intangible cultural heritage.

In the far east of Ladakh live the Buddhist Brokpa people of Da- Hanu. They are identified by their Indo-Aryan appearance and their clothing, especially their head ornaments. For hundreds of years Ladakh has been at the crossroads with invaders and traders from China, Central Asia, Tibet and India. Its recent isolation after the border closure with China has assisted the survival of its traditional multi-ethnic and multi-faith society. Ladakhi open mindedness, acceptance and strong sense of identity has in general stood them on good stead to manage and survive the Western influences of tourists and NGOs that have come in increasing numbers since the early 1970s. Climate change however will sorely test this resilience

The Intangible Cultural Heritage Committee is working successfully to increase the membership of the ICHC. It is developing the Representative List, overseeing the implementing of State Party's national and regional safeguarding directives and responding to problems in the ICHC processes such as weaknesses in complying with the nomination application criteria of 'ensuring the widest possible participation of communities , groups and where appropriate individuals in safeguarding activities and management'. But despite the Committee's immense knowledge dissemination potential, the overarching problem of climate change is not, as yet, well recognised. The probability is that the threat will massively impact its good works.

UNESCO and the Intangible Cultural Heritage Committee are working to right the imbalances as discussed in Chapter 19 across regions but with emphasis on 'knowledge dissemination and sharing' operations in Small Island Developing States (SIDS), Pacific Century Premium Developments (PCDP) and Africa.

Feedback from heritage workers indicates the need for more resources to enable the implementation of the ICHC, for fostering and re-fostering greater awareness of the importance of safeguarding intangible heritage and for assistance to States Parties to enhance capacity building particularly in the task of building national inventory programmes. The Intangible Cultural Heritage Committee, after having reviewed many nominations, has identified that more community participation in the nomination process is needed in order to stay true to the intention of the ICHC. Through these improvements may come more 'bottom-up' observations on the impact of climate change on the cultural lives of communities.

One of the acknowledged positive and progressive initiatives in knowledge sharing is the new role of UNESCO Category 2 Centres in the area of intangible heritage. The **Intangible Cultural Heritage Centre for Asia and the Pacific (ICHCAP)** (UNESCO 2010) (Galla 2010; Galla 2009a) was established in 2009 in the Republic of Korea. The Centre specialises in networking and information utilisation. It aims to promote the ICHC, contribute to its implementation in the Asia-Pacific region, increase community participation in safeguarding, raising awareness of intangible cultural heritage, enhance the capacity for safeguarding intangible cultural heritage through coordination and dissemination of information, and to foster regional and international cooperation for the safeguarding of intangible cultural heritage.

UNESCO is constantly reviewing its responsiveness in meeting the needs of States Parties for knowledge about the ICHC and its implementation. An example, amongst many, of its practices is the series of **Intangible Cultural Heritage workshops** in Abu Dhabi (United Arab Emirates), Havana (Cuba), Sofia (Bulgaria), Libreville (Gabon), Harare (Zimbabwe) and Beijing (China). The workshops were devoted to the training of trainers in different regions from January to March 2011 to create a network of expert trainers to further strengthen national capacities for safeguarding intangible cultural heritage.

An opportunity to introduce climate change policy to safeguard intangible heritage comes from a reasonably longstanding move to bring together the WHC and the ICHC. Since 2004 strategists have been discussing the **synergy that could be built between the two cultural conventions** (Smeets 2004).

> "In the case of interdependent sites that are inscribed on the World Heritage List and the intangible heritage elements that are going to be safeguarded under the 2003 Convention, common and integrated approaches will have to be developed. Integrated approaches should start from an inventory of both the tangible and the intangible heritage present within the specific setting of a region or a community. One group of such cases was already mentioned, that of

the cultural spaces, i.e. built or natural environments that are essential for the enactment of one or more Intangible Cultural Heritage elements. If a cultural space appears to be inscribed on the World Heritage List, and if the intangible heritage element that is narrowly associated to it, is protected under the 2003 Convention, then the two, Intangible Cultural Heritage and its link to Tangible Cultural and Natural Heritage Committees concerned will have to cooperate" (Smeets 2004 p149).

More recently this potential collaboration has been seen to have applications through "programmes involving bearer communities and relating to educational activities, dissemination, protection, safeguarding and social ownership" (UNESCO 2010n). Relevant in this context is that established World Heritage climate change policies and management practices which already cover the 'intangible heritage' at World Heritage Sites could be shared with the intangible cultural heritage fraternity, a prospect that is in the nature of the ICHC. By this means intangible heritage workers could work alongside World Heritage workers trialling and refining climate change policies.

Further initiatives in information sharing include preliminary proposals to use indigenous knowledge systems for the safeguarding of intangible cultural heritage through the sustainable conservation of natural heritage sites, and the **Indigenous Cultural Mapping Initiative** (UNESCO 2010n). A further opportunity exists teaming with LINKS. In 2002 UNESCO launched **The Local and Indigenous Knowledge Systems (LINKS),** a project which aims to empower local and indigenous peoples in biodiversity governance by advocating full recognition of their unique knowledge, know-how and practices.

Yet there are still missed opportunities for important knowledge sharing. In October 2011 the United Nations University, IPCCC, the Secretariat of the Convention on Biological Diversity, UNDP and UNESCO co-organised two workshops, under the theme of **Indigenous Peoples, Marginalized Populations and Climate Change**. The workshops brought together indigenous peoples and marginalized populations, natural and social scientists, and other experts. The goal was to identify, compile and analyse relevant indigenous and local observations, knowledge and practices related to understanding climate change impacts, adaptation and mitigation. The workshops provided an opportunity to ensure that experience, sources of information and knowledge (scientific, indigenous and local), along with data and literature (scientific and grey), focusing on vulnerable and marginalized regions of the world, be made available to the authors of the **IPCC 5th Assessment Report** and the global community. The theme of the first workshop was *Climate Change Vulnerability, Adaptation and Traditional Knowledge*. Selected on recommendation from an International Panel of Experts, 50 participants met in Mexico City to present their research. The second workshop will be held early in 2012 and will focus on climate change mitigation (UNESCO 2011). This represents a splendid opportunity for the Intangible Cultural Heritage Committee to be involved in a meeting mainstream to its interests. Unfortunately UNESCO has advertised these events under the Natural Sciences Sector with *no* link to either the WHC or ICHC websites.

Recommendations

23.7 Refer to the ICHC's well developed **knowledge sharing mechanisms** and their potential role in climate change capacity building in the proposed *Intangible Cultural Heritage Climate Change Response and Policy Document*.

23.8 Propose that Category 2 Centres, e.g. the Intangible Cultural Heritage Centre for Asia and the Pacific (ICHCAP), act to safeguard **regional ICH** against climate change impacts by accessing and contributing to the **WH:ICH:CC Database & Network**; mediating regional climate change science for local communities, interpreting climate and weather predictive data, assisting with assessments of climate change vulnerability and risk, and facilitating cross disciplinary collaborative climate change programmes with other projects such as Local and Indigenous Knowledge Systems (LINKS).

23.9 Propose that **climate change vulnerable States Parties** including SIDS, PCDP and LDCs be assisted in nominating their ICH or **collectively nominating** their shared ICH such as for instance their knowledge of navigation in the Pacific.

23.10 Propose that the accreditations of NGOs (*Operational Directives* paragraph 88) include consideration of **the NGO's climate change qualifications**.

23.11 Endorse recommendation 23.3 in relation to the importance of **museums** in climate change education extending this role to include threats to ICH.

Stakna Gompa (monastery) positioned commandingly beside the Indus River 25km upstream of Leh

Chapter 24

Can the Conventions Address Climate Justice, Human Rights and Cultural Rights Issues Relating to Climate Change Impacts on Cultural Heritage?

19-year-old Swar is survivor of the May 2008 cyclone Nargis that devastated Myanmar. He recounted his families experience during the storm. His family clung to a coconut tree as the cyclone hit together with another man who was forced to relinquish his youngest daughter to the storm's fury, never to be seen alive again. "On just the first day my country had 100,000 casualties," said Linn Kyaw Swar. "Never before has there been such a devastating catastrophe." Uprooted trees meant no shade in a hotter than- usual summer. He initiated a tree planting project, to plant thousands in that area. "I can plant 100 trees in a day with 10 colleagues," he said. "But this is no long-term solution to climate change. We're the foot soldiers in this war; we know what happens if we don't act," he said. "But now we all need to step up."

Can the World Heritage Convention and the Intangible Cultural Heritage Convention address climate justice, human rights and cultural rights issues relating to climate change impacts on cultural heritage?
What roles do human rights law and cultural rights law have in relation to climate change-impacted cultural heritage in these Conventions? Consider references made to 'existing international human rights instruments' (ICHC Art.2) and associated Conventions, Recommendations, Covenants and Declarations.

The World Heritage Convention

Legal Matters

The legal history of the World Heritage Committee's response to climate change is detailed in Chapter 26. Suffice here is to draw attention to some main actions that so far typify the nature of this response.

Initially, as described in chapter 15, the World Heritage Committee's public acknowledgement of the climate change threat was triggered by petitioners. There followed the development and implementation of a policy which concentrated on site based mainly adaptive efforts. Addressing the broader scale of the climate change dilemma arose when the World Heritage Committee recognised that there is uncertainty as to the legal status of a World Heritage site under threat or anticipated threat from climate change. For instance: "Should a site be inscribed on the World Heritage List while knowing that its potential Outstanding Universal Value (OUV) may disappear due to climate change impacts?" and, "Should a site be inscribed on the List of World Heritage in Danger or deleted from the World Heritage List due to impacts beyond the control of the concerned State Party in circumstances where these impacts have resulted in serious deterioration of or loss of OUV (UNESCO 2008a p12).

In answer to the first question, it found that unless all the criteria (under which it was nominated) were destroyed by climate change, under which circumstance the site would be delisted (OP 192 – 198), the site could remain listed. The criteria that were affected by climate change may be able to be modified or altered following an assessment by the World Heritage Centre and Advisory Bodies (under the *Operational Guidelines* 163-165). Re-nomination would need to take place if the criteria were to be altered (*Operational Guidelines* 166). In answer to the second question, as has been discussed in Chapter 15, the World Heritage Committee makes a case-by-case judgement as regards the 'serious and specific' threat that qualifies an inscribed site to be transferred to the List of World Heritage in Danger. In the case of a site impacted by climate change induced threat, the process is the same.

These legal responses are reactive in nature, in that they do not normally implicate the underlying cause of the threat, for instance they do not identify and apportion blame to the companies, industries and nations that in the case of climate change related damage, emitted the greenhouse gases. Rather attribution of damage is directed to the proximal cause for example to the storm, flood or other event that was immediately involved in the destruction with climate change perhaps being mentioned secondarily but with no mention of those that were responsible for the causative emissions.

A 2010 Report by **The World Commission on the Ethics of Scientific Knowledge and Technology (COMEST)** titled ***The Ethical Implications of global climate change*** lists one of the categories of vulnerability as: "Threats to cultural heritage, mainly to traditional ways of living, or to architectural masterpieces of various kinds, particularly in the case of sudden irreversible submergence of inhabited land" (COMEST 2010 p13). The report makes it clear that global climate change "poses a clear and present threat to the well-being of the community of life on Earth, which includes non-human life, but also the social and cultural dimensions of human existence". In 2006, 86 States Parties identified damage to their World Heritage properties from climate change. It is now becoming possible to determine the probability of an extreme weather event having been intensified by climate change and in some cases even quantify the likely contribution of climate change to the adverse event (that may be, for example, responsible for the damage to a World Heritage monument). Sea level rise, as distinct from subsidence can also be measured (for example, at submerging World Heritage coastal sites such as the **Ruins of Kilwa Kisiwani and Ruins of Songo Mnara, Tanzania**) thus enabling apportion of damage between geological events and human-induced climate change. The important legal question of whether a State Party could legally challenge carbon polluting States Parties for defying Article 6.3 "Each State Party to this Convention undertakes not to take any deliberate measures which might damage directly or indirectly the cultural and natural heritage referred to in Articles 1 and 2 on the territory of other States Parties to this Convention" has yet to be addressed.

Final Statement of the participants of the Conference on European and Pacific Responses to Climate Change in the Pacific
Berlin, November 21st 2010

Point 7 Respect *for the traditions and cultures of people displaced by climate change, which must be a precondition to all actions undertaken. As such the principle of free, prior and informed consent must prevail. Indigenous peoples' link to the land is fundamental for most of the Pacific communities; the decision to resettle needs to be well-founded and unavoidable. Furthermore, climate change migrants need to be empowered to become agents of their own resettlement process, not victims of it. Capacity building, transparent consultations, the provision of relevant information and the rehabilitation at resettlement destinations are to be guidelines of the required help from outside. The preservation of their way of life (right to culture), as reflected in traditional knowledge and local culture, including all aspects of personal, family, social, political and spiritual life, needs to be strongly supported to help affected migrants maintain and keep alive their cultures, languages and national identities.(On the Run Final Statement of the participants of the Conference on European and Pacific Responses to Climate Change in the Pacific.)*

Insert 24.1 On the Run

Other obvious legal questions hark back to the original petitions. How much temperature rise, caused by human induced greenhouse gas (GHG) pollution, is acceptable before the WHC article 4 is breached? Can the WHC use Article 4 to slow black carbon pollution (anthropogenically generated) endangering World Heritage sites threatened by glacial melt and sea level rise? Article 4 states, "Each State Party to this Convention recognizes that the duty of ensuring the identification, protection, conservation, presentation and transmission to future generations of the cultural and natural heritage referred to in Articles 1 and 2 and situated on its territory, belongs primarily to that State. It will do all it can to this end, to the utmost of its own resources and, where appropriate, with any international assistance and co-operation, in particular, financial, artistic, scientific and technical, which it may be able to obtain." These legal questions have not been widely tested and no guidelines have been incorporated into the *Policy document on the impacts of climate change on World Heritage properties*.

As has been discussed, climate change impacts humanity, and the consequent 'human impacts' may affect World Heritage. Given the unprecedented climate driven displacement of people, the question of the rights of climate change displaced populations and their cultures to a form of protection from recipient countries and GHG polluting countries (could be one and the same) will inevitably arise. In December 2010 **the Conference on European and Pacific Responses to Climate Change in the Pacific** drew up a 10 point final statement *On the Run* which appealed for "Consideration of the human rights aspect of climate change. The dignity of people displaced by climate change has to be a guiding principle of all plans and actions taken. In the same way the human rights of the communities receiving displaced people need to be respected." Importantly it also articulated the issue of cultural rights of the displaced, see Insert 24.1. To encompass the new issues arising in the climate change context, the **Australian Human Rights and Equal Opportunity Commission** claims that a new international agreement is needed "one that equitably shares the emerging burden of climate-induced displacement flows across the world and upholds the human rights of the individuals affected" (HREOC 2008). This is just one of the 'human impacts' among others such as health and sustainable development, which has ethical implications for World Heritage protection of cultural sites and of intangible cultural heritage.

Ethics

COMEST is an advisory body to UNESCO. It develops ethical principles that could provide decision-makers with criteria that extend beyond purely economic considerations. COMEST's report ***The Ethical Implications of Global Climate Change*** concludes that ethics is "...not something added on top of other issues related to climate change, but rather a constitutive part of all of the reasonably justifiable responses to the challenges of climate change. Therefore, it can be stated unequivocally that climate change cannot be dealt with adequately and properly if the ethical dimensions discussed in this report are not highlighted, well understood, and taken into account in decisions about responses. The purpose of this report was therefore not to make climate change a (new) theme of ethics, but rather to make ethics a core and necessary element of any debate about climate change and its challenges (COMEST 2010 p38)."

COMEST has argued that there is an urgent need to establish 'universal ethical principles' that can guide responses to the challenges of climate change at global, regional, national and local levels.

- We need to understand climate change as a constellation of extremely complex phenomena in order to lay out coherent and credible scenarios for its possible development. This calls for a concerted scientific effort, focusing on the most urgent needs; in recognition of the universal right "to enjoy the benefits of scientific progress and its applications" (article 27.1 of the **Universal Declaration of Human Rights**).
- We need also to reduce greenhouse gas emissions on the basis of fair burden sharing that does not impede legitimate expectations of development.
- We need, finally, to soften the impact of climate change to enable States and populations to adapt without damaging their vital interests.

In other words, at every level of action – scientific knowledge, mitigation, and adaptation – the key, inherently ethical issue is responsibility (UNESCO 2010ll).

Many of the key principles required by an ethical approach to climate change are already enshrined or referenced in international conventions and declarations. The equity principle, UNFCCC's common but differentiated responsibilities (CBDR) and respective capabilities has been discussed in Chapter 13.

Other important and well-established principles include:

- the precautionary principle, see Insert 16.2;
- the right to share in scientific advancement and its benefits;(ICHRP 2008; UN 1948)
- the principle of sustainability;
- the principle of integrity as applied to ecosystems; and
- the principle of safeguarding and promoting the interests of future generations.

COMEST emphasises, "There is therefore an implicit basis for international consensus to provide international debate on climate change action with a stronger and clearer ethical underpinning (UNESCO 2010ll)."

Insert 24.2

UNESCO CLIMATE CHANGE INITIATIVE: **Core Program:**
Climate Change, Ethics, Social and Human Sciences Dimensions: Research Programme.

The design and implementation of appropriate climate change adaptation actions, based on the MOST and environmental ethics programmes, benefiting the most vulnerable related to the cross-cutting issues of energy, water and biosphere management, as well as improve understanding of gender equality issues related to climate change.

The Initiative will maximize the synergies among its main components and will be carried-out from the global down to the regional, national and local levels. (UNESCO 2009h)

Universal Declaration of Ethical Principles in Relation to Climate Change, Draft

After discussion, the UNESCO Executive Board recommended that the General Conference investigate the *advisability* of preparing a **draft *Universal Declaration of Ethical Principles in relation to Climate Change*.** This recommendation was vigorously discussed at the 35th General Conference of UNESCO in October 2009 where a Resolution was adopted to launch a process that could lead to the development of this declaration. The Director-General was requested to submit a report at its 185th session in October 2010 on the *desirability of preparing a draft* of a declaration of ethical principles in relation to climate change. After wide consultation including 8 regional meetings and expert workshops, a survey, analyses of submissions, meetings with agencies and a report by COMEST, UNESCO Executive Board found : that although "ethics has been recognized as having an essential contribution to make," there was "no consensus that preparation of such a declaration would make a useful contribution to the international response to climate change and the opinions of Member States remain divided (UNESCO 2010s p3)."

Insert 24.3 Shifting ethics 'to the background'

> ... far-reaching ethical questions can be asked about the continuation of human actions that not only cause climate change, but also contribute to its intensification and acceleration.
> The ethical stakes surrounding climate change cannot be avoided or reduced. Failure to act could have catastrophic implications, but responses to climate change that are not thought through carefully, with ethical implications in mind, have the potential to devastate entire communities, create new paradigms of inequity and maldistribution, and render even more vulnerable those peoples who have already found themselves uprooted by other man-made political and ideological struggles. Moreover, it is well known that global climate change has the potential to bring about conflict mobilized by the quest for scarce resources. The need for an ethical approach is therefore compelling.
> The formulation of an ethics of response to climate change will have profound implications for the immediate and future well-being of vast numbers of people who are the immediate victims of global climate change, or fall into the vague category of those causing it. However, **ethical concerns are in fact rarely made explicit in discussions about climate change, and therefore are not adequately scrutinized or debated.** Climate change discussions predominantly take place on a factual and technical level, i.e. they focus on the causes, the impacts and the effects of climate change, or on technical policy issues regarding responses to its challenges. As Ten Have (Ten Have 2006) has pointed out with regard to responses to environmental problems in general, there seems to be a tendency to move directly from concerns about climate change to climate change action, without self-consciously and critically reflecting on the aims, the nature, the extent and the justification of these actions. Thus, the ethics already embedded in concerns about and responses to global climate change are shifted to the background, and effectively taken off the agenda of matters that need to be seriously considered (COMEST 2010).

The Board found that the "Response to global environmental change requires assessment tools that can take adequate account of extended and still to be understood causal chains across time and space, including the conflicting bases of differentiated responsibility in mitigation and adaptation, fundamental uncertainties relating to the knowledge bases required for the successful elaboration of effective policies, and the challenge of responsible management of collective risk at the global scale. In addition, there are major ethical issues with respect to the scope, focus, availability and accessibility of basic science, monitoring information and early-warning capacity that should be taken into account in the development and management of environmental knowledge (UNESCO 2010ll p1)."

"While there was broad support from workshop participants for the idea of a declaration in general terms, there were significant differences in opinion regarding the availability, relevance, and universal acceptability of currently recognized ethical principles as they apply to climate change." The Board "Requests the Director-General to take into account recent scientific and scholarly findings on the ethics of climate change and submit to the Executive Board at its 187th session a report reviewing the issues relevant to assessment of the desirability of preparing a declaration of ethical principles in relation to climate change, with particular reference to the outcomes of United Nations Framework Convention on Climate Change (UNFCCC) Conference of the Parties (COP)-16 and further work by COMEST on the key controversies surrounding such ethical principles in relation to climate change and their implications on policy (UNESCO 2010s p5)."

COMEST explains that the 'reservations' that for the past 15 years have hampered the establishment of an agreed international framework for action that might rise to the challenge the planet faces: are "...key unresolved ethical questions." "For example, with reference to the UNFCCC, how can the interests of present and future generations be balanced? What type of response to the challenges of climate change would be truly "equitable"? Which responsibilities are truly "common" and which are "differentiated"? Do those who have the "capacity" to act have a duty to do so, regardless of their historical contribution to greenhouse gas emissions? (UNESCO 2010ll)"

The evolution of this issue has been described here in some detail to underline the high sensitivity surrounding any admission of culpability for GHG emissions and any acknowledgement of the imperative that they be heavily cut. Insert 24. 4 indicates the number of World Heritage sites held by the identified 20 highest carbon dioxide emitting States Parties and in total by the remaining 167 lower emitting States Parties. The key to the top 20 emitters is ranked in diminishing order in relation to recent annual emission status. It will be noted that the top 20 emitting States Parties possess almost half of the 936 World Heritage properties. It is likely that the many if not all of these high emitters are not meeting their undertakings to do their utmost to protect their large holdings of world heritage and are not fulfilling their obligations under the WHC to do their utmost to protect all World Heritage.

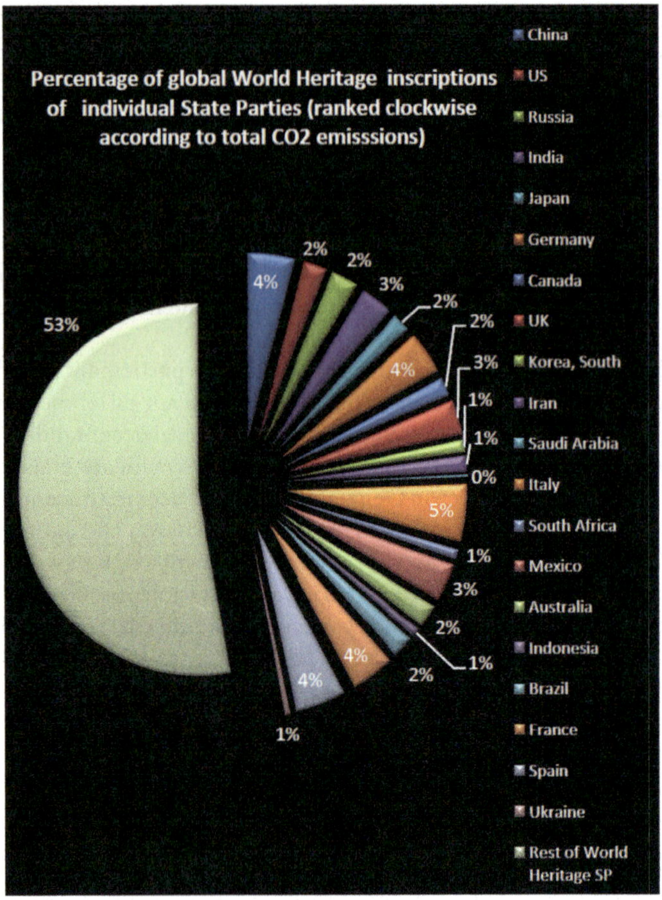

Insert 24.4 States Parties ranked clockwise in descending CO_2 emission order. Size of segments and associated percentages refer to proportion of world's WH inscriptions held by each State Party

UNESCO addresses the ethics of climate change both directly and indirectly through reference to other branches of the UN, its parent body, (e.g. International Panel on Climate Change (IPCC). The ethics of climate change are addressed directly through advice from COMEST, in education through the UNESCO Ethics Education Programme started in 2004 (UNESCO 2004c), and in society through the UNESCO Intersectoral Platform and **World Heritage Strategy for Action to Address Climate Change**. In 2009 UNESCO included 'ethics' in its medium term strategy, 2008-2013, as part of the **UNESCO Climate Change Initiative core program: Climate Change, Ethics, Social and Human Sciences Dimension: Research Programme**, (see Insert 24.2).Across this range of programmes the question of climate change, ethics and culture is as yet undeveloped as is any application of an ethical stance in relation to climate change and the protection of cultural World Heritage properties. (See Insert 24.3 Shifting ethics 'to the background').

To reiterate, the World Heritage Committee's attitude on climate change is clearly expressed in the 2006 comment in the report *Predicting and Managing the Effects of Climate Change on World Heritage* paragraph 35: climate change will alter "the way people live, work, worship and socialise in buildings, sites and landscapes with heritage values." Climate change-induced socio-economic changes "will have a greater possible impact on the conservation of cultural heritage than climate change alone" (UNESCO 2006a p.29). However the implications of climate change have evolved significantly since 2006 and the World Heritage Centre has not been clear as to where it stands about meeting its responsibilities in the light of current knowledge and expectations. In fact its post- report *(Predicting and Managing the Effects of Climate Change on World Heritage)* climate change publications retreat from these initial perspectives. The World Heritage Centre appears uncertain of the extent of its commitment to an ethical stance, intergenerational responsibilities, mitigation commitment, and full climate change knowledge dissemination and human rights. Of relevance here is UNESCO's careful stance to act in its 'fields of competence'. As shown in the discussions concerning the Director-General's preliminary proposals for priority fields of action (UNESCO 2010) UNESCO often qualifies its proposals e.g. UNESCO proposes "Emphasis on the normative aspect of the promotion of human rights in all of UNESCO's fields of competence" and "The importance of climate change was recognized by several Member States, not only to respond to environmental challenges but also to address social and development implications in UNESCO's fields of competence" (UNESCO 2010 n).

Unlike the Intangible Cultural Heritage Convention (ICHC), the World Heritage Convention (WHC) does not recognise previous standard setting *instruments by name* in the text. The World Heritage climate change publications do recognise other conventions *(Predicting and Managing the Effects of Climate Change on World Heritage (UNESCO 2006a p36-40); Policy Document on the Impacts of Climate Change on World Heritage Properties* (UNESCO 2008a p4) and *A Strategy to Assist States Parties to Implement Appropriate Management Responses,* (UNESCO 2006g p5)) as does the WHC Preamble (paragraph 6). These references extend to including the benefits of developing "synergies and partnerships with other multilateral environmental agreements" (UNESCO 2006a p36), and seeking "international co-operation with other Conventions" (UNESCO 2006g p5).

The instruments, agreements and other organisations most commonly referred to in relation to 'cultural' World Heritage include the **UNFCCC, IPCC, ICHC, Convention on the Protection and Promotion of the Diversity of Cultural Expressions, the International Human Dimensions Programme on Global Environmental Change (IHDP), the International Committee of the BlueShield, the Organisation of World**

Heritage Cities and others in the financial and technical realm. The thrust of the anticipated collaborations is research, information sharing, and exchange of best practice, education, training, awareness raising and capacity building. Human Rights are not specified in the development of these "synergies and partnerships".

Human rights concerns have remained relatively marginal to the WHC but there are opportunities for their consideration, in part, through some of the Conventions and organisations named above. The UNFCCC (1994) makes no mention of human rights per se but importantly in Article 3 has enshrined the equity principle 'common but differentiated responsibility' (CBDR) discussed in Chapter 6 which was introduced precisely to acknowledge the justice claims of developing countries, and in particular to balance the differences of contribution and capacity (Rajamani 2006).

The **IHDP** sees the Earth moving into the 'Anthropocene' era. Its activities focus on developing and sustaining cutting-edge research; developing world-wide capacity to understand and deal with these challenges; and promoting interaction between scientists and World Heritage climate change policymakers on these topics. It interprets climate change through a social justice and human rights framework.

The monumental IPCC 4th Assessment Report, 2007 has reference to human rights throughout. It links climate change with principles and practices applicable to *policy* formulation such as: equal access to the global atmosphere being consistent with the UN Human rights declaration underlining the equality of all human beings; climate change litigation by indigenous Arctic dwellers; climate change and ethics; climate justice, and 'rights' based equity.

By extending the interpretation of WHC's preamble concerning "international conventions", the following Human Rights agreements can be considered to inform and influence World Heritage: The **Universal Declaration of Human Rights (UDHR), 1948,** the **Declaration of Human Duties and Responsibilities (DHDR) 1998** and **the United Nations Committee on Economic, Social and Cultural Rights, 2003.**

The Universal Declaration of Human Rights (UDHR)

Climate change is likely to affect several rights enshrined in the UDHR. First and foremost is Article 19 *the right to know*, the right to knowledge, the right that has the potential to equip people to face the inevitable changes due to climate change and to empower them. The **Advisory Council of Jurists of the Asia-Pacific Forum on National Human Rights Institutions** recently endorsed the idea that (as climate change will affect the right to an environment of a particular quality) the protection of the environment is "a vital part of contemporary human rights doctrine and a *sine qua non* for numerous human rights, such as the right to health and the right to life (HREOC 2008)."

Monastery mural depicting the Tibetan Buddhist deity Namgyalma Blazing Crown

Climate change is already undermining a broad range of human rights, such as rights to food, water and shelter; rights associated with livelihood, culture, migration and resettlement; gender equality; and personal security and the rights of indigenous peoples. The worst effects of climate change are likely to be felt by individuals and groups whose rights protections are already precarious (GHF 2009b) and those who are very young and yet unborn. As Kyung-wha Kang, the UN Deputy High Commissioner for Human Rights has started, "Global warming and extreme weather conditions may have calamitous consequences for the human rights of millions of people...ultimately climate change may affect the very right to life of various individuals... [countries] have an obligation to prevent and address some of the direst consequences that climate change may reap on human rights" (MacInnis 2008).

It is in this setting that Human Rights and Ethics have the greatest relevance to climate change impacts on World Heritage. World Heritage properties both support and are supported by local communities. As climate change impacts intensify, human rights will come under pressure and under increased scrutiny to change to meet the 'new' questions of justice and distribution. As suffering increases and new solutions are grappled for to answer questions such as, 'is climate change relevant to the responsibility to protect World Heritage?' stakeholders are arguing for human rights to be the resolution framework. Human rights provide clarity and direction by recognising the moral link between local causes and distant effects, and "a shared and legally codified moral language" (ICHRP 2008 p11). Human rights considerations are "clearly relevant to adaptation policies", "to technology transfer" from rich to poor nations, to setting thresholds when assessing threats, (ICHRP 2008 p12) to claiming the right to information and the right to participation in climate change management.

The Declaration of Human Duties and Responsibilities (DHDR)

This declaration is less well recognised than the UDHR but deserves consideration for its intrinsic value and relevance to the emerging Gaian perspective of responsibilities to Mother Earth. It was written to reinforce the implementation of human rights (under the auspices of UNESCO) so helping to address the lack of political will for enforcing the UDHR. The preamble states "The effective enjoyment and implementation of human rights and fundamental freedoms is inextricably linked to the assumption of the duties and responsibilities implicit in those rights....." (Foundation and UNESCO 2009). The Declaration is long and wide ranging with 41 Articles organised into 12 Chapters. Of significance to the WHC is Chapter 2 which lists the responsibilities which accompany the right to life and human security for present and future generations in the awareness that for the first time in human history our survival is in self inflicted peril. Article 3 and 4 address intergenerational responsibility, the right to peace and the right to live in a balanced ecological environment. Article 8 describes the duty and responsibility of humanitarian assistance and intervention to those in need including the millions of displaced people. Article 9 enunciates the duty and responsibility to protect and promote a safe, stable and healthy environment, promoting respect, protection and preservation of the uniqueness and diversity of all forms of life. An adequate use of resources avoiding excessive exploitation and consumption, and collaborative scientific research and exchange of information are required. This article promotes an international and legally binding agreement to reduce GHG emissions worldwide and an urgent change of attitude towards the environment, now reinforced by the reality of climate change and the human responsibility. UNESCO's Patricia

Morales comments "Recent history including disrespect by so many of the efforts to address climate change demonstrates that disregard of duties and responsibilities such as those proposed by the DHDR constitutes a tragedy for our human condition today" (Morales 1998 p14).

The International Covenant on Economic, Social and Cultural Rights, 1966, (ICESCR)

ICESCR Article 15 and UDHR Article 27 recognize the right of everyone to *culture* and to take part in cultural life without any discrimination. Parties to these instruments also work to promote the conservation, development and diffusion of science and culture, and encourage international contacts and cooperation in these fields. Interestingly the Covenant contains a principle similar to CBDR which implicitly acknowledges differences in capacity when it says that each state is required to take steps "individually and through international assistance and cooperation" with a view to a "progressive realisation" of the rights in the Covenant (Article 2(1) (UNHCHR 1966). If the World Heritage *cultural* sites and those participating in the *life* of the World Heritage sites were threatened by climate change, there would be means by which claims might be brought against polluting nations.

United Nations Declaration on the Rights of Indigenous Peoples (UNDRIP)

Under the **UNDRIP** 'first' peoples have the right to practice and revitalise their cultural practices, customs and institutions. There is an intrinsic link between indigenous culture and land. According to one expert, "Indigenous people don't see the land as distinct from themselves in the same way as maybe society in the south-east (of Australia) would. If they feel that the ecosystem has changed it's a mental anxiety to them. They feel like they've lost control of their 'country' — they're responsible for looking after it" (FOE 2007 p6). For this reason, the right to participate in and to strengthen indigenous cultural life is directly threatened by climate change. The rights of indigenous people are relevant to this question as a number of cultural World Heritage sites have been inscribed because of their indigenous cultural OUV which is maintained and 'embodied' by indigenous custodians.

Universal Declaration of the Rights of Mother Earth

The **Universal Declaration of the Rights of Mother Earth** was formulated at the UN sanctioned **World People's Conference on Climate Change and the Rights of Mother Earth** in Bolivia in 2010. It grew out of frustration with the powerful international community for ignoring the real threat of climate change to developing nations, for not making a meaningful international agreement to cut GHG emissions and for not taking an ecological 'Mother Earth' perspective to secure a sustainable future. The 'rights' that implicate climate change are listed in Insert 24.5

Insert 24.5

Preamble: "We, the peoples and nations of Earth: are: conscious of the urgency of taking decisive, collective action to transform structures and systems that cause *climate change* and other threats to Mother Earth;
Article 1: Mother Earth includes "The rights of each being are limited by the rights of other beings and any conflict between their rights must be resolved in a way that maintains the integrity, balance and health of Mother Earth;
Article 2: Inherent Rights of Mother Earth include " the right to regenerate its bio-capacity and to continue its vital cycles and processes free from human disruptions" and " the right to full and prompt restoration of the violation of the rights recognized in this Declaration caused by human activities;
Article 3: Obligations of human beings to Mother Earth include "Every human being is responsible for respecting and living in harmony with Mother Earth and "Human beings, all States, and all public and private institutions must guarantee that the damages caused by human violations of the inherent rights recognized in this Declaration are rectified and that those responsible are held accountable for restoring the integrity and health of Mother Earth (World People's Conference on Climate Change and the Rights of Mother Earth 2010)

The appeal of this movement is understandable. This is an extract from the Honduras Weekly in March 2010 *before* the "World People's Conference on Climate Change and the Rights of Mother Earth" conference in April 20-22 in Cochabamba, Bolivia. As the journalist begins,

> "This (Conference) is an opportunity to make up for the lack of conformity and action on targets and commitments at the Copenhagen Climate Summit of 2009". Honduras is "one of the nations that seem to be suffering the effects of climate change as a result of the actions of other countries, destructive multi-national business practices ..., and a low climate change priority in country.
>
> Southern Honduras is currently experiencing a severe drought leaving tens of thousands with little food or water, and many surviving on rations donated by international organizations. Concurrently, as Honduras lies within the hurricane belt, its residents are at risk from ever more frequent and catastrophic hurricanes. Rebuilding physical infrastructure is of great concern, but in the future it will become increasingly difficult to treat the sick when illnesses such as malaria and cholera reach acutely high levels after extreme rain and flooding — intrinsic symptoms of hurricanes.
>
> So, with a bleak prognosis for Honduras and a present preview of what is to come, it's time for Latin America, along with the rest of the world, to use this second chance to discuss and repair some of the Earth's immeasurable damage and prevent it from occurring further" (Taylor 2010).

A Human Rights Approach

Despite the fact that to date the social and human rights implications of climate change have received little publicised attention, the environmental dimension of Human Rights has not been extensively articulated, and the precise connection between climate change and the international human rights law system is as yet undeveloped, a human rights-based approach is widely recognised as the most effective way to respond to climate change (HREOC 2008; GHF 2009a; ICHRP 2008). Climate change is an issue of social justice.

A Human Rights approach provides a conceptual framework for climate change World Heritage policies and legislative responses; a framework which is *normatively* based on international human rights standards and which is *practically* directed to promoting and protecting human rights. In applying a human rights-based approach, decision-makers should be guided by the core minimum human rights standards when weighing competing demands on limited resources.

A human rights-based approach addresses equity issues. By focusing on individuals as *rights-holders* responsibility is placed on government to allow for participation and input from affected members of society into the development of adaptation policies. A human rights-based approach to the implementation of a policy such as a revised *Policy Document on the Impacts of Climate Change on World Heritage Properties* would require that decision-makers engage in a thorough and proper consultation with those affected to minimise the disproportionate impact on vulnerable groups. Such an approach applies **the principles of *non-discrimination* and *substantive equality*.** This could be achieved by requiring that all new legislative-based policies concerning climate-change adaptation be accompanied by a human rights compliance statement. Recognising this, the UNFCCC places international obligations on Member Parties to help developing nations meet the costs of climate change adaptation and to develop regional mitigation and adaptation programs.

Through a human rights-based approach in relation to World Heritage, aid should focus on strengthening communities from the bottom up, building on their own coping strategies to live with climate change and empowering them to participate in the development of climate change policies which protect their World Heritage's OUV. Aid needs to be locally grounded and culturally appropriate. To this end, adaptation assistance should be part of mainstream strategies. This could mean, for example, incorporating the traditional cultural practices of indigenous communities into climate change responses.

The same approach applies to climate change induced-disaster risk management. Adopting a human rights-based approach addresses inequities by linking World Heritage risk policy to international human rights law encompassing all relevant guarantees—civil and political as well as economic, social and cultural rights.

The WHC itself has been criticised for not being as equity based as it could. Arguably Article 24, which deals with large scale assistance for threatened sites, is prejudicial against developing States Parties because the more threatened the property, in question, the more investment is needed to apply for detailed scientific, economic and technical studies and comply with assistance and thus the more it becomes unaffordable. Article 24 continues: "the studies" shall also "seek means of making *rational use* (my emphasis) of the resources available in the State concerned" (Article 24). This could be seen as interference in the workings of what would usually be a developing country.

It is worth noting that the World Heritage Committee can refuse nominations from States Parties who have "incompatible practices" relating to the nomination. These are usually related to a threat such as a dam, mining operations or an industrial development. However this power could conceivably be used to refuse a nomination to a State Party that nominates a potentially climate change threatened site but which has no national GHG mitigation policy and has not signed any international agreement such as the Kyoto Protocol.

Recommendations (as for ICHC Recommendations)

The Intangible Cultural Heritage Convention

When we talk about climate justice we are talking about looking at climate catastrophe through the injustice lens.
Either we get this right together or we all sink together. Every little [action] counts.
Kumi Naidoo, Global Campaign for Climate Action and Global Call to Action Against Poverty

Human rights instruments are referred to in the ICHC's Preamble and in Article 2, "For the purposes of this Convention, consideration will be given solely to such intangible cultural heritage *as is compatible with existing international human rights instruments*, as well as with the requirements of mutual respect among communities, groups and individuals, and of sustainable development." In this respect, unlike the WHC, the ICHC provides direct opportunities for interpreting its safeguarding role through a human rights and cultural rights framework.

A Brokpa dancer from Dha-Hanu at the Leh Festival

As discussed in chapter 13, climate change poses a direct threat to a wide range of universally recognized human rights. One of the most relevant human rights is enshrined in the UDHR in Article 27, "Everyone has the right to freely participate in the cultural life of the community, to enjoy the arts and to share in scientific advancement and its benefits." Cultural rights are referred to in the preamble and Article 3 of the 1976 **International Covenant on Economic, Social and Cultural Rights** drawn up by the Office of the United Nations High Commissioner for Human Rights. The Preamble sets the context. "Recognizing that, in accordance with the UDHR, the ideal of free human beings enjoying freedom from fear and want can only be achieved if conditions are created whereby everyone may enjoy his economic, social and cultural rights, as well as his civil and political rights." Article 3 states "The States Parties to the present Covenant undertake to ensure the equal right of men and women to the enjoyment of all economic, social and cultural rights set forth in the present Covenant."

This places States Parties, especially of poor and vulnerable nations in a strong position to claim safeguarding assistance for their intangible cultural heritage in the face of climate change caused through the actions of polluting nations. This would be in the form of compensation under CBDR and should be in the form of access to expertise, technical know-how and financial support. "Accordingly, the developed country Parties should take the lead in combating climate change and the adverse effects thereof" Article 3.1 UNFCCC (UNHCHR 1966).

These claims are founded on the belief that indigenous peoples' cultures and worldviews, and also the enjoyment of their human rights, are linked to the preservation of their eco- and livelihood systems, both of which will come under increased threat as climate change intensifies.

"The consequences of calamitous weather conditions are already visible in many parts of the world. A human rights approach compels us to look at the people whose lives are most adversely affected," the High Commissioner for Human Rights told the **20th Forum on Global Issues** in October, 2009.

"It provides the legal rationale and grounds to advocate the integration of human rights obligations into policies and programmes countering the negative effects of environmental challenges," (OHCHR 2009). Climate change is already affecting the well-being and security of people and their effective enjoyment of human rights in both developed and developing countries and across all cultures and boundaries. While climate change affects everyone, it will hit the poorest and the most marginalized groups the hardest. Those who are already in a situation of vulnerability because of factors such as age, gender, and socio-economic status, will be disproportionally affected. This underlines the importance of effective human rights guarantees to reduce vulnerability in the face of climate change. Climate change therefore should be addressed in a way that is fair and just.

The Intangible Cultural Heritage Committee could adopt a human rights approach to address climate change including pressuring polluting States Parties to reduce GHG emissions to levels that would not further interfere with the human rights of those States Parties most vulnerable to the effects of climate change, and requesting polluting States Parties to assist those States Parties needing to adapt to changes that can't be avoided and which will impact their intangible cultural heritage and infringe their human and cultural rights.

Inserts 24.6 shows a similar relationship to that demonstrated for the World Heritage properties and indicates the number of intangible cultural heritage elements possessed by the identified 14 highest carbon dioxide emitting States Parties and in total by the remaining 128 lowest emitting States Parties. The key to the 12 identified States Parties is ranked in diminishing order in relation to recent annual CO2 emission status. It will be noted that the top 12 emitters possess almost half of the 267 intangible cultural heritage elements. The big polluting nations of the USA, Canada, Australia and Germany have not ratified the ICHC and hence are not represented in the diagram.

Article 13 of the ICHC says: "States may adopt legal, technical, administrative and financial measures aimed at ensuring access to the intangible cultural heritage while respecting customary practices governing access to specific aspects of such heritage". It could be argued that if access to intangible cultural heritage practises is denied by climate change induced impacts (such as crop failure leading to forced migration thus separating a community from the homeland context of their customary rituals) then States Parties may adopt legal measures. In such a case it would be a human rights legal claim against those States Parties responsible for polluting the atmosphere, thereby exacerbating desertification leading to crop failure. Similarly it could be argued that if climate change impacts caused a break in "transmission" of a listed element, or undermined the element's "viability" this would amount to a breach of the human right "to participate in the cultural life..."

As has been discussed UNESCO is assessing support for a draft Universal Declaration of Ethical Principles in relation to Climate Change. For selfish reasons such support may be difficult or impossible to obtain from those States Parties with the highest per-capita emissions. The dilemma of climate justice is controversial, crucial and may be the defining factor for a sustainable future.

One of the outcomes from a human rights approach to climate justice in relation to the safeguarding of intangible cultural heritage might be 'bottom up' community strengthening. Communities might build on their coping strategies to climate change and become empowered to participate in the development of climate change policies.

Again support for safeguarding actions needs to be locally grounded and culturally appropriate. To this end, adaptation assistance should be part of mainstream poverty reduction strategies and budget planning in developing countries, rather than one-off 'special initiatives'. Modelling international aid delivery on the human rights-based approach allows for this, as it emphasises the importance of local knowledge and seeks the active participation and consultation of local communities in working out how best to adapt to climate change. This could mean incorporating the traditional intangible cultural practices of indigenous communities into climate change responses.

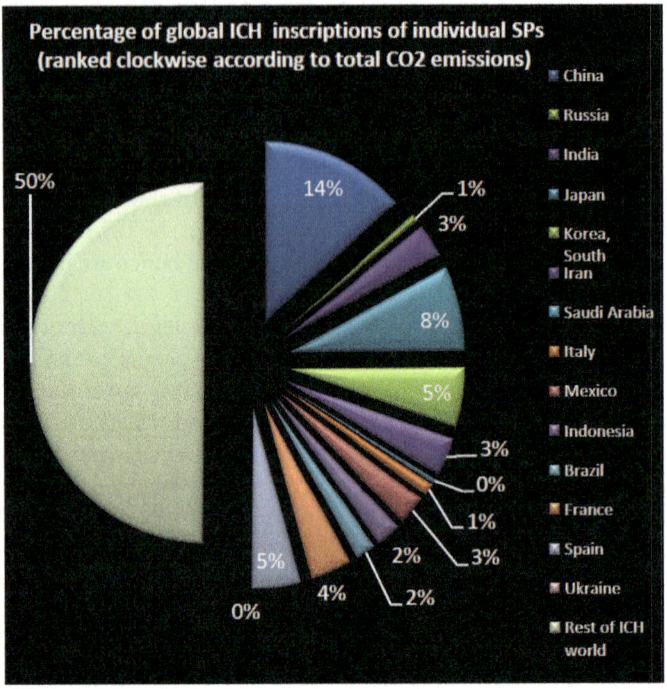

Insert 24.6 States Parties ranked clockwise in descending CO2 emission order. Size of segments and associated percentages refer to proportion of world's Intangible Cultural Heritage inscriptions held by each State Party

The same approach applies to risk management for climate change induced disaster. Affected populations are most often forced to leave their homes because of extreme weather events, storm surges, torrential downpours, wild fires and desert storms. Adopting a human rights-based approach addresses inequities by linking the proposed *Intangible Cultural Heritage Climate Change Response and Policy Document* to international human rights law.

Recommendations

24.1 Progress the development of a **Universal Declaration of Ethical Principles in Relation to Climate Change.** Inform heritage workers about the threats of climate change to tangible and intangible cultural heritage in order to gain their support in lobbying UNESCO leaders to progress the above Declaration (COMEST 2010; UNESCO 2010ll).

24.2 Support the development of a **human-rights approach** to the impact of climate change on tangible and ICH which would include the precautionary principle (Insert 16.2); the need to preserve culture in order to enable participation in such; the principle of sustainability; the principle of integrity as applied to ecosystems and the principle of safeguarding and promoting the interests of **future generations**.

24.3 Encourage the incorporation of the proposed human rights approach into the revision of the current *Policy Document on the Impacts of Climate Change on World Heritage Properties* (15.2), and in the proposed *Intangible Cultural Heritage Climate Change Response and Policy Document* (15.14) and the provision of **human rights criteria** to be taken into account when States Parties with inscribed World Heritage or ICH seek resource allocations.

24.4 Ensure that, once reviewed, the current *Policy Document on the Impacts of Climate Change on World Heritage Properties* (15.2) and the proposed *Intangible Cultural Heritage Climate Change Response and Policy Document* (15.14) **champion information regarding the need for climate change mitigation as a human right-to-know** as enshrined in the UDHR Article 19.

24.5 Propose that the World Heritage and the Intangible Cultural Heritage Committees adopt stronger official positions **recognising that 'anthropogenic' climate change is threatening accelerating damage to World Heritage sites and some ICH practices** including those in the least polluting and most poorly resourced States Parties. Further recognise that GHG pollution has proven and well known consequences that increasingly make such pollution a "**deliberate measure** which might damage directly or indirectly the cultural and natural heritage... situated on the territory of other States Parties" (WHC Article 6.3) or "**threatens**" the purposes of the ICHC (Article 1) and the various measures of "safeguarding" (Article 2, 12 and 13) at national (ICHC Article 11) and international (ICHC Article 16, 17, and 18) level and that it is beholden on the polluting and resourced States Parties through "international co-operation and assistance" (WHC Article 7; ICHC Article 1) and "collective assistance" (WHC Preamble par.5) or "universal will" (ICHC Preamble paragraph 5) to support the inscription of World Heritage and ICH in these least resourced States Parties as a means to protect, "conserve, and identify that heritage" (WHC Article 4) and "safeguard" intangible cultural heritage through "the identification, documentation, research, preservation, protection, promotion, enhancement, transmission particularly through formal and non-formal education, as well as (through) the revitalisation of the various aspects of such heritage" (ICHC Article 2.3).

24.6 Propose that the "assistance" referred to above should recognise **common but differentiated responsibilities** and include the **transfer of relevant technology and adaptive projects** that "protect" from future impacts, recognising that the "protection" in this context includes the "safeguarding" of the community's culture and cultural heritage.

24.7 Propose that States Parties be encouraged to develop **mutual obligations** mindful of the "development" and "human rights" needs of World Heritage custodians and intangible cultural heritage practitioners.

24.8 Adopt a human rights approach with respect to **disaster risk management** (DRM). This should address inequities by linking the World Heritage/ Intangible Cultural Heritage risk policies to international human rights law encompassing all relevant guarantees-civil and political as well as economic, social and cultural rights.

24.9 Integrate the **rights of Mother Earth** - living in harmony with nature - into the revised *Policy Document on the Impacts of Climate Change on World Heritage Properties* and into the proposed *Intangible Cultural Heritage Climate Change Response and Policy Document.*

24.10 Encourage more countries to ratify the two UNESCO Conventions and in so doing further implement appropriate just and ethical - legal, technical, administrative and financial measures and policies to ensure the safeguarding, the development and the promotion of WH and ICH at national and local level, with the participation of their communities, groups and individuals.

24.11 Build a **body of knowledge** by research and information sharing on human rights, ethics and culture in relation to the protection of cultural World Heritage and Intangible Cultural Heritage from climate change.

Chapter 25

How do the Conventions Address Climate Change-Induced Threats to the Security of Cultural Heritage?

The heat in Northern India is killing more people every year. Climate refugees are created by water needs. In Bhopal, lakes and wells have dried up. While the rich safeguard their water supplies with security guards, the poor walk miles for contaminated water. "Murder over water is not rare anymore," she said. "Surely this is not the future you want to give us." Children are passionate and want revolution. Treaties and laws aren't the answer. "We need to mobilize everyone to preserve water and create local solutions. Should we be doing this alone, or should you be joining us?" Lavanya Julaniya (India).17 speaking about climate refugees created by water.

We sink or swim together. Climate change can be a threat to peace and stability.
There is no part of the globe that can be immune to the security threat.
Rajendra K. Pauchauri, IPPC Chairman

How do the World Heritage and Intangible Cultural Heritage Conventions address climate change-induced threats to the security of World Heritage and Intangible Cultural Heritage? For example how do the Conventions address the threats of abuse to cultural heritage as a result of societal breakdown and conflict due to climate change-exacerbated poverty?

World Heritage Convention and the Intangible Cultural Heritage Convention

The World Heritage Committee is diligent in addressing climate change threatened World Heritage sites with 'in Danger' listing on a case-by-case basis. Sites on the in Danger list are ones confronted by 'serious and specific dangers' which "are amenable to correction by human action"... "noting that the emphasis of the corrective measures to be recommended should be on 'adaptation' rather than on 'mitigation'" (Decision 32COM7A.32). For almost four decades the World Heritage Committee has been complying under the World Heritage Convention (WHC) and adding and removing properties threatened by such 'dangers' to the List of World Heritage in Danger. The transference process to the in Danger List sets in train a sequence of support and repair actions to address the 'danger.' Of growing concern to the World Heritage Committee has been the increased threat of 'dangers' in the form of both human-induced and 'natural' disasters. The World Heritage Committee has responded carefully and supportively with the release in 2010 of a new manual *Managing Disaster Risks for World Heritage* and directives for States Parties to address disaster risk management. The *Policy Document on the Impacts of Climate Change on World Heritage Properties* on addressing the global threat of GHG emissions, has taken the fall back position of requesting States Parties to protect properties from the adverse effects of Climate Change, "to the extent possible and within the available resources, recognizing that there are other international instruments for coordinating the response to this challenge" (Decision 30COM7.1) and later to urge States Parties to participate in the UN Climate Change conferences with a view to achieving a comprehensive post-Kyoto agreement (Decision 31COM7.1). This means to defer mainly to the UNFCCC the role of tackling the catastrophic trajectory of global warming the planet is presently on.

Longer term security threats include conflict such as might arise from economic and societal collapse and competition for land, food and water. The **United Nations Department of Economic and Social Affairs (UNDESA)** reports that climate change has the potential to create humanitarian disasters, ecological collapse and knock-on socio-economic effects through mass environmental displacement, the loss of livelihoods, rising hunger, and water shortages leading to pandemics thereby having the potential to unleash national, regional and global security threats(UN-DESA 2009). Furthermore, UNDESA claims, if the countries that carry primary responsibility for the problem are perceived to turn a blind-eye to the consequences, the resentment and anger that will follow could foster conditions for political extremism (UNDP 2007b; Campbell 2008).

Viewed from another perspective these longer term threats are addressed by World Heritage driven efforts to introduce and maintain adaptive interventions, capacity building and education programmes and culturally sustainable development, thus contributing to a more secure society.

For instance, such actions can be compared in a smaller way to what is happening in the Pacific. Here climate change impacts are being felt more directly than anywhere else with up to 6 billion expected to be affected in the coming years. Development of these islands is dependent on natural resources, and land is scarce leading to conflicts over its access and use. Climate change impacts like floods, storms and salt-water intrusion further increase pressure on limited land resources. Effective land-use planning, including that based on traditional knowledge, is of the utmost importance for agriculture, infrastructure and settlements. Up-to-date climate data and projections are being introduced and existing plans will be updated based on the information that will be supported through technical advice. This is expected to reduce risks to the health and livelihoods of people and thereby increase security (Schmitt 2009).

The WH Committee acts as a force for social cohesion and in some cases, renewal and restoration. Through its primary work of implementing protection for heritage at all levels and of managing and increasing the World Heritage Lists (not necessarily increasing the 'in Danger List') it acts as security insurance. As it proclaims, the WHC is "not only 'words on paper' but is above all a useful instrument for concrete action in preserving threatened sites and endangered species(UNESCO 2008d)." "By recognizing the Outstanding Universal Value (OUV) of a site, States Parties commit to its preservation and strive to find solutions for its protection." The WHC is a force for positive achievement that is relevant at site level and globally as a 'very powerful tool to rally international attention and actions, through international safeguarding campaigns'. The World Heritage Committee increases the WHC's influence and promotes social solidarity by adding to the variety of its engagement processes. This it does by initiating celebrations and thematic activities such as the **Thematic Initiative on Religious and Sacred Heritage, World Heritage Education,** and **Youth Forums** e.g. 2nd Ibero-American Youth Forum and many more.

As attested by the success of the book *Case Studies on Climate Change and World Heritage* (Colette 2007) the iconic standing of a number of World Heritage sites is being used in selling the climate change message generally. In addition World Heritage sites are being embraced for their potential as priority reference sites for the understanding of the impacts of climate change on human societies and cultural diversity, biodiversity and ecosystems services, the world's natural and cultural heritage and possible adaptation and mitigation strategies (see Insert 15.2). The 'security' implication is that the use of World Heritage sites in this context along with the many culturally sustainable development projects at World Heritage sites contributes to generating a culture of global responsibility through informed local care. There is the potential to generate a sense of identity with the site, to its OUV and to its contribution to the heritage of the planet. As 'priority reference sites for research', World Heritage sites will be appreciated as evolving systems, rather than grounded in the perspective 'of keeping sites preserved'. As a living concept realised though nearly 1000 sites inscribed worldwide, World Heritage is insurance for social stability and security. It has a rightful place in the philosophy of Mother Earth and represents a hopeful opportunity for a Gaian future.

Involvement with World Heritage sites supports UNESCO's vision as articulated in the preamble of the UNESCO Constitution: "to construct the defences of peace in the minds of men". To this end, UNESCO asserts, it shall remain unstinting in seeking to bring about a culture of peace and to develop and deepen mutual understanding, reconciliation and dialogue (UNESCO 2007f, 2008c). As has been noted, longer term climate driven security threats are not addressed directly in the publication, *A Strategy to Assist States Parties to Implement Appropriate Management Responses,* or the *Policy document on the impacts of climate change on World Heritage properties.* Heritage strategists may be drawn to the wisdom of Elinor Ostrom, Jared Diamond and traditional elders or to the Gaia idea and the Mother Earth movement to find creative ways to combine World Heritage advantages with alternative ways to build a positive attitude to mitigation. Without reductions in GHG pollution we will have forfeited the wealth of 40 plus years of cultural and natural heritage protection. The WHC, as a peak body, has a unique opportunity to create grass root opportunities for carbon neutral practices while using all means 'to the utmost' of its influence to create the universal perspective necessary for a sustainable future. Perhaps the Dalai Lama could become the first World Heritage climate change 'ambassador' giving teachings on climate change and heritage? He has been one of the few leaders who have prioritised climate change over national political interest (in his case Tibet, a formerly held but now occupied, national estate). The Dalai Lama's promotion of global interdependence and protection of the environment was one of the reasons that he was awarded the Nobel Peace Prize in 1989. In his acceptance speech, he said: "Both science and the teachings of the Buddha tell us of the fundamental unity of all things. This understanding is crucial if we are to take positive and decisive action on the pressing global concern with the environment" (Dalai Lama 1989).

The Intangible Cultural Heritage Convention (ICHC), its *Operational Directives* and Periodic Reporting requirements make no specific reference to direct threats to the security of intangible cultural heritage through abuse or conflict, which can be attributable to climate change. However, as is the case with its filial Convention, the WHC, the ICHC addresses direct threats to inscribed heritage by urging States Parties to nominate the element for inscription on the List of Intangible Cultural Heritage in need of Urgent Safeguarding. The threats are not detailed other than to state that the heritage's "viability is at risk despite the efforts of the community, group or, if applicable, individuals and State(s) Party (ies) concerned" or that "The element is in extremely urgent need of safeguarding because it is facing grave threats as a result of which it cannot be expected to survive without immediate safeguarding" (*Operational Directives* 1.1u2). Once listed, the safeguarding potential of the intangible cultural heritage machine is galvanised to assist State Party/States Parties to restore the threatened practice, custom, performance etc. Climate change threatened listed intangible cultural heritage could be protected under this process. For example a storm surge affected *intangible cultural heritage*-owning community group may need relocating as a group to safeguard their particular heritage practice. The Intangible Cultural Heritage Committee could assist if the State Party was not able to do so and if the 'practice' qualified to be on the List of Intangible Cultural Heritage in need of Urgent Safeguarding. The threat to an intangible cultural heritage-owning community of having its ceremonial relationship with a significant site severed through climate change induced forced migration presents an intractable problem for the ICHC.

Both Conventions stall at this hurdle. They share this inability to protect and safeguard under the present *Guidelines* and *Directives*. But both Conventions which do have far reaching, widespread and in depth potential and are graced with Committees that put self interest and national politics behind them and have the capacity to meet the challenge of making membership conditional on meaningful mitigation undertakings particularly GHG emissions reductions.

The Periodic Reporting process for listed intangible cultural heritage could serve as a regular feedback on climate change impact. The 'specific conditions' clause, would allow the reporting of climate change induced threats to listed elements such as the imminent abandonment of a festival or ritual due to a severe weather event such as a cyclone. Despite the lack of an intangible cultural heritage climate change policy and the absence of climate change being singled out as a potential threat to be responded to, the Reporting Forms do require respondents to describe threats to the viability of the representative element, if it is, for example, threatened by disappearance. For 'in Urgent Need' elements, a more detailed report on the severity and immediacy of the threat is required. Periodic Reporting for inscriptions on the Representative List is every 6 years, and every 4 years for 'Urgent need' listed elements.

Additionally under the ICHC States Parties are mandated to draw up one or more inventories of the country's intangible cultural heritage on which they also report to the Intangible Cultural Heritage Committee periodically (Article 12). This reporting could provide another feedback to the Intangible Cultural Heritage Committee on climate related threats to intangible cultural heritage, so providing further insight into element- based national climate change threat to intangible cultural heritage. Such knowledge could be a valuable cultural barometer to measure and gauge societal cohesion and identify security problems.

UNESCO has proposed to complement efforts to protect heritage in 2012-2013. Special emphasis will be on intangible cultural heritage, movable cultural property and action to combat trafficking in such property (UNESCO 2010n). Its Medium-Term Strategy, 2008 -2013, is built around the following mission statement: "As a specialized agency of the United Nations, UNESCO contributes to the building of peace, the eradication of poverty, sustainable developmentand intercultural dialogue through education, the sciences, culture, communication and information (UNESCO

2010b)." In respect of the commitment to create stability through building peace and eradicating poverty, intangible cultural heritage practice is a security enhancer, and by safeguarding its transmission against climate change, it is a long term security enhancer.

In the medium term, UNESCO's specific role will be twofold. It will 'aim to build decision-makers' and the general public's awareness of the importance of the heritage, especially the Intangible Cultural Heritage, to development and to the establishment of a pluralist society inclusive of marginalized communities and groups in particular, and capable of being open to their social practices, rituals and festive events. Moreover, it will aim to instil understanding of the continuity between cultural objects and that which has made it possible to produce and to continue producing them, namely the values, talents and skills that belong to the creators of the heritage and that are protected by the 2003 Convention on the intangible heritage" (UNESCO 2008c). With these intentions in mind, it is the opportune time for UNESCO and the Intangible Cultural Heritage Committee to include climate change in their strategies.

Potential 'climate change related' security threats such as armed conflict, illicit trade, looting and other forms of heritage theft fall outside the WHC except when they pose a 'serious and specific danger' to a World Heritage site. In the latter case the site could be moved to the 'List of World Heritage in Danger'.

As discussed in Chapter 6, these security threats also fall outside the ICHC. Though the thrust of the ICHC is to safeguard 'intangible' heritage, the practices, performances, rituals and customs may be accompanied by tangible attributes; 'instruments, objects, artefacts' (Article 2.1), that attract thieves aware of their value on the international black market. States Parties to both Conventions are encouraged to ratify the following three legal instruments and to note the UN Draft Resolution to address these potential threats:

- 1954 Hague Convention for the Protection of Cultural Property in the Event of Armed Conflict and its
- 1954 (First) Protocol, involving the conservation of monuments and sites and their protection from looting and the illicit trafficking of movable property and the
- Second Protocol (1999) which establishes an enhanced system of protection for specifically designated cultural property, and which came into force in 2004 and
- UN's Draft Resolution on Return (or) Restitution of Cultural Property to the country of origin (UN 2009a), cultural property that would have been originally stolen or looted.

Recommendations

25.1 Appreciate and learn from the WHC's 40 years of contributions to conservation and to awareness raising for the World's natural and cultural treasures, reflecting on its role in **social cohesion and stability**, and as a vehicle for raising national **identity and pride**.

25.2 Exploit the WHC's reputation of stability and security enhancement (25.1) by supporting the **heritage site vulnerability campaign** proposed at 15.4 to draw attention to the vulnerability of World Heritage to climate change.

25.3 Appreciate that ICH contributes to **community strength and resilience** as well as international understanding and approbation making it a force for peace.

25.4 Recognise UNESCO, the WHC and the ICHC as global **forces for developing collective responsibility**.

Chapter 26

How have the Legal Powers of the World Heritage Convention Protected World Heritage from the Impacts of Climate Change?

For a highly vulnerable country like Tuvalu, we cannot just sit back and watch our homeland slowly disappear. If necessary, we will use whatever legal means available to us to seek the necessary restitution for all damages created by climate change. Hopefully, the international community will respond before such action is necessary. But time is running out fast. Climate change could well be the greatest challenge that humanity has ever known. I make a very strong plea to all to act quickly and responsibly, to ensure that countries like Tuvalu do not disappear. – Tuvalu Prime Minister Apisai Ielemia (OXFAM 2009)

How have the legal powers of the World Heritage Convention protected world heritage from the impacts of climate change?

The World Heritage Convention

Since 1972 the World Heritage Convention (WHC) has wrestled with difficult issues but none as complex and difficult as that of climate change. Legally the issue was a non-starter in the 1970s and 1980s. Now with many World Heritage properties under threat and with the prospect of what is 'in the pipeline' with business as usual (BAU), the task is daunting, see Insert 26.2. Progress to date has been interspersed with episodes of altruism, power play and frustration. It is well worth summarising the story in some detail to appreciate the conservative path travelled to date and gain a feel for the difficult journey ahead. It also records how the perception of climate change has evolved.

The following table ***World Heritage Convention Decisions involving 'climate change' – summary table*** (Insert 26.2) only sources those documents and Decisions which specifically mention the terms 'climate change'. It is a selective overview of the increasing pressure being brought to bear on the World Heritage Committee by concerned members, individuals, organizations and advisory groups and of the earnest, mostly comprehensive and diplomatic responses of the World Heritage Committee. The table centres on the 'Decisions' and does not attempt to cover all documents which may have some bearing on climate change. The Table is dominated by references to climate change impacts on natural sites, there being few references, as yet, to climate change impacts on cultural sites. However it is instructive to consider the former in that they do chart the 'means by which' the WHC has been brought to the defence of all State Party's climate change impacted heritage. The reality that international law is often driven by national politics and self-interest that limit its effectiveness must be balanced against an appreciation of the significant positive contribution the law has made to addressing heritage problems. There are glimpses of national positioning, national pride, cultural sensitivity, economic motivation, high handedness, lack of scientific conviction and legal wrangling but overall a pervading sense of global responsibility to protect heritage.

World Heritage Convention Decisions involving 'climate change' – summary table

Initiating event	Response	Decision	Decision conditions mentioning 'climate change' and Comments	Comments
Nomination of Virgin Komi Forests, Russian Fed.	Inscription as a natural site, 1995	19COM VIII.A.1	OUV '... has pristine boreal forests and is an important site for scientific research including climate change'.	Climate change considered as a subject for scientific research.
Operational Guidelines recognised as out of date, 1996	Operational Guidelines revised. The WH Committee confirmed as the decision-making body of WH statutory organs and is not bound by recommendations by the Bureau and advisory bodies	**20COM XVII.1-9** Revision of the Operational Guidelines for the implementation of the WHC	Revision of Form for the submission of nominations of cultural and natural properties: 'Factors Affecting the Site. a. Development Pressures (e.g. encroachment, adaptation, agriculture, mining).b. Environmental Pressures (e.g. pollution, climate change). Natural disasters and preparedness (earthquakes, floods, fires, etc.). d. Visitor/tourism pressures. eg. Number of inhabitants within site, buffer zone. f. Other.	The confirmation of the authority of the WH Committee became important in the contested issue of whether World Heritage Committee can place a WH site on the 'in danger' list without approval of the State Party responsible for that site (see 32COM 7A.32). Climate change identified as a 'pressure' affecting a site.
Progress Report on Regional Actions for the Implementation of the Global Strategy for a Representative and Balanced WH List	Discussion, 2001	25COM IX.1-19 - IX	Australia referred to a number of partnerships in support of the World Heritage Global Strategy in the Asia-Pacific region ... 'that proposed' IUCN and the WH Centre look at the impact of climate change in the region with reference to WH sites...'	Climate change identified as an 'impact', not pressure.
Nomination of **Cerrado Protected Areas: Chapada dos Veadeiros and Emas NPs, Brazil**	Inscription as a natural site, 2001	25COM X.A	The site ... has acted as a relatively stable species refuge when climate change has caused the 'Cerrado' to move north-south or east-west. This role is ongoing as Earth enters another period of climate change.	Climate change seen as a periodic phenomenon.
Nomination of Jungfrau of Aletsch-Bietschhorn, Switzerland	Inscription as a natural site, 2001	**25COM X.A**	Site is of significant scientific interest in the context of glacial history...and ongoing processes...related to climate change. The global phenomenon of climatic change is ...reflected in... rates of retreat of the different glaciers.	Climate change considered an ongoing global phenomenon.
Nomination of Mapungubwe Cultural Landscape, Sth Africa	Inscription as a cultural site, 2003	**27COM 8C.30**	Site graphically illustrates the impact of climate change and records the growth and then decline of the Kingdom of Mapungubwe as a clear record of a culture that became vulnerable to irreversible change.	Climate change is recognised as a threat historically.

Nomination of **Ilulissat Icefjord, Denmark**	Inscription as a natural site, 2004	**28COM 14B.8**	Site is...example of ...last ice age...; ice-stream is one of the fastest and most active. The glacier has been... of scientific attention for 250 years ...and has ... added to the understanding of ice-cap glaciology, (and) climate change...'	Climate change is reflected in ice-cap glaciology.
WH Committee received ***4 Petitions in 2004 to place WH sites of Sagarmatha NP, Nepal; Huascaran NP, Peru; Belize Barrier Reef Reserve System, Belize and Great Barrier Reef (GBRWHA), Australia- the planet's largest WH Area and the first designated Particularly Sensitive Sea Area, on List of WH in Danger on the basis of climate change impact.***	In 2005 WH Centre responds by proposing concerned States Parties and petitioners might wish to collaborate with WH Centre, Advisory Bodies and other interested States Parties and partners, to organize a workshop to consider the impacts of climate change on WH properties,	Draft Decision 29COM 7B.a Rev, 2006 Threats to WH properties	Notes petitions. Appreciates the ...concerns relating to threats to natural World Heritage properties that are or may be the result of climate change; Notes that the impacts of climate change are affecting many and are likely to affect many more World Heritage properties, both natural and cultural in the years to come; Encourages State Party's to seriously consider the potential impacts of climate change within their management planning, in particular with monitoring and risk preparedness strategies and take early action in response to these potential impacts; Requests the World Heritage Centre...consider...an international experts workshop on reacting to the threat of climate change on World Heritage properties; Decides not to include the 4 properties...on the List of World Heritage in Danger Contested interpretations and issues. Co-Director, Climate Justice, P. Roderick: "We are extremely angry that the WH Committee has not taken any meaningful action to protect some of the most important sites on Earth from climate change. They are good at drawing up wonderfully drafted documents, but the idea of actually doing anything seems to pose a problem. Moreover, ducking the issue of why climate change is affecting these sites will make their efforts to adapt to the impacts largely futile. The world is entitled to expect better from the Committee. Bending over backwards as a result of fear of the US and Canada will tarnish the Committee's reputation"(Climate Justice 2006).	The WH Committee describes climate change as an "external, global and indirect threat impacting on property but with origins outside the property itself". The WH Committee considers climate change less of an immediate threat than other site specific threats. The human contribution to climate change is not addressed. The WH Committee couches response to Petitions cautiously, questioning some of the science implications proposed by the petitioners, on the one hand recognising the global nature of the threat but resisting the argument for a 'global' response from a global (world) convention. The WH Committee is immediate in recognising the opportunities for it to facilitate information sharing. The petitioners suggest a growing consensus among the various stakeholders for the need for more concerted efforts in conserving sites threatened by climate change. The WH Committee responds that climate change is affecting other reefs so is not specific to the GBRWHA. The WH Committee claims too little site specific evidence is provided. The Petitions are substantial. The WH Committee takes an inappropriate position when implying that as some States Parties are acting independently to address climate change and that it should not act in response to climate change by listing sites on the 'In Danger' list.

Petitioners claimed 'Serious and specific, ascertained and potential dangers have arisen or are likely to arise from the impacts of climate change' and these need 'an adaptive programme of corrective measures' WHC Articles 4,5,6.3 (UNESCO 2005b)	appropriate adaptive management strategies and explore options of improved collaboration between the States Parties of affected properties		US claimed that a State Party's consent is needed to propose a site for WH 'in Danger' listing. It is not (O'Keefe and Prott 2011 p78). US claimed that only a State Party can submit info. re: 'endangeredness' of a WH site. Any person or group can. The WH Committee refuted the claim and accepted the Petitions. Is climate change 'serious and specific' enough to warrant WHC efforts? State Party's differed in opinion. International lawyers and scientists strongly agreed that the threat was profoundly dangerous. Do GHG emissions equate to 'deliberate measures which might damage directly or indirectly' (Art 6.3) WH sites? Yes, considering today's level of climate change science knowledge but powerful States Parties would not agree as this would have far reaching ethical, moral and economic implications. Are State Party's obliged to sign Kyoto Protocol to avoid 'deliberate measures'? No. Whether a State Party is doing its utmost to mitigate climate change is the question. Is there sufficient evidence to distinguish human-induced climate change from natural variability? Yes. Should a State Party's obligation to reduce GHG emissions apply equally to all States Parties or should the accepted principle of common but differentiated responsibilities apply (CBDR)? The only ethical position is to embrace CBDR and while this equity principle is generally accepted, it is not formalised. How does the WH Committee deal with States Parties that make incorrect statements about the WHC and ignore climate change science? (Wold and Thorson 2006). The WH Committee goes to great lengths to facilitate, catalyse, share, disseminate and promote over considerable time the most authoritative knowledge.	Belize Barrier Reef was placed on 'In Danger' List in 2009, but curiously, not for climate change impacts, although evidence solidly supported climate change as the serious threat in the petition (Burns 2009p154). "History may well mark this decision as a lost opportunity to take a small step to better preserve WH sites for future generations, the whole idea behind the Convention. Those who failed to act will have their names and actions remembered for all time, as the fate of these sites remains in danger" (Climate Justice 2006). The Sydney Centre for International and Global Law argues that given the obligations under WHC, the WH Committee consider placing the GBRWHA on the 'in Danger' List as the Aust.Gov.(a State Party) has not taken adequate protective steps to address all of the threats to the WH property (Operational Guidelines 83(ii)(d)) by not making substantial reductions in GHG emissions (Rothwell 2004) and that States Parties must endeavour, both individually and in co-operation with other State Party, to ensure that emissions of greenhouse gases and other pollutants are controlled so as to minimise the potential deterioration of the GBRWHA.

In all, the **WH Centre** received ***6 petitions from more than 37 NGOs and individuals asking the WH Committee to inscribe properties on the List of WH In Danger due to the threats posed by climate change.*** Two further petitions were: **Glacier NP, 2006, The Greater Blue Mts. WH Area, 2007**. The latter calls for both on-site management measures and actions to reduce GHG emissions by Australia which is cited as the world's largest exporter of coal (CANA et al. 2007).	As above	29COM 7B.a **Threats to WH properties, 2006**	Recognizing ...UNFCCC and need for a proper coordination of (its) work with WHC. Takes note of the four petitions...; Appreciates (as in last 'draft' decision, above); Further notes (as above); Encourages all State Party (as above). Requests World Heritage Centre...to establish....group of experts to: a) review the nature and scale of the risks posed to World Heritage properties arising specifically from climate change; and b) jointly develop a strategy to assist States Parties to implement appropriate management responses; Requests...working group... prepare ...report on *Predicting and managing the effects of climate change on World Heritage* Strongly encourages State Party's...to use the network of World Heritage properties to highlight the threats posed by climate change to natural and cultural heritage, start identifying the properties under most serious threats, and also use the network to demonstrate management actions that need to be taken to meet such threats, both within the properties and in their wider context; Also encourages UNESCO to do its utmost to ensure that the results about climate change affecting World Heritage properties reach the public at large, in order to mobilize political support for activities against climate change and to safeguard in this way the livelihood of the poorest people of our planet.	**The decision not to place sites on the 'In Danger' list is a "....disheartening response because other 'in danger' sites have no more compelling cases (Burns 2009)."** The WH Committee changes the draft decision and mainly makes site specific climate change directives. The WH Committee seems to want to shift the climate change debate towards existing WH Committee processes at individual sites eg monitoring and reporting. WH Committee takes a measured stance without any reference to human induced causes of climate change focussing on site generated identification with solutions to be shared. "Despite the overwhelming scientific evidence of the various threats that climate change poses to these sites,... the WH Committee has not yet added any of these properties to the List of WH in Danger"(ACJP and Earthjustice 2009). "Any site-specific mitigation occurring within Glacier NP's boundaries, while commendable, is inevitably inadequate to address the devastating consequences of climate change within the park. Even a total ban on GHG emissions within the park would not slow and could never reverse the climate change effects on glacial melt within the Park. Yet this type of mitigation is all that the Joint report (*Predicting and Managing the Effects of Climate Change on World Heritage*) and the *A Strategy to Assist States Parties to Implement Appropriate Management Responses*, suggest should occur-a wholly inadequate response to the threat of climate change because it will not protect the OUVs of the Park" (Thorson 2008 p21).

Recommendations of the Kobe Thematic Session on ***Risk Management for Cultural Heritage***	Discussion on strategy for risk preparedness generally termed Disaster Risk Management (DRM).	**29 COM 7B.b** **Threats to WH properties, 2005**	'...taken note of the serious threat posed by disasters on... World Heritage and SD and poverty eradication of communities living around affected World Heritage properties; Strongly encourages States Parties to act swiftly...to integrate concern for heritage....within...policies and operational mechanisms for disaster mitigation and to develop...risk-sensitive Management Plans for the World Heritage sites Requests World Heritage Centre...takes into account the recommendations of the Kobe Thematic Session on risk preparedness to be examined by the World Heritage Committee at 30th session.	Climate change is not identified as a 'factor' in DRM.
A **meeting of 50 experts** was held in March 2006 to prepare a *report Predicting and Managing the Effects of Climate Change on World Heritage* (UNESCO 2006) and	The publications: *Predicting and Managing the Effects of Climate Change on World Heritage* and *A Strategy to Assist States Parties to Implement Appropriate Management Responses*, developed.	30COM 7.1	Endorses the *Strategy to assist States Parties to implement appropriate management responses and* requests....World Heritage Centre to lead the implementation of the "Global level actions" ... Takes note of the *Report- Predicting and managing the effects of climate change on World Heritage.* Encourages UNESCO...to disseminate widely this strategy, the report. ; Requests States Parties...to protect OUV, integrity and authenticity of World Heritage properties from the adverse effects of climate change, to the extent possible and within the available resources, recognizing that there are other international instruments for coordinating the response to this challenge; Invites States Parties and WHCentre... to build on existing Conventions and programmes... in their implementation of Climate Change related activities; Also requests States Parties, WHCentre...to seek ways to integrate, to the extent possible and within the available resources, this *strategy* into all the relevant processes of the WHC including: nominations, reactive monitoring, periodic reporting, international assistance,	The Decision misses opportunity to strongly urge States Parties to reduce GHG emissions. States Parties could through Operational Guidelines, report mitigation efforts incl. GHG emissions and policies to reduce or offset them The publication *A Strategy to Assist States Parties to Implement Appropriate Management Responses was* criticised by IUCN as not being stronger and explicit. "Environmentally sound choices" (as in *A Strategy to Assist States Parties to Implement Appropriate Management Responses)* should have explicitly emphasised reducing GHG emissions and enhancing carbon sinks. The strategy publication was regarded as limp because it had no binding components (Burns 2009). The report *Predicting and Managing the Effects of Climate Change on World Heritage* summarises IPCC but this information is marginalised in the subsequent *A Strategy to Assist States Parties to Implement Appropriate*

the publication, A Strategy to Assist States Parties to Implement Appropriate Management Responses, (UNESCO 2007) for States Parties in addressing **Climate change**.		**The Impacts of climate change on World Heritage properties, 2007**	capacity building, other training programmes, as well as with the *Strategy for reducing risks from disasters at World Heritage Properties.* Strongly encourages World Heritage Centre…in collaboration with States Parties…to develop proposals for the implementation of pilot projects at specific World Heritage properties especially in developing countries, with a balance between natural and cultural properties as well as appropriate regional proposals, with the objective of developing best practices for implementing this *Strategy* including preventive actions, corrective actions and sharing knowledge… Further requests States Parties and World Heritage Centre to work with IPCC,…including a specific chapter on World Heritage in future IPCC assessment reports; Requests World Heritage Centre to prepare a *policy* document on the impacts of climate change on World Heritage properties… with relevant climate change experts… a draft to be presented to the 31st session in 2007 for comments. This draft should include considerations on: a) Synergies between conventions on this issue, b) Identification of future research needs in this area, c) Legal questions on the role of the WHC with regard to suitable responses to Climate Change, d) Linkages to other UN and international bodies dealing with the issues of climate change, e) Alternative mechanisms, other than the List of World Heritage in Danger, to address concerns of international implication, such as climatic change; Considers that the decisions to include properties on the List of World Heritage in Danger because of threats resulting from climate change are to be made by the World Heritage Committee, on a case-by-case basis, in consultation and cooperation with States Parties, taking into account the input from Advisory Bodies and NGOs, and consistent with the *Operational Guidelines*.	*Management Responses* and later in the *Policy*. The geographical **differential impact** of climate change is omitted, as are the most impacted regions and heritage therein: namely mountains, polar regions and ocean environment. Cooperation with UNFCCC is only tentatively expressed and does not appear to have been acted upon. The *report Predicting and Managing the Effects of Climate Change on World Heritage* and the *World Heritage Climate Change Strategy* reinforce the position to only address climate change mitigation at WH site level. The WH Committee expresses interest in requesting the IPCC to include a chapter on WH and climate change In the 2014 5th Assessment Report. To date there does not appear to be any preliminary meetings planned to achieve this. Experts endorse the importance of 'monitoring changes' as a strategy in climate change mitigation. An Australian lawyer advocates assisting mitigation efforts on WH sites by reducing stress from unsustainable activities by using the WHC and through it, to use States Parties' laws relating to across the board factors – pollution, land use, planning, construction, conservation and Environment Impact law (Gruber 2008).

Request for WH climate change strategy on Disaster Risk Management incorporating recommendations of the Kobe Thematic Session on ***Risk Management for Cultural Heritage***	Development of the *Strategy for Reducing Risks from Disasters at World Heritage Sites*	**30 COM 7.2** **Strategy for Reducing the Risks from Disasters at World Heritage Properties**	Takes note of, and endorses the objectives of the *Strategy for Reducing Risks from Disasters at World Heritage properties, and* requests World Heritage Centre and...to work together...to prioritise the proposed actions... Requests World Heritage Centre disseminates it widely through its web-site...; Calls upon States Parties and...to give more consideration to the impacts of disasters on cultural and natural heritage when designing...strategic goals and plans; Further encourages States Parties...to integrate concern for World Heritage into wider national disaster reduction plans...and to develop management plans that include a risk-analysis and management component for World Heritage properties...; Requests WHCentre and...to develop... resource material to build-capacity on disaster reduction at World Heritage properties... as well as a training module...; Also requests WHCentre and...to prepare a rev. draft...for Emergency Assistance requests incl. requirement to clarify what is the specific serious threat/danger affecting the property, how it might affect its overall OUV, and how the proposed activity intends to mitigate/prevent it...;	Climate change still not linked with DRM. Decision uses same descriptors 'serious' and 'dangerous' as the criteria for inscribing a site on the 'in danger' WH List.
Nomination of River Is. of Majuli in midstream Brahmaputra River in Assam, India	Inscription as a cultural site, 2006	**30COM 8B.40**	'Refers the nomination back to the State Party in order to ... undertake an appraisal of the overall river basin in which Majuli lies, and the potential impact of climate change, in order to ascertain the chances of the island surviving in the medium term..."	Climate change as a potential impact on the survival of a proposed WH site is recognised.
2007 **Provisional Agenda of the 31st Session of the WH Committee**	Discussion and adoption	**30COM 17 for 31COM**	The examination of State of conservation reports of WH properties stimulates an 'Issues' discussion on ... the impacts of Climate Change ...' Governor-General of New Zealand, Anand Satyanand, spoke of the need to protect WH properties in the face of serious challenges including "climate change...while the Pacific Region covers one third of the globe; it is	Climate change termed an 'impact', 'serious challenge', 'threat' and a 'risk' to WH. Recognition of the particularly vulnerable position of indigenous peoples' heritage under

			underrepresented on the World Heritage List". Rep. of Korea advises that the WH Committee should take note of the preservation by indigenous peoples of their heritage under threat from climate change. IUCN noted that a recent monitoring mission to Macchu Picchu had highlighted the need to keep climate change and the risks it posed to WH properties high on the agenda.	threat of climate change, and of under-representedness of the Pacific on WH List.
State of conservation reports on **Chan Chan Archaeological Zone, Peru; Air; Tenere Natural Reserves, Niger; Keoladeo NP, India**	Discussions about serious climate change threats to the one cultural site and two natural sites, 2007	**31 COM 7A.30** **31 COM 7A.10** **31 COM 7B.17**	Climate change was not included it in any of the Decisions arising out of the State of conservation reports.	Climate change is recognised as threatening OUV, but it is not the subject of directives in the Decisions. Climate change is addressed 'to the extent possible and within the available resources,' (30COM7.1) which may mean it is unaddressed at site level because the action (reduction in GHG emissions) needed is at international and at State Party level.
Meeting: Working Gp. of 40 to develop *Policy*, 2007. Australia submitted that WHCom should follow **4 principles**: 1. '...actions by WHC... must focus upon site-level adaptation and should not address issues eg mitigation of GHG. 2. must avoid overlap or duplication with other international conventions. 3. climate change is one of a no. of pressures facing WH properties and any response must	Development of draft *Policy* is a watershed climate change position document which undergoes revision and Gen. Assembly discussion before Decision. Contestable issues: Australia is against using WHC to address climate change beyond site level, incl. by obliging reduction in GHG. It argues States Parties, under WHC, can address climate change almost any		Endorses the *Policy Document on the Impacts of Climate Change on World Heritage Properties* decides to authorize the Chairperson of the World Heritage Committee to vet the Policy Document, incorporating views expressed at the 31st session, and, as appropriate, to consult World Heritage Committee members by email and other means; Decides to transmit the revised Policy Document for discussion and adoption at the 16th General Assembly of States Parties in 2007; Recommends that the *Policy*...be read in conjunction with the *Report and* the *Strategy* together with other relevant conventions...; Urges the World Heritage community to integrate actions pertaining to climate change in risk preparedness policies and action plans, making use...of the *Policy*...and *Strategy for Risk Reduction* at World Heritage properties, so as to protect their outstanding universal value, authenticity and/or integrity. Also urges States Parties to participate in the UN Climate Change conferences with a view to achieving a comprehensive post-Kyoto agreement and...support the research.... as identified in... the *Policy*...	The Decision and the *Policy document on the impacts of climate change on World Heritage properties and A Strategy to Assist States Parties to Implement Appropriate Management Responses* are flexible, discretionary and collaborative in nature but weak in protecting WH from climate change reflecting the risk adverse stance of a few dominant high GHG emitting States Parties and the "specious" interpretation of the WHC (Thorson 2008).The many States Parties with low carbon footprints requesting a global and regional response were overridden in discussions. Legal questions e.g. development of WH enforceable climate change mitigation measures including GHG emissions reduction at State Party level that were brought up at the working group were sidelined. Realpolitic influences are evident in this stance and the sidelining of proposals from Peru and Canada for alternate Lists for climate change threatened WH sites or sites that have lost OUV. The climate change sensitivity of the

address all pressures in an integrated manner. 4. States Parties must not rely solely upon the WH process to integrate approaches to WH and those on climate change, but have an obligation to work domestically with climate change policy-makers and other relevant areas of govt. and society. Canada twice asserts that WH Committee should be careful to avoid being taken over by a preoccupation with climate change at the expense of other issues. WH Secretariat stated "WHC does not have a leading role to play in terms of mitigation efforts. States Parties should be encouraged to address these issues through UNFCCC" (UNESCO 2007e). Argentina raised CBDR	way they see fit. It argues 'monitoring and reporting' under WHC and Operational Guidelines are sufficient to address climate change. Canada asks "Why should climate change be treated any differently from other issues facing WH sites? Given the *irreversible* nature of climate change, it seems unlikely that the 'In Danger' list is an appropriate response". Many other States Parties requested regional and international, as well as site based, policy on climate change. The IUCN Environmental Law Programme argued for urgency of research in climate change and heritage law and that Operational Guidelines "could	31COM 7.1 **The Impacts of climate change on World Heritage properties** **Policy Document**	Encourages World Heritage Centre to sensitize States Parties, as appropriate, to the need to establish inter-disciplinary mechanisms to deal with policy and governance issues relating to the effect of climate change on World Heritage properties; Recommends World Heritage Centre strengthen its relations with all organizations working on climate change, particularly with the UNFCCC and IPCC... and specifically with regard to the effect of climate change on World Heritage properties; Welcomes book *Case Studies on Climate Change and World Heritage* Encourages UNESCO...to disseminate widely the *Policy...and...*related publications Adopts... specific research priorities...in *Policy*...and recommends to the UNESCO Forum...to prioritize these subjects and to open discussions on the effects of climate change on World Heritage properties; Requests World Heritage Centre...to develop in consultation with States Parties criteria for the inclusion of those properties which are most threatened by climate change on the List of World Heritage in Danger, for use in prioritizing vulnerability assessment, mitigation and adaptation activities; Decides, for future sessions...to add to the State of conservation working document a section on those properties most affected by climate change; Adopts a carbon neutral policy for all future sessions, to the extent feasible. **Contested interpretations and issues** Lawyers have vehemently argued there is not a legal case for abdicating climate change global and regional mitigation to UNFCCC (Thorson 2008) (Burns 2009). "There is a very real threat that if the World Heritage Committee waits for the UNFCCC to 'solve' this problem, much of the world's cultural and natural heritage may be lost" (Burns 2009p160). The most judicious approach	Netherlands, Africa, glaciated States Parties and Pacific Islands and the expressed need for 'precautionary measures' and to confront 'this all-encompassing threat', although raised, were passed over. States Parties that supported participation in UN climate change meetings to achieving post Kyoto agreement successfully argued against States Parties that stressed that the WHCom's own forums should be used. The WHCom sidelined the legal proposals to 'confront head-on and with specific legal obligations mitigation measures to protect WH from effects of climate change (UNESCO 2007e). The IUCN Law group. asserted that unless WHC could oblige compliance with legally binding GHG emissions conditions, States Parties would not take climate change mitigation 'seriously'. Does the abdication of the WHCom from national and global mitigation action reflect the criticism that the WHC is a lopsided convention, weighted to education, publicity and status; popular because it generates national prestige and tourism; but with moderate demonstrable protective obligations and no legal teeth? Technology transfer was recommended in other conventions as a means of coping with climate change, and was supported by some States Parties, but was not considered appropriate for the WHC by WHCom. to make policy. Carbon offsets opportunities were dropped. The lack of 'alternative mechanisms' in dealing with climatic change impacted sites reveals a retreat to BAU processes and is a reminder that climate change WHC policy, ineffective as it is for mid and long term protection, has been

but no other moral perspective was reported. Canada said WHC "does not give WH Committee authority to take a leading role internationally on climate change... the WHC cannot and should not be seen to lead the international response to...climate change"(UNESCO 2007e).	set specific processes and criteria to ensure States Parties are doing 'to the utmost of their resources' (Art 4) and meeting international duty by ref. to climate change obligations under Kyoto Protocol, UNFCCC, and nationally to protect WH and reduce GHG emissions". Does the fact that WHC has no specific compliance and enforcement mechanisms (teeth) prevent it from explicitly encouraging, if not requiring States Parties to reduce GHG emissions to address climate change damaged WH? In fact High Court of Aust. found that Art 4 and 5 **DO** impose legally binding obligations. (Thorson 2008p24).		would be for WHC concomitantly to address climate change in the context of sites it is committed to protect. The WH Committee's non-confrontational stance, unlike that of Australia and Canada, set the tone of this disappointing Policy...which struggles to deem itself as 'legitimate'? The WHC preamble discredits the Policy that the WHC authorises site level mitigation action only (Burns 2009). "Every treaty in force is binding upon the parties to it and must be performed in good faith." (under *Vienna Convention of the Law of Treaties)* A question that arises in this context is whether the failure of a State Party to carry out its obligations under the UNFCCC by not achieving its prescribed targets under the Kyoto Protocol results not only in the breach of the UNFCCC but also a breach of the obligations under Article 4 of the WHC by failing to do to "do all it can", "to the utmost of its resources" to ensure that WH properties are protected (UNESCO 2007ep34). WH sites threatened by climate change may be i) Listed as 'in danger' regardless of whether the State Party is responsible for the climate change concerned, or ii) deleted. The *Policy* does not suggest which of these strategies to choose, or discuss their ramifications. Should the Policy have a revision clause that ensures it remains relevant to Climate change science? See **Resolution 16 GA 10 A,** recommendation 15.1. The indecisiveness of the 'In Danger' listing process is confusing e.g. Aust's Great Barrier Reef, the WH's largest site, qualifies for such listing on every legal argument as The Sydney Centre for International and Global Law found,(Rothwell 2004) and has been submitted twice in Petitions for such listing but remains unlisted despite accelerating deterioration.	driven by petitions, not by proactive 'insiders'. Australia's arguments beggar the question when it added: "...the In Danger' list is ill-equipped to deal with a challenge that faces most, if not ultimately all WH 'sites. Although some properties have been the focus for attention in relation to climate change, there has been no discussion of whether these 'sites' are more vulnerable to climate change than others..." nor any "indication of how 'In Danger' listing may address a challenge of the nature of climate change" (UNESCO 2007e). 'In Danger' listing may have forced an obligation for State Party/s to reduce GHG emissions with added possibility that powerful US would have ignored this (Burns 2009p161). The insistence to avoid overlap and duplication with outside bodies and conventions; and the dictum not to create ' 'new processes of research' may be counterproductive to addressing the big picture of climate change and not sympathetic with the WHC's thrust for cooperation in meeting the climate change challenge? The Decision does not address 'human impacts' of climate change on WH especially on cultural sites, their communities and intangible heritage; Decision terminology: In the knowledge of the urgency for climate action, 'Burns suggests 'urges the WHCen strengthen its relations with... UNFCCC and IPCC' would seem better than 'recommends'(Burns 2009).

	Australia acknowledges *possibility* of gaps in existing research re: climate change science, impacts and adaptation and with Canada, opted for collaboration with outside groups stating "WH process should not establish any climate change specific research process" (UNESCO 2007e).			
Need to 'build public and political support' by better 'info. distribution and communication' ' to address the multiple challenges posed by **climate change** to the world's irreplaceable and fragile cultural and natural heritage' (Case Studies)	UNESCO compiled book of case studies 'using the iconic character of WH sites as an important asset for raising public concern and enthusiasm and therefore, building up support to take preventive and precautionary measures for adapting to climate change'.	**Published 2007 by UNESCO**	***Case Studies on Climate Change and World Heritage*** **report that the following sites are affected by climate change: Sagarmatha NP,** Nepal; **Huascarán NP,** Peru; **Ilulissat Icefjord,** Denmark; **Kilimanjaro NP,** United Rep. of Tanzania; **Jungfrau-Aletsch-Bietschhorn,** Switzerland; **Great Barrier Reef,** Australia; **Sundarbans,** India, Bangladesh; **Komodo NP,** Indonesia; **Cape Floral Region Protected Areas,** South Africa; **Greater Blue Mts. WH Area,** Australia; **Ichkeul NP,** Tunisia; **Wet Tropics of Queensland,** Australia; **Area de Conservación Guanacaste,** Costa Rica; **Chan Chan Archaeological Zone,** Peru; **Ivvavik /Vuntut/Herschel Is.** Canada; **Chavín Archaeological Site,** Peru; **Golden Mts. of Altai,** Russian Fed; **WH sites of the City of London, UK; Venice and its Lagoon,** Italy; **Historic Centres of Cesky Krumlov and Prague,** Czech Rep; **Timbuktu,** Mali; **Ouadi Qadisha (the Holy Valley) & Forest of the Cedars of God (Horsh Arz el-Rab),** Lebanon.	These WH Sites have documented seriously threatening climate change impacts. As of 2011 only 1 is listed as 'in Danger.' Maybe the sensitivity and controversial nature of 'in Danger' listing brings more publicity, raises more awareness, and creates more debate than the case would be if all endangered sites were routinely listed?(A danger list in danger 2010)

Request for a prioritised list of **DRM actions** from WH Centre and others (Decision 30 COM 7.2)	WH Centre requests WH Committee to endorse revised *Strategy* incl. Action: Support risk identification and assessment at WH sites incl. climate change impacts. The *Policy document on the impacts of climate change on World Heritage properties* links climate change and risk management or reduction.	**31COM 7.2 Revision of the Strategy for Reducing Risks at World Heritage Properties**	Approves the *rev. Strategy for Risk Reduction at World Heritage Properties* with its prioritized list of actions; Encourages States Parties....and WH Centre to implement the *Strategy* within their spheres of activities; Requests WH Centre...to integrate policies and strategies est. by the WH Committee on climate change in the implementation of the *Strategy*; Recommends a risk management component be incorporated in the Management Plan for WH properties in accordance with Operational Guidelines 118.	Climate change threats are included in DRM. The need for WH policies and strategies on Climate Change and Risk Reduction to be 'consistent and complementary' recognised. It is relevant to note that climate change *'in this context'* (in this *Strategy for Risk Reduction at WH properties)*, should be considered as one of the factors that...can threaten the State of conservation of WH.
Nomination of **Rainforests of the Atsinanana, Madagascar.**	Inscription of this natural site of outstanding biodiversity, 2007	31COM 8B.9	OUV '... important refuge for species during past periods of climate change and will be essential for the adaptation and survival of species in the light of future climate change'.	Climate change recognised as a long term phenomenon.
Nomination **Primeval Beech Forests of Carpathians, Slovakia, Ukraine.**	Inscription of natural property, 2007.	31COM 8B.16	Recommends: 'Give priority in Integrated Management Plan to research and monitoring as this ... can provide a valuable contribution to understanding the potential impact of global climate change.	This outstanding example of complex, undisturbed temperate forest is recognised for its research potential into climate change.
2007 Nomination of **Teide NP, Spain**	Inscription of a natural property with buffer zone.	31COM 8B.17	Key issues incl... potential impact of climate change...' Recommends: 'Encourage improved research and monitoring of the potential impact of global climate change and the need for adaptive management strategies.'	A rich and diverse assemblage of volcanic features and landscapes suitable for climate change research.
2007 Nomination of *extension* of **Jungfrau-Aletsch-Bietschhorn, Switzerland**	Approval for this natural site extension.	31COM 8B.18	' is of significant scientific interest in the context of glacial history and ongoing processes, particularly related to climate change. '... climate change is well-illustrated....in the varying rates of retreat of the different glaciers...'Issues incl. potential impact from climate change...'	Ongoing, global and potential threat of climate change recognised.

2008 State of conservation report **Galapagos Is, Ecuador,** - a natural property threatened by invasive species, unbridled tourism and overfishing	Requests compliance to develop a comprehensive climate change plan as present threats are being exacerbated by global warming.	**32 COM 7A.13**	Climate change not mentioned in Decision.	Omitting the state of conservation recommendation (for climate change to be addressed) from the Decision, raises the question of inconsistency in prioritising action to comply despite 'In Danger' Listing from 2007 to 2010. Delisting from 'In Danger' list was considered precipitous and a sign of appeasement to a sensitive State Party.
State of conservation report on **Ruins of Kilwa Kisiwani** and **Ruins of Songo Mnara, United Rep. Tanzania**	Requests compliance with State of conservation recommendations for this cultural site to continue to be listed as 'in danger'	**32COM 7A.14**	'<u>Notes with concern</u> ...challenges faced by the property from climate change, leading to among others beach erosion...' <u>Notes</u> the danger posed to heritage by these challenges and their overwhelming nature; '<u>Adopts</u> ...management and... conservation plan... halted sea-wave action...' '<u>Also requests</u> States Parties to invite a Reactive Monitoring mission...'	Retained on WH in Danger List. Kenya said climate change would have a huge impact (eg sea level rise, storms) on the site and asked how the WH Centre could help. WH Committee does address difficulties of 'overwhelming' nature and thus has precedence for addressing climate change.
Request (in **Decision 31COM 7.1)** for criteria for the inclusion of those properties which are most threatened by climate change on the List of WH in Danger? Submission from Aust. Climate Justice	Requests approval for criteria IUCN Env. Law. Programme had argued that Operational Guidelines for the inscription of properties on the List of WH in Danger which require " the threats and/or their detrimental impacts	32COM 7A.32 – **Amendments to Operational Guidelines setting criteria**	<u>Noting</u> the real danger from **climate change** faced by many orld properties, <u>Decides</u> to adopt the criteria...for assessing properties which are most threatened by climate change for inclusion on the List of World Heritage in Danger, noting that the emphasis of the corrective measures to be recommended should be on "adaptation" rather than on "mitigation"; <u>Approves</u> the following amendments to the Operational Guidelines: a) Amendment to Paragraph 179 (b) (vi): *threatening impacts of climatic, geological or other environmental factors.* ~~gradual changes due to geological, climatic or other environmental factors.~~ b) New Paragraph : Paragraph 180 (b)(v): *threatening impacts of climatic, geological or other environmental factors.* c) Amendment to Paragraph 181: In addition, the ~~factor or factors which are threatening~~	The existing Operational Guidelines are minimally modified to address climate change threats for WH 'in danger' determination. The contested issue of inscribing sites on List of WH in Danger: Climate change will eventually threaten all WH sites, and clearly it is impacting on many now differentially. Perusal of this table shows sites repeatedly that are so identified but are not Listed as 'in Danger'. How will the WH Committee explain what 'in Danger' means as more sites are threatened? Is the WH image/brand being undermined by the contested and confusing categorisation of WH sites as eligible to be listed as 'in Danger'? How can the WHCom overcome the 'stigma'

Program (ACJP) and Int. Env. Law Project (IELP) sought amendments to Operational Guidelines to strengthen WHC to incorporate adaptation and mitigation issues into nomination process to enable emergency nominations for climate change risk sites and enhance reporting and monitoring management.(ACIP 2008)	on the integrity of the property must be those which are amenable to correction by human action" could be met because climate change can be specificied as a threat against which 'clearly identifiable human actions...can be taken to correct it' thus meeting Operational Guidelines requirements (UNESCO 2007e).	**for assessing properties that are most threatened by climate change for inclusion on the List of World Heritage In Danger 2008**	*threats and/or their deleterious impacts* on the integrity of the property must be those which are amenable to correction by human action. In the case of cultural properties, both natural factors and man-made factors may be threatening, while in the case of natural properties, most threats will be man-made and only very rarely a natural factor (such as an epidemic disease) will threaten the integrity of the property. In some cases, ~~the factor or factors which are threatening~~ *threats and/or their deleterious impacts* on the integrity of the property may be corrected by **administrative or legislative action**, such as the cancelling of a major public works project or the improvement of legal status.' Contested interpretations and issues: WCG Burns states: The WH Committee should consider expanding the scope of its guideline revisions to include the following components: • specification of measures that parties should take to mitigate their GHG emissions to protect WH sites, including adoption of the Kyoto Protocol and its successor; • a requirement that the Convention's parties specify activities that are being taken to reduce GHG emissions, including pertinent legislation and policies; • specification of adaptation protocols for WH sites that may be threatened by climate change, including vulnerability assessments and methods to improve site resilience(Burns 2009 p163).	associated with 'in Danger' listing and turn it to an advantage in urging States Parties to address the global issue as well as the regional and site-based issue? How can the WH Committee address/meet the threat of losing credibility by concentrating on local actions to List and de-List from the 'in Danger' List? The carefully qualified conditions for 'In Danger' listing undermine the urgency and perilous nature of the threat some WH sites are suffering from climate change. The identified 'shame' associated with 'in Danger' Listing is linked to cultural sensitivities which may have an impact on access to funding and technical help, a dilemma which indicates that alternate mechanisms are still needed. If all WH sites had vulnerability scales for climate change damage (with Crisis code symbols) 'with what are State Party and global responsibilities', this may lessen the shame and facilitate application for funding and scientific help, see recommendation 15.2, 15.3 and 15.4.
Reactive monitoring mission to **Shiretoko, Japan** found OUV is related to...sea ice...(which)... influences... marine and terrestrial ecosystem.	Requests implementation of recommendations for this natural site where the impacts of long term climate change could have a significant impact in relation to	**32COM 7B.16**	Requests development of a climate change Strategy which incl. i) a monitoring programme; and ii) adaptive management strategies to minimise any impacts of climate change on its values;	Climate change impacts are not identified e.g. temperature rise, warming sea, sea ice melting, SLR, loss of biodiversity. 'Long term climate change' impact recognised.

	marine and salmon management. 2008.			
State of conservation report on **Yellowstone NP, USA**	Requests State Party compliance with this and previous State of conservation reports for this natural site, 2008.	**32COM 7B.29**	Requests 'i) ...review ... programme to remove the lake trout invasive species; investigate the effects of reduced lake levels and drought on the cutthroat trout and consider the potential role of climate change in further affecting the recovery of this species;.. '.	Climate change recognised as having a potential contributing impact on biodiversity.
State of conservation report on Everglades **NP, USA** – a natural site, 2008	Requests State Party compliance with State of conservation report recommendations.	**32COM 7B.30**	Also encourages State Party to do a vulnerability assessment and develop a risk reduction strategy for climate change, including effective solutions to restoring water flow and functioning of the Everglades ecosystem that will allow it to adapt to projected SLR.	Gentle wording 'encourages' and does not suggest the urgency of the problem caused by climate change. Climate change action should be qualified 'strongly' and 'urged.' This site was added to List of WH in Danger, 2010.
Reactive monitoring mission to **Tasmanian Wilderness, Australia** found climate **change** threats- SLR, storms, eroding archaeological sites, drought, and fires, temp. and precipitation changes.	Requests State Party implement recommendations for this mixed site. Consider incorporating the implications of deforestation in light of the discussions on successor to Kyoto Protocol emission commitments. 2008.	**32COM 7B.41**	Requests State Party 'Prepare and implement ...management plan....to reduce risks, particularly from fires and climate change'. 'Establish...programme for monitoring the impacts of climate change....'	Addressing climate change linked with Kyoto Protocol commitments in pre-Decision discussion.
Nomination of **Mt. Sanqingshan NP, China, 2008**	Inscription of this natural site.	32COM 8B.6	Recommends State Party 'establish research and monitoring programmes to assess and adapt to impacts of climate change including the potentially adverse impact of fire and invasive alien species on the park's aesthetic and natural values;'	Climate change linked to fire frequency and invasive species.
	Inscription of this cultural site includes Roi Mata's			

Nomination of **Chief Roi Mata's Domain, Vanuatu** of 3 early C17th sites on islands – Efate, Lelepa and Artok associated with the life and death of the last paramount chief, Roi Mata, of Central Vanuatu. 2008	residence, the site of his death and Roi Mata's mass burial site. It is closely associated with the oral traditions surrounding the chief and the moral values he espoused. The site reflects the convergence between oral tradition and archaeology and bears witness to the persistence of Roi Mata's social reforms and conflict resolution, still relevant to the people of the region.	**32COM 8B.27**	Decision makes no mention of climate change despite the following references in the Nomination file: The two principal impacts that have been identified are coral bleaching of the reefs and SLR caused by global climate change. The effect of SLR will "depend partly on the response of governments worldwide to respond to global climate change." Sea level rise carries with it the risk of increased flooding of low-lying areas during storm surges and increased erosion. Given that most of the archaeological features... are located in low-lying coastal areas, SLR will pose a significant threat to the whole site. The recommended course of action is to continually assess sea level data for the area and to consult specialists on this issue to find practical solutions to the problem. The State Party is...establishing monitoring points on Artok Is. to provide sea level change data for the property as a whole. These monitoring points will also measure tectonic uplift of the islands that may counter the effects of SLR at various points in time (UNESCO 2008f).	Climate change may not have been considered in the Decision as the threat is not imminent. This position may be inconsistent with other decisions which take into account the longer term threat. Gruber notes that the responses available to the impact of climate letechange on cultural heritage sites are very limited in many cases, such as in Arctic regions or when areas become permanently or periodically inundated. Cultural heritage is irreplaceable, mostly immoveable, and will be hugely affected by climate change esp. along coasts, deltas, low-lying riverine flood prone areas and in the Arctic and Antarctic (Gruber 2008).
Nomination of **Lagoons of New Caledonia,** 2008	Inscription of this natural site. Climate change identified in nomination discussions	**32COM 8B.10**	<u>Requirement</u> State Party 'protecting and managing large areas in the form of no-take zones and proactive management of water quality and fisheries regulations will help maintain reef resilience in the face of climate change. <u>Requests</u> ' Full protection should be given,...to all herbivorous fish species as these species are critical in the face of climate change to maintain reef health and ensure the most rapid recovery from bleaching events;'	**Climate change** impacts are not elaborated e.g. acidification of sea, warming sea and sea level rise.
Nomination of **Wooden Churches of the Slovak part of Carpathian Mt. Area, Slovakia**	Inscription of cultural site. 2008	**32COM 8B.37**	<u>Recommends</u> State Party 'define and implement a common integrated management plan ....taking into account the increased risk of forest fires due to global climate change'	**Climate change** linked with forest wild fires.

Thirty-third session Seville, Spain 22-30 June 2009		**World Heritage 33 COM WHCom 7B climate change a challenge for WHC** and **for SIDS in particular**	Pilot climate change projects in Madagascar, Indonesia and Peru are under negotiation with the donors. They seek to promote biological connectivity and adaptive forest landscape management at specific WH properties to increase their resilience to and reduce risks from climate change impacts. Regional capacity building workshop for Asia- Pacific on vulnerability assessment of WH properties to disasters and climate change December 2009, Beijing.	The WHC contains fifteen references to 'international assistance', emphasising the centrality of global co-operation and partnerships to the protection of WH (Baxter 2006p70).
2009 State of conservation reports **on Galapagos Is, Ecuador, Ruins of Kilwa Kisiwani** and **Ruins of Songo Mnara, United Rep. of Tanzania,; Banc d'Arguin NP, Mauritania**	Requests compliance of States Parties with State of conservation recommendations for these 2 natural sites and 1 cultural site	**33 COM 7A. 13** **33 COM 7A.14** **33 COM 7B.11**	'Galapagos' site: climate change discussed and one State Party said climate change should be addressed but this was not reflected in the Decision. 'Ruins' site: climate change impacted but not reflected in the decision. 'Banc' site: climate change claimed as a critical issue for this property but wording changed from climate change to 'other environmental phenomena'. One State Party claimed it 'unfair' for Mauritania to be asked to address climate change impacts without assistance.	During decision making one State Party called for consistency regarding reporting the impacts of climate change. With reference to 'Banc' site, one State Party repeated that when referring to climate change, the WH Committee should refer to "adaptation" and not "mitigation", especially for climate change affecting cultural and natural properties. WH Committee is divided on emphasis put on addressing climate change. WH Committee is persistent in downplaying 'mitigation' and opting to omit climate change from decisions.
Received **Petition from 3 Australian environmental/law groups** which states that the obligations under the WHC and the UNFCCC are consistent with each other. Hence, States must	Discussed this 2008 petition with UNFCCC.	**World Heritage 33 COM (World Heritage Committee session 7B)**	The petition: *States Parties Responsibilities under the World Heritage Convention in the Context of Climate Change – Absolute Minimum Temperature Rise Necessary for Compliance with the World Heritage Convention addressed to the Ad Hoc Working Group on Further Commitments for Annex 1 Parties under the Kyoto Protocol (AWG-KP)* was referred on to the UNFCCC Secretariat which reported that the petitioners had been advised to re-submit it to the Ad Hoc Working Group on	The reworked petition was resubmitted as *State Parties Responsibilities to Reduce Emissions to Ensure the Protection, Conservation and Transmission of World Heritage to Future Generations to the Ad Hoc Working Group on Long-term Cooperative Action (AWG-LCA)under the UNFCCC* in September, 2009, supported by 12 conservation organisations. It stated that Annex 1 Parties (States Parties with developed economies) must ensure decisions concerning the reduction of GHG emissions and emissions reduction targets take the obligations under the WHC into consideration during negotiations under the UNFCCC as a matter of high priority

reduce GHG emissions and meet the targets decided under the UNFCCC and AWG-KP to protect WH.			Long-term Cooperative Action 'as it (the petition) raises a broader and more general concern, which relates to both the UNFCCC and the Kyoto Protocol'	and seriousness and ensure consistency bet. the respective Conventions; stabilize GHG concentrations in the atmosphere below 350ppm CO2e; ensure the global temp. increase is limited to well below 2 degrees C above the pre industrial level; reduce GHG emissions by at least 40 per cent below 1990 levels by 2020; reduce GHG emissions by at least 80 per cent below 1990 levels by 2050.
Reactive monitoring mission to **Mt Kenya, Kenya** and **2009 Petition on 'The Role of Black Carbon in Endangering WH Properties Threatened by Glacial Melt and SLR change**	Compliance with 2008 mission recommendations and submission of State of conservation report. including report of glacier retreat. Letter sent to Kenya about the 'Black Carbon' Petition although Mt Kenya not named in the petition.	33COM 7B.3	Notes with concern the reported impacts of climate change on the site and recommends the State Party to exchange experience with other States Parties and experts working on mountain WH conservation, climate change and other environmental phenomena to explore appropriate and practical strategies for maintaining OUV of site in the long term; Welcome efforts to 'enlarge the NP'. Ensure ...monitoring of climate change...and implement management practices that support ecosystem adaptation.'	In the Decision wording "adaption and mitigation" replaced by climate change. Then "other environmental phenomena" was added. No mention was made of Black Carbon in Decision. Development of wildlife migration corridors was advocated.
2009 State of conservation report on **Cape Floral Region Protected Areas, Sth. Africa**	Requests State Party include a buffer zone for wildlife corridors to increase resilience to climate change in this natural site.	33COM 7B.6	Encourages State Party to...continue and enhance its programmes for fire management, control of invasive species and mitigation of climate change impacts; Requests State Party to submit...a report on progress made in fire management, control of invasive species, mitigation of climate change impacts'	IUCN notes... reports of a range of other stresses to the property, in particular on water use and pollution, livestock and infrastructure. The WHCen and IUCN note that the removal of other stresses ... is a key strategy to maximise its resilience. for climate change. Mitigation projects advised. Climate change linked with wildfires and invasive species.
	Requests State Party compliance with State of		Invites State Party to exchange experience with other States Parties and experts...on World Heritage	

State of conservation report on Rwenzori Mts. NP, Uganda	conservation recommendations regarding glacial retreat, illegal logging, mining and poaching.	**33COM 7B.7**	conservation and climate change, to explore appropriate and practical adaptation and mitigation strategies for maintaining OUV and integrity of the site in the long term.	Long term mitigation and adaptation strategies are advised.
State of conservation report on Banc d'Arguin NP, Mauritania of threats: less rainfall; poaching, wood gathering, illegal fishing; oil exploration.	Requests State Party complies with recommendations in this and in previous reports which included need for management capacity and resources.	33COM 7B.11	'incorporate World Heritage Committee's Climate Change and Disaster Risk Reduction Strategies. Proposal to use satellite imagery for monitoring and management. Urges State Party to complete the POLMAR (management plan) Notes with concern that threats from ongoing low rainfall, which are contributing to a decline in terrestrial habitat and wildlife; and encourages State Party to assess adaptation measures to respond to climate change;	Multiple impacts of climate change: physical, biological and 'human' impacts are addressed.
State of conservation report on The Sundarbans, Bangladesh of risk of increased flooding & increased salinity from **SLR.** A 25cm SLR could result in loss of 40% of mangroves. Cyclone 'Sidr' damage led to loss of management capacity. Black carbon is impacting WH by SLR.	State Party requested, re: Petition, to call on WH Committee to take action to protect the OUV of WH properties most vulnerable to climate change. Letter sent to Bangladesh about Carbon Petition Restoration of community nurseries and green belts recommended by State of conservation	33COM 7B.12	Requests State Party to develop a programme of ecological monitoring... documenting the impact of climate change...;' 'The World Heritage Centre and IUCN encourage the international community to provide the assistance requested by the State Party to help to 'understand, mitigate and adapt to the impacts of black carbon.' The State Party is encouraged to closely monitor the changes in sea level in the property and the potential impact of climate change. The State Party may also benefit from engagement with other States Parties with properties whose OUV and integrity are at risk from the impact of climate change in coastal areas.'	'On 22 April 2009 the WH Centre received a response to the Black Carbon letter from Bangladesh advising that the WH Committee "...should try to influence the UNFCCC and Kyoto Protocol to explore the possibility of incl. black carbon as an active agent for climate change". The letter also states that in the present scenario "Sundarbans WH site may be included in the 'List of WH in Danger' by black carbon". IUCN considers that the WH Committee has an important role to bring to the attention of UNFCCC the threat to the OUV and integrity of WH properties from the impact of climate change, and to recognize the need for and encourage action to reduce emissions, including of black carbon.'
	The 2009 mission again considered 'in Danger' listing as			

State of conservation Mission to **Tropical Rainforest Heritage of Sumatra, Indonesia, 2009**	threats remain critical. State Party expressed the strong opinion that such listing would create a negative perception and could hinder efforts to restore the integrity and effective protection and management of the site. The mission considered that the benefits of 'in Danger' listing were outweighed by the possibility that listing could reduce political will to act in relation to the conservation concerns. Request State Party comply with mission's recommendations to add site to 'in Danger' List	33COM 7B.15	Urges State Party to establish an effective and prioritised monitoring system to assess the status and trends of key factors affecting the OUV of property, incl. wildlife populations, invasive species, deforestation, poaching, wildlife trade and any anticipated climate change impacts. "The plan is a response to a pledge last year by Norway to invest up to $1 billion in Indonesian conservation efforts under REDD scheme if Indonesia would get serious about halting the many devastations of deforestation. The Norwegian offer is part of an ongoing international effort to put forest protection front and centre in trying to overcome the serial disappointments of recent worldwide climate summits and begin to put together an effective global architecture to combat climate change"(Bacchus 2011).	WH Committee prioritises the State Party's sensitivity to having the site listed as 'in danger' above repeated recommendations from WH missions. 'In Danger' listing perceived as a mark of public failure. The site was not added to the 'in Danger' Listing. WH Centre regrets that State Party continues to regard the 'in Danger' inscription as a criticism, rather than a means to strengthen international support for the property as WHC intended.
State of conservation report on Sagarmatha NP, Nepal incl. **climate change** induced melting of glaciers, lake formation, threat of violent flooding and glacial lake outbursts disastrous for villagers,	Request Nepal adopt State of conservation recommendations and responds to the 2009 Petition on black carbon which by accelerating impacts of climate	33COM 7B.17	Invites State Party to exchange experience with other States Parties and experts...working on mountain World Heritage conservation and climate change, to explore... adaptation and mitigation strategies to maintain the OUV and integrity of the property in the long term and to develop a DRM plan. 'The World Heritage Centre and IUCN encourage the State Party to implement adaptive management	Despite substantial research on the profound climate change threats, Sagarmatha NP, an iconic natural site with cultural and spiritual values, remains unlisted as a site 'In Danger'. Adaptative strategies include early warning systems, monitoring using remote sensing such

loss of mt. biodiversity, & collapse of mt. ecosystems potentially leading to famine & disease; tourism pressure, firewood collection & the role of black carbon.	change threatens properties protecting glaciers including Sagarmatha NP. The WH Centre sent Petition to all States Parties whose properties were mentioned.		measures to optimise the ability of the ecosystem and resident wildlife to adapt to changing conditions. Resilience should be maintained by ensuring ecosystem connectivity and genetic diversity and reducing threats and pressures that could increase vulnerability to these rapid changes. Pressures that can reduce the ability for ecosystems to adapt include pollution and habitat fragmentation.	as Landsat and the draining of glacial lakes. Does globalization value heritage only to its greater devaluation?
State of conservation report on Tubbataha Reef Marine Park, Philippines	Requests State Party complies with recommendations.	33COM 7B.18	Requests State Party put in place a programme of ecological monitoring... particularly the effect of climatic events on sea surface temperature and coral bleaching, storm frequency and other factors...related to climate change.	Ecological monitoring, a complex practice, is becoming dominated by climate change.
State of conservation report **Waterton Glacier International Peace Park, Canada/ US** incl. glacial retreat and Petition re: Role of Black Carbon in endangering World Heritage sites Threatened by Glacial Melt and SLR.	Requests State Party complies with State of conservation recommendations re: 2009 Petition which calls on WH Committee to take action to protect the OUV of WH properties most vulnerable to global warming (33COM 7C).	33COM 7B.22	Notes... high level of public concern regarding the protection of the property from threats from both mining and climate change; Requests the two States Parties... collaborate on the monitoring of climate change and the development of adaptive management strategies to maximise the resilience of the property to climate change... Requests State Party submits... report...on progress in developing climate change adaptation strategies..., taking into account the broader ecosystem, together with a State of conservation report.	James Hansen has posited that this NP, the first International Peace Park may have to be more properly called the 'Glacierless NP' in the not too distant future. As the climate is warming the glaciers are rapidly retreating.
State of conservation report on **Ilulissat**	Requests State Party complies with State of conservation to collaborate with Arctic States			Downplaying climate change impacts with

Icefjord, Denmark incl. the role of black carbon and mitigation strategies to reduce it esp. from ship fuel and diesel. IUCN recommends adaptive management to optimise the ability of ecosystem and resident wildlife to adapt.	Parties, indigenous peoples and Arctic communities as a follow-up to *World Heritage and the Arctic* '07 meeting and UNESCO '09 Meeting on *Climate Change and Arctic SD.* Resilience enhanced with ecosystem connectivity and reducing threats that increase vulnerability to climate change.	33COM 7B.23	Requests State Party collaborate with other States Parties whose World Heritage properties contain glaciers to monitor the impacts on those properties of global climate change, and to develop adaptive management strategies to ensure the long-term protection of the OUV integrity of the properties in response to climate and other environmental change.	carefully chosen words e.g. "The State Party notes that monitoring of climate, glaciers and permafrost areas show that site is responding to climatic change". The impacts of climate change are multiple, positively feeding back so increasing the momentum of the impacts especially in the polar regions.
State of conservation report on **Belovezhskaya Pushcha / Białowieża Forest, Belarus** and **Poland**	Request State Party addresses climate change and increased risk due to lack of coordination with State Forest Enterprise.	**Draft Decision: 33 COM 7B.24**	No mention of the climate change threat to biodiversity in decision.	
Repeated request by WH Committee for evidence of Govt. approval for **"Wrangel Is" Reserve, Russian Fed.** management plan; need for a more comprehensive monitoring plan,	Request compliance from State Party to meet repeated requests incl. increased attention to climate change monitoring and planning. Climate change impacts on northern Russia's ecosystem, is now	33COM 7B.30	Requests State Party confirm to World Heritage Centre that the necessary ministerial approval and adequate finance are in place for the implementation of the management plan, including in relation to infrastructure, increased security and inspection officers and an effective monitoring system, considering climate change impacts on the property;	Climate change is recognised as an accelerating

considering climate change and which should be based... (on) potential changes due to climate change, such as shifts in species composition, 2008.	an urgent issue and is probably the most serious threat to the values of the site ... with increasing evidence that the rate and scale of change may be greater than even relatively recent predictions.		Also requests State Party to submit to the World Heritage Centre, a report on the State of conservation of site, including a report on the status of its ecosystems and including an assessment of the impacts of climate change...'.	phenomenon.
2009 Petition on the role of black carbon ... calling on WH Committee...to protect...WH sites **most** vulnerable to global warming in high latitude and altitude glaciers and low-lying sites threatened by SLR. Request States Parties to find sources of black carbon and recommend measures... to reduce emissions; Place 17 WH sites on 'in Danger' List and develop... with...State Party... corrective measures ...; Coordinate with UN bodies working on climate issues to educate...on the	Petition lists 17 WH sites threatened by black carbon impacts: Glacier and Montane: **Ilulissat Icefjord,' Denmark, Sagarmatha NP, Nepal, Three Parallel Rivers of Yunnan Protected Areas ,China; Swiss Alps Jungfrau-Aletsch Switzerland; Kilimanjaro NP, United Rep. of Tanzania, Waterton Glacier International Peace Park, Canada/USA; Huascaran NP, Peru; Pyrenees – Mont Perdu, Spain/France.**	**33COM 7c General Decision on the State of conservation of World Heritage properties 2009**	Noting the increasing number of natural disasters affecting World Heritage properties, Requests World Heritage Centre...prepares a report on...progress made in... implementation of the *Strategy for Disaster Risk Reduction* at World Heritage properties. Notes... petition on *the Role of Black Carbon in the endangering of World Heritage properties Threatened by Glacial Melt and Sea Level Rise* and encourages States Parties to exchange information on existing national policies, regulations and opportunities for immediate voluntary action to control the generation of black carbon that can affect World Heritage. Requests World Heritage Centre...adopt a consistent approach to reporting on the impact of climate change on World Heritage properties and to ensure that future decisions... are based on World Heritage Committee's Strategy to assist States Parties to implement appropriate management responses to climate change.	Black carbon is the second most powerful contributor to global warming after CO2, and because of positive feedback effects on snow, warms the planet X 3 more than CO_2. Petition "suggests that reducing these emissions may be among the most effective near-term strategies for slowing the amplified climate warming experienced at high latitudes thus reducing glacial melt and SLR. Because black carbon is a short-lived climate forcing agent... reducing emissions has an immediate effect... slowing near-term global warming", without which it may be impossible "to protect many World Heritage properties from damage or destruction caused by climate change". "...black carbon is released ... during the inefficient burning of fossil fuels, biofuels and biomass. Because the dark-coloured particles absorb sunlight and when deposited on ice and snow, the reflectivity (Albedo) of these surfaces decreases and the rate of melting increases. As previous petitions and reports have not addressed this situation, the WH Committee has been unnecessarily limited in its role to protect WH. Because climate change is already adversely affecting much WH, if no

impacts that climate change and black carbon, are having on WH and to encourage... to mitigate the impacts of black carbon; Encourage and fund transfer of technologies to help mitigate impacts of black carbon emissions and assessing impacts of black carbon WH sites through Reactive Monitoring and Periodic Reporting.	Low-lying: **Great Barrier Reef, Australia; Belize Barrier Reef, Belize; Tubbataha Reef Marine Park, Philippines; Aldabra Atoll, Seychelles; Donana NP, Spain; Komodo NP, Indonesia; Kakadu NP, Australia, Sundarbans NP, India; The Sundarbans, Bangladesh.** The WH Centre forwarded a copy of this petition to the Secretariat of the UNFCCC for their comments, as well as to all the States Parties whose properties are listed and to Kenya see **Mt Kenya**, Decision 33COM 7B.3		Contested interpretations and issues: Petitioners request the WH Committee to take action that is specific and that falls squarely within the WH Committee's mandate under the WHC. *27 glaciers, out of the 150 recorded in 1850, remain*	action is taken in the short-term, many WH sites will not survive until the long-term GHG reduction measures take effect. In *Policy* discussions, two States Parties called for WH Committee not to create its own research agenda or program of research specifically addressed to climate change but cooperate with others. In the case of Black Carbon (not addressed by UNFCCC) and the undeveloped area of impact of climate change on cultural heritage (especially indigenous cultural heritage) the WH Committee may have a responsibility to lead with its own research agenda in order to protect WH. WH Committee *Policy* says that mitigation strategies at global and State Party level to be addressed by UNFCCC. But as UNFCCC and Kyoto do not regulate black carbon and nor is it likely to be regulated under the successor to Kyoto potentially undercutting the argument that the WHC should defer to the UNFCCC (Burns 2009p163). Note The petitioners found the17 WH properties met the criterion for the **List of World Heritage in Danger**: (b) Potential Threat (v) Threatening effects of climatic, geological or other environmental factors. Petitioners propose 5 WH tools: 1 'in Danger listing; 2WF Fund to support protective studies, 3 co-operation with expert organisations to adapt & mitigate, 4. WH studies of Black Carbon impact on WH sites, 5. Reactive monitoring and periodic reporting....noting that any adaptation measure should seek ways in which to mitigate.(UNESCO 2006).

Nomination of *extension* for **Tubbataha Reef Marine Park, Philippines**	Approval for extension to this natural site	33COM 8B.3	Site could be a demonstration property to study a natural reef system in relation to the impacts of climate change. Requests State Party to put in place a programme of ecological monitoring…, particularly the effect of climatic events on sea surface temperature and coral bleaching, storm frequency and other factors that could be related to climate change;'	The number of studies of climate change at WH sites is increasing. The multiple impacts are becoming better understood. The grasp of the enormity of the growing impact is being realised but the global protective measures necessary are high jacked by BAU GHG emissions.
Nomination of **Wadden Sea, Germany/Netherlands**	Inscription of this natural site	33COM 8B.4	Key threats incl. threatened fisheries, industrial facilities, maritime traffic, residential and tourism and climate change.	
2009 mission report on Simien Mts. NP Ethiopia	Decides to recommend retaining this natural site on List of World Heritage in Danger.	34COM 7A.9	Recommends State Party implement management planning, (for agricultural encroachment), tourism planning and climate change adaptation to drying of ponds and invasive plants.	
State of conservation report for **Serengeti NP, United Rep. of Tanzania**	'… impacts, combined with… effects of climate change, could potentially lead to… droughts and in the worst case scenario, stop the Mara River's water flow and compromise the Serengeti's iconic migration.	**Draft Decision: 34 COM 7B.5**	No mention of climate change in Decision.	
2009 monitoring mission report on the **Tropical Rain forest Heritage of Sumatra, Indonesia.**	Requests State Party complies with recommendations from 3 monitoring missions incl. that this natural property be put on the 'In Danger' list.	34COM 7B.14	Requests State Party develop and implement effective and prioritized monitoring system for agricultural encroachment, illegal logging, mining, poaching, wildlife trade, invasive species, and any anticipated climate change impacts. Strongly encourages State Party to consider alternative approaches to addressing threats…by making explicit provision within the REDD strategy and specifically the FIP…'	Site placed on 'In Danger' list in 2011.

State of conservation report on **Sagarmatha NP, Nepal** notes that **climate change** listed as a threat in previous reports.	Requests State Party complies with previous recommendations.	**Draft Decision: 34 COM 7B.16**	No mention of climate change in decision.	This site is most seriously impacted by climate change (see petitions 2006, 2009) and UNESCO's 'Case Studies….' book through increase in temperature, glacier retreat, flooding, glacier lake formation, biodiversity loss… but these impacts are not reported on in State of conservation decisions, nor is site listed as 'in danger'.
Reactive monitoring mission State of conservation report on Waterton-Glacier International Peace Park, Canada /USA says partnerships related to climate change issues are commitments and incl. initiatives… to address climate change.	Requests State Party complies with recommendations: management, monitoring and research should be developed to combat climate change impacts with further promotion of trans-border co-operation as in the new MOU which incl. mitigation of, and adaptation to climate change.	34COM 7B.20	Encourages State Party to share their experiences in the development of climate change mitigation and adaptation strategies with other World Heritage properties; Requests State Party to keep the World Heritage Centre informed regarding significant developments with respect to the above issues, (inc. climate change) and to give particular attention to these issues in their contribution to the periodic reporting process.	Climate change impacts of temperature rise, glacial retreat, introduced species spread and fragmentation of habitat may be proving to be outside the WHC's sphere of competency to address to protect the OUV.
State of conservation report on Lake Baikal, Russian Fed. strongly recommends monitoring of seals and impacts of climate change eg reduction in ice cover,	Requests State Party complies with recommendations re: climate change. By 2100 lake's ice cover, upon which its endemic plankton and Baikal seal depend, is likely to significantly recede. Melting permafrost may	34COM 7B.22	Also requests State Party to submit to World Heritage Centre a report on State of conservation of the property and progress made in preventing the discharge of untreated wastewater into Lake Baikal, addressing continuing high levels of pollution…, developing a	Climate change impacts are becoming increasingly more obvious. The Convention's **Decision 30COM7.1**

development of adequate mitigation measures and implementation based on early detection of emerging trends.	exacerbate the effects of current industrial pollution and accelerate the release of stored toxic chemicals, such as polychlorinated biphenyl (PCB) and dioxins, into Lake.		comprehensive tourism strategy and monitoring the Baikal seal population and the impacts of climate change on the property.	requesting 'States Parties...to protect OUV, integrity and authenticity of World Heritage properties from the adverse effects of climate change, to the extent possible and within the available resources, recognizing that there are other international instruments for coordinating the response to this challenge' is not proving to be a very effective protective policy.
State of conservation on **Doñana NP, Spain, Yellowstone NP, US, Alejandro de Humboldt NP, Cuba; Tasmanian Wilderness, Australia**	Climate change discussed for each site.	**Draft Decisions 34 COM 7B.26** **34 COM 7B.28** **34 COM 7B.33** **34 COM 7B.38**	Climate change not mentioned in draft decisions and later decisions	
State of conservation report on **Everglades NP, US,** 2010	Request State Party restores ecosystem of this natural site and Re –inscribe. Listed as 'In Danger.'	34COM 7B.29	Considers that 'the single most effective strategy to preserve the Everglades aquatic ecosystem in the face of climate change and SLR is the rapid implementation of the additional proposed restoration projects...'	To date, efforts to implement the adaptative measures have resulted in minimal improvements in water volumes and flow distribution. Prone to hurricanes and flooding. State Party recognises that 'in Danger' Listing appropriate.
State of conservation response by State Party, 2010, about **Madriu - Perafita - Claror Valley, Andorra**	Requests compliance of State Party for further reports on the WH management plan.	34COM 7B.75	Requests State Party does fauna study and pursue the work targeted in this study, concerning the «supramountain» fauna, and the potential conservation options relating to climate change; Recommends that this study be used to monitor 'impacts of climate change'.	WH site specifically recommended to be used to monitor the impacts of climate change as heralded by the UNESCO Climate Change Initiative.
Nomination of **Phoenix Islands Protected Area,**	Inscription as a natural site of '...largely pristine,'	34COM 8B.2	Site is of crucial scientific importance in identifying and monitoring the processes of sea level change, growth rates and age of reefs and reef builders and in evaluating	Decision makes no reference to the prospect that the State Party responsible for protecting the WH site is undergoing a profound transformation, including a migration program due to climate change and may not be able to manage the site as SLR submerges land, undermines the economy and will alter the lives of the people, the custodians of the site. Thus far, no international law claims have been

Kiribati	replete with largely intact...atolls...'		effects from climate change...'	brought by low-lying nations likely to be inundated by the SLR predicted to accompany climate change. Recent scientific findings asserts that SLR is likely to be larger than previously predicted, affecting as many as 6 billion people on low-lying Pacific islands and southeast Asia delta areas.
Nomination of **Putorana Plateau, Russian Fed.** 2010	Inscription as a natural site	34COM 8B.8	The site 'may provide valuable evidence on the impacts of climate change to large-scale natural arctic ecosystems if proper monitoring and research take place.' Recommends setting up a long-term scientific research and monitoring program to document and better understand the impacts of climate change..'	Recognition that Arctic sites are particularly vulnerable to climate change.
Nomination of Papahānaumokuākea, US - includes a significant portion of the Hawai'i-Emperor hotspot trail. 2010.	Inscription as a mixed site with islands, archaeological sites, sea habitats, reefs, lagoons, with deep cosmological and traditional significance.	34COM 8B.10	Also recommends State Party...develop response plans for the property related to climate change... to harmonize existing agency plans and activities in a coherent framework that can further strengthen conservation and management efforts.... generate information...beyond the property itself.	Physical and biological impacts eg SLR, ocean acidification, warming waters are not specificied thereby disallowing the chance to educate about the range and ubiquitous impacts of climate change on this fascinating mixed site.
Nomination of **Bikini Atoll Nuclear Test Site, Marshall Is.** 2010	Inscription of cultural site which is most threatened by climate change.	34COM 8B.20	Requests State Party provide details on site's marine surveillance system; strengthen visitor reception and the presentation of the property's cultural values in connection with the Peace Museum project...'	The site's biggest threat, climate change, is not mentioned in the decision to inscribe.
2010 State of conservation report on **Port, Fortresses** and **Gp. of Monuments, Cartagena, Colombia.** Climate change impacts: 'tide rise', warming and	Requests State Party complies with State of conservation finding that climate change should be in management plan. Notes State Party proposes to restrict settlements, construct locks and	**Draft Decision 34 COM 7B.107**	No mention of climate change despite recommendations.	State of conservation report notes Management and Protection Plan has not been finalized since it was requested in 2006 and 'timeframe for implementation is not ensured due to financial constraints.' Underlines the problematic nature of addressing climate change physical and biological impacts in developing States Parties with scarce funds to

humidity.	install pumps and valves in areas at risk.			spend on public works.
2011 joint WH Centre/IUCN monitoring mission **Manas Wildlife Sanctuary, India**	Recommendations State Party to set up an integrated monitoring system and a swamp deer recovery plan.	**35 COM 7A.13**	Encourages the State Party to consider the extension of the property in three stages. Conduct a joint feasibility study with the State Party of Bhutan on a possible transboundary extension of the property, in order to increase its ability to adapt to climate change	Manas Wildlife Sanctuary is removed from the List of WH in Danger
Nomination of **Ningaloo Coast, Australia, 2011**	Inscription of natural site	**35 COM 8B.7**	Sea level rise and increases in seawater temperatures associated with climate change have had comparatively little effect on the property. ... Still, careful monitoring is highly recommended.	Climate change impacts are being anticipated and adaptive measures are being adopted in advance.
Nomination of **Ogasawara Islands, Japan, 2011**	Inscription of 'natural' serial property	**35 COM 8B.11**	Strongly encourages the State Party to develop and implement a research and monitoring programme to assess and adapt to the impacts of climate change on the property.	
Nomination of **Prehistoric Pile Dwellings around the Alps, Switzerland, Austria, France, Germany, Italy, Slovenia**	Inscription of an almost complete underwater archaeological cultural site.	**35 COM 8B.35**	The revealed archaeological evidence allows a unique understanding of the way these societies interacted with their environment, in response to new technologies, and also to the impact of climate change.	It is recommended that the State Party develop an over-arching presentation framework that allows coordination between museums and an agreed standard of archaeological data to ensure understanding of the value of the whole property and how individual sites contribute to that whole.
2011 WH Centre/IUCN reactive monitoring mission to **Everglades National Park, USA**	Mission concluded that the OUV of the property continues to degrade due to an inadequate level of water flow and quality into the property.	**35 COM 7A.14**	Requests State Party to address the delays in the implementation of the Modified Water Deliveries (MWD), C-111 and Comprehensive Everglades Restoration Plan (CERP) projects, and related water quality initiatives which will result in continued degradation of the property and likely reduce the resilience of the Everglades ecosystem in the face of climate change	Everglades National Park remains on the List of World Heritage in Danger

2008 mission to **Mount Kenya NP/Natural Forest** Kenya	Requested adoption of Mt. Kenya Management Plan 2010-2020 and the completion of the Environmental Impact Assessment.	**35 COM 7B.2**	Remains concerned about the long-term impacts of climate change on the property, and encourages the State Party to resubmit a proposal for its extension in order to preserve as much lower altitude undisturbed forest and wildlife corridors as possible, and increase its resilience against climate change.	Long term impacts of climate change are raising concerns.
Report of the WH Centre and the Advisory Bodies on **Great Barrier Reef Australia**	Notes with extreme concern the approval of Liquefied Natural Gas processing and port facilities on Curtis Island within the property	**35 COM 7B.10**	Welcomes the State Party's commitment to improve the property's resilience and its ability to adapt to climate change and other forms of environmental degradation following the extreme weather events Regrets that the State Party did not inform the Committee as per paragraph 172 of the *Operational Guidelines* and requests the State Party to report, in accordance with paragraph 172, its intention to undertake or to authorize any new development that may affect the Outstanding Universal Value of the property before making decisions that would be difficult to reverse.	Link between climate change and extreme weather events implied.
Report of the WH Centre and the Advisory Bodies on **The Sundarbans Bangladesh**	Recommends ecological monitoring data for the property	**35 COM 7B.11**	Notes with satisfaction the initiation of the Sundarbans Environmental and Livelihoods Security project, which includes support for ecological monitoring and documenting the impacts of climate change on the OUV of the property, and welcomes the State Party's commitment to expand its coastal greenbelt zone through mangrove afforestation as a mitigation measure to climate change.	The State Party is requested to forward a state of conservation report on progress achieved with regards to post-cyclone restoration, as well as of the results from the ecological monitoring programme.

2011 IUCN monitoring mission to **Lagoons of New Caledonia: Reef Diversity and Associated Ecosystems France**	Mission recommends among other requests: Ensure timely response to threats identified and concerns rose relating to risks from mining exploration and non-compliance of regulations for the protection of the property.	**35 COM 7B.22**	Also requests the State Party to Facilitate the finalization and implementation of the co-management plans, and incorporate appropriate climate change considerations with particular attention to planning, monitoring and disaster risk reduction.	

Recommendations

26.1 Assuming the acceptance of **recommendations 15.1, 15.2, 16.3, 20.1, 20.3, 21.6, 23.8, 24.2, 24.8, 28.8, and 25.2** and

- Appreciating the leading role the WHC plays in World Heritage **site based protection** and
- Informed by the **adaptability, capacity and integrity** of the WHC and the ICHC and
- Reflecting on more than 60 **World Heritage Committee Decisions related to climate change** tabled in this chapter and
- Noting with concern the **present limitations** to protecting World Heritage properties from climate change impacts and
- Recognising the **strong public interest** in World Heritage, intangible cultural heritage and in their protection from future threats and
- Witnessing the **progressive momentum** building in anticipation of Rio+20,

Propose the development of a **joint *World Heritage and Intangible Cultural Heritage Climate Change Policy* with a strong position on mitigation.**

26.2 Urge the World Heritage and ICH leaders to move beyond site specific mitigation actions **to national and international level action** by developing a **leading position on global climate change mitigation.**

26.3 Support the adoption of the **mitigation rationale** that: to ensure protection and safeguarding of heritage and the possibility of a sustainable, equitable future for all, concrete action is necessary to stabilise the climate by moving to renewables and reducing carbon emissions to no more than 350 ppm. 350 parts per million (ppm) of CO_2 is what many scientists, climate experts, and over 112 national governments say is close to the safe upper limit for CO_2 in our atmosphere (350org 2011).

26.4 Urge the consideration of the proposed joint *World Heritage and Intangible Cultural Heritage Climate Change Policy's* mitigation position to oblige States Parties to make **clear commitments** to move to clean energy **by reducing GHG emissions according to the CBDR equity principle**, to make improvements in energy efficiency, and to contribute to and/or benefit from universal access to low or no carbon energy services.

26.5 Advocate that the proposed joint *World Heritage and Intangible Cultural Heritage Climate Change Policy* co-ordinates with the proposed **heritage site vulnerability campaign** (using the Climate Change Vulnerability Scale with climate change crisis symbols for display at World Heritage sites as recommended in 15.4).

26.6 Uphold the objectives of the ICHC (Preamble paragraph 11, Article 2, 13,) and the WHC (Article 4, 6,) that seek to prevent the devaluation of heritage. Failure of wealthy States Parties to make ambitious cuts to their greenhouse emissions constitute a deliberate avoidance of **responsibility that discounts the future** and runs counter to their undertakings as signatories to protect both their and other States Parties' tangible and ICH from the impacts of climate change for the benefit of future generations.

26.7 Encourage a greater willingness by the WHC to **engage openly, inclusively and constructively** in the protection of climate change threatened heritage.

26.8 Assuming the acceptance of recommendation 15.2 (proposed revision of *Policy Document on the Impacts of Climate Change on World Heritage Properties)*, **re-examine the petitions** that call for the listing of threatened WH properties to be inscribed on the 'in Danger' list due to the impacts of climate change.

26.9 Encourage concerned **citizens to petition** the Conventions about the need to address climate change, the need for increased listing of World Heritage sites on the List of World Heritage in Danger and climate change threatened practices on the list of Intangible Cultural Heritage in Need of Urgent Safeguarding. This would help combat political influence (especially from powerful polluting States Parties) which has constrained the process to date.

26.10 Assuming acceptance of recommendations 15.6 and 20.1 and WHC Decisions 29COM 7B a and 31COM 7.1 continue to seek **stronger ties with the UNFCCC**. It appears that the UNFCCC has made very little progress in negotiations relating to the Petition *State Parties Responsibilities to Reduce Emissions to Ensure the Protection, Conservation and Transmission of World Heritage to Future Generations* (September, 2009) and to the relevance it has for the WHC.

26.11 Noting recommendation 26.10 propose mandatory consideration of **UNFCCC advice** in deciding whether to list World Heritage sites on the **'in Danger' List** because of climate change. The advice would arise from a formal review of the science presented in petitions to the WH Committee as well as from World Heritage State of Conservation reports. This process would further depoliticise decision making. It would also recognise that climate change judgements increasingly necessitate local, regional and global climate change science knowledge.

As climate change impacts intensify, differentiating the part played by anthropogenic climate change becomes more important if compensation is linked to climate change threat or damage.

26.12 Invite a representative from the UNFCCC and an international environmental lawyer (as permitted by both Conventions) to advise on the latest **climate change science and climate change litigation** at all World Heritage and Intangible Cultural Heritage Committee, Conference and Assembly meetings.

26.13 Endorse action upon point 12 *"Further requests the States Parties and the World Heritage Centre to work with the Intergovernmental Panel on Climate Change (IPCC), with the objective of including a specific* ***chapter on World Heritage in future IPCC assessment reports"*** of the WHC decision 30COM7.1 re: *The Impacts of climate change on World Heritage properties 2007.*

Chapter 27

The Conventions Compared

We know the earth's resilience and resource base cannot be stretched infinitely. Moreover we are now uncomfortably aware that "business as usual" is not an option anymore. Our societies and economies are integral parts of the biosphere and it is time for the leaders of the world to act as stewards of nature's invaluable and inescapable contribution to human livelihoods, health, security and culture.
Professor Johan Rockström, Symposium Chair and Director of Stockholm Resilience Centre at Stockholm University, and Stockholm Environment Institute. (3rd Nobel Laureate Symposium 2011b)

The World Heritage Convention and the Intangible Cultural Heritage Convention

The following table is a side by side summary of how the two Conventions compare in relation to their capacity to protect and safeguard climate change threatened cultural heritage.

Lamas in training making carpet cleaning at 15th century Thikse Gompa (Monastery) an occasion for fun. (Leh, Ladakh). Buddhist monasteries are central to the spiritual and political lives of their followers who comprise half of the population of Ladakh. Many families have a son who will become a monk. The Dalai Lama has emphasised the importance of climate change and encouraged teaching Lamas to do likewise. The welcoming and largely open minded nature of Ladakhi Buddhists has led to a more enlightened acceptance of the reality of climate change.

UNESCO WHC and ICHC and their capacity to protect/safeguard cultural heritage from climate change impacts								
Questions	***World Heritage Convention 1972***	*Operational Guidelines*	*Reports, meetings, Decisions, strategies*	*Recommendations (Quotations included derive from the WHC and ICHC. See Recommendations at the end of Chapters for 'article' reference for these quotations)*	***Intangible Cultural Heritage Convention 2003***	*Operational Directives*	*Reports, meetings, Decisions*	*Recommendations*
*What do the **statistics** concerning the WHC and ICHC indicate about the state of protection offered to world cultural heritage against damage caused by climate change?*	***Number of States Parties per world nations*** *188/201*		Encourage ratification through **Global Strategies:** Target Pacific 1997(Fiji), 1999 (Vanuatu), Mauritius Strategy. **Programmes:** '5 Year WH– Pacific 2009'; SIDS Intersectoral Platform on SIDS. **Activities:** WH Natural Strategy Workshop.	**14.1**Ensure that the WH Centre continues its analyses of statistics thus supporting the development of evidence based WH **protection programmes** that implement the proposed *revised Policy Document on the Impacts of Climate Change on World Heritage Properties*, see recommendation 15.2	***Number of States Parties per world nations*** *142/201*		Encourage ratifications through activities of UNESCO Category 2 Centres, worldwide workshops and capacity building programmes: 'Focus on Africa.'	**14.5** Continue to initiate programmes that **encourage ratification** of the ICHC; nominations from 'developing' SPs which have one inscription or no inscriptions; and promotion of opportunities for multinational, serial and trans-boundary inscriptions. **14.6** Propose addressing **marginality** by formalising a strategy to address imbalances within and between the 'Lists'. It should take into consideration human development, climate change vulnerability, size, geographic placement, remoteness, (plays a role in multinational nominations), educational attainment and information technology capacity.
	Number of Inscriptions on Represent-ative List *936* *(725cultural, 183 natural, 28 mixed)*	OGs for **nominations** now address climate change. OGs paragraph 54,57.	**Global Strategy for Representative, Balanced & Credible WH List,** 1994 to address imbalance of: category (natural /cultural), geographic and human developmental representation.	**14.2** Ensure that the WH Committee continues to encourage and **assist nominations** especially from SPs with zero inscriptions as inscribed properties have the best chance to benefit from WH protection	***Number of Inscriptions on Represent-ative List*** *232* *Article 16*	SPs submit Periodic reports on status of ICH elements every 6 years.	ICH Committee notes imbalance in inscriptions per List and in inscriptions per SP and in inscriptions per region. Addresses by limiting 3 nominations per SP per round and to favour 1st time inscribers. Decision 4COM19, 4COM1SUB/6 to limit 100/round.	**14.7** Ensure UNESCO continues **capacity building** programmes for disadvantaged States Parties. This should assist listing which is a proven safeguarding action, and in time may prove to be a powerful protective action against the impact of climate change. **14.8** Continue to provide technical support and expertise to **assist ratification**, implementation and translation of the ICHC into national policies and guidelines.
	Number on	OGs 179-181 Endangered	**New categories** introduced: cultural landscapes, itineraries &		***Number on 'In Urgent***	States Parties submit Periodic	Committee claims this List as 'the most important list'.	**Gap in knowledge** relating to

	'In Danger' List *35*	status due to climate change can be recognised.	industrial heritage. New Science and technology category proposed.		***need of safeguard-ing' List*** *27* *Article 17*	reports on status of 'in urgent need' ICH elements per 4 yrs.	Work thru. Cat. 2 Centres to promote programmes.	innovative ways to address imbalances especially between the Lists.
	States Parties with Tentative Lists *167*	OGs Annex 2. OGs Paragraph 54, 57.	Promote through Intersectoral Platform on SIDS.		***No tentative list***	n/a	n/a	
	States Parties with inscriptions *148*			**14.3** Support the WH Committee's continued facilitation of nominations. Adopt Cherif Khaznadar's recommendation of tying the nomination of an element from a country of the North for inclusion in the Representative List to that country's sponsorship of the inclusion of an element from a country of the South on the Urgent Safeguarding List.	***States Parties with inscriptions*** *87*		Disappointment in representative and geographic imbalance and lack of *urgent* nominations. Will target SIDS /Africa	**14.9** Use the **language**(s) native to the State Party in assisting the State Party to implement the ICHC. **14.10** Consider instigating a **'Tentative Urgent Safeguarding List'** for ICH to provide easier access for inexperienced States Parties.
	States Parties with zero inscriptions *35*		Programmes initiated to address zero status: 2003 SIDS, WH Intersectoral Platform- SIDS 2008. *Manual: Preparing World Heritage Nominations* published online 2010 to upgrade assistance.	It is recommended that Cherif Khaznadar's suggestion for **North-South and South-South partnerships** be applied to WH nominations, taking into account the similar views of UNESCO experts about increasing the nominations from underrepresented States Parties, directed assistance to SIDS and LDCs and the pivotal need for information exchange and cooperation in the face of climate change.	***States Parties with zero inscriptions*** *55*		4th ICH Committee said no time to encourage zero inscription status SPs because of large number of nominations. Suggestion for a north SP with nomination 'adopts' south SP which has an 'in urgent need' nomination.	**14.11** Promote Cherif Khaznadar's recommendation of "tying the nomination of an element from a country of the **North** for inclusion in the Representative List to that country's sponsorship of the inclusion of an element from a country of the **South** on the Urgent Safeguarding List". This **'tied' relationship** may lead to greater empathy with and responsibility for the insecurity of ICH of climate change vulnerable States Parties. **14.12** Propose making **universal ratification** by the 10th anniversary of the ICHC in 2013 a goal.
	Number of inscriptions limit per round *45*			**14.4** Encourage WH authorities to use	***Number of inscriptions limit per round*** *100*		Preference given to SPs with zero inscriptions, deferred nominations and multinational	**14.13** Include programmes to encourage the nomination of ICH elements in the UNESCO **longer term strategies** such as 'Priority Africa', SIDS and Pacific Century

				their statistics to support the proposed worldwide **heritage site vulnerability campaign** (see 15.4).			nominations.	Premium Development (PCPD) countries.
	Number of nominat-ions per SP per round=2 (at least 1 to be a 'natural' property)		'Suzhou Decision' offers better chance to underrepresented SPs to achieve successful nominations. **New manual** for writing Nominations.		***Number of nominat-ions per SP per round*** *3*	Operational Directives (Paragraph 20 amendment) .	Decision 4.COM 19; 4COM 1SUB/6.	**14.14** Encourage SPs to **use the Practices and Programmes** from the ICH List 3.
	List of Good Practices	*not offered*	n/a	n/a	***List of Good Practices*** *8*		Disappointment at lack of nominations.	
To what extent are WH cultural properties and ICH elements protected and safeguarded by the WHC and ICHC from the ***physical and biological impacts of climate change?*** *Impacts include temperature rise and increased humidity; extreme weather events (storms, tropical cyclones, floods, storm surges, coastal erosion, drought, desertification,*	*Articles 11.4 used to list 'WH in danger' for 'serious and specific dangers'* *Articles 4, 5, 29 offer opportun-ities to introduce wider climate change protective*	*Use Reactive Monitoring to monitor climate change through 'state of conser-vation' reports. OGs Paragraph 169-176.* *Could use climate change information from Reinforced Monitoring, Decision 31COM 5.2, and 6 yearly Periodic Reporting,*	Decision 29: Survey of States Parties 2005 re: assessing climate change damage to WH sites. Decision 29 7B, a, Decision 30COM7.1, Decision 31COM7.1 re: climate change management. Decision 30COM 7.1, 14 re: 'case by case basis' for deciding climate change inclusion 'in Danger' listing. Decision 32 COM7A.32 re: alterations to OGs to upgrade climate change threat. Decision 33COM 7C and	**15.1** Urgently adopt **Resolution 16 GA 10 A**. requesting "**…review and update** *the Policy Document on the Impacts of Climate Change on WH Properties* & related documents, so as to make available **the most current knowledge and technology** on the subject to guide the Decisions/ actions of the WH community." **15.2 Strengthen and broaden the scope of the 2007 *Policy Document on the Impacts of Climate Change on WH Properties*** (see 15.1) to address predicted climate change impacts on WH over the short, medium and long term and include positions on mitigation, adaption, climate justice and ethics. It should also advise on **amendments to the OGs** in order to implement the 'revised' *Policy document* especially in relation to the criteria for listing on the **List of World Heritage in Danger**. **15.3** Propose developing a **climate change vulnerability scale** with a	*Article 17 ICH in urgent need of safe-guarding. Could use Article 8.3 to appoint a science advisor and to form a* ***committee*** *to evaluate WH-climate change policy, survey SPs re: climate*	Use ODs 1. 33-34.89. 102. 105. 107.160-164 for ICH in 'Urgent need of safeguarding.'	**Gap in addressing** physical and biological impacts on ICH. Consider adding 'climate change' to the **criteria for selection**, (ODs 52, a, i) on safeguarding and in subsection (i), 'The programme, project or activity is primarily applicable to the particular needs of developing countries'.	**15.9** Propose writing an ***Intangible Cultural Heritage climate change response and policy document*** compatible with the Climate Change Initiative. It should adopt a HR approach, include a climate change strategy and DRM and policy implementation options. **15.10** Consider the following to achieve this: **study the WHC's response to the threat of climate change** e.g. the WHC's 'case-by-case' approach to 'in Danger' Listing; the WH position regarding CBDR; lack of formal recognition of differential impact of climate change on States Parties due to geographic position; appropriateness or otherwise of the WHC's lack of obligatory requirements supporting climate change mitigation. Gain a greater understanding of the impacts of climate change on ICH by reading reports, by surveying holders of ICH and by reference to academic articles eg. the IJIH. Propose initiating **engagement** with Mali

heatwaves, bushfires); fresh water stress (shortage , contamination); poleward shift of climate zones; acidification of oceans; extinctions including coral bleaching	*measures.*	*OGs 199-207, Annex 7, to determine best protection manage-ment, plan.*	Decision 34 COM 7C re: 'state of conservation' reports and remote monitoring of WH sites for climate change-induced impacts. 2009 UNESCO 'Climate Change Initiative' re: WH sites as reference for climate change monitoring, impacts, adaptation and mitigation strategies.	matching set of **climate change crisis symbols** to raise awareness of the link between ineffective GHG emissions control and the survival of the OUV of WH Sites. **15.4** Propose launching a worldwide campaign **– a heritage site vulnerability campaign** employing the climate change vulnerability scale and climate change crisis symbols - to introduce **the 'reviewed and updated' World Heritage position on climate change** (refer 15.1) **15.5** Propose developing a ***WH:ICH:CC Database & Network*** to be a clearinghouse for climate change information in relation to both natural/ cultural heritage and inclusive of impacts on custodians. Link to UCCAF, UNFCCC, IPCC, ICOMOS, ICCROM, UNDP. It should have a **searchable database that accepts and integrates multiple input variables to provide a multifaceted response.**	*change impacts on ICH and oversee 'expert meeting' Use Article 1, 7,13, 19, 29 to support evaluation, survey, analysis, policy, manage-ment reporting and monitoring.*	**Gap in directives** about how to address the physical and biological impacts' on ICH through ODs.	Consider amending ODs in the nominating examiners' report where 'environmental transformation' (OD 7) could be elaborated and climate change could be detailed as a risk factor; and in paragraph 30 relating to the transfer of elements from one list to another for safeguarding reasons.	Voi and other disaffected ICH owners. Encourage involvement, consultation, partnerships and **participation** in developing *an **Intangible Cultural Heritage climate change response and policy document.*** Consider initiating a **questionnaire survey** of States Parties incl. effects of climate change on poverty, ill-health and migration on cultural custodians, their communities, the element's practice, and its transmission and inventorying. **15.11**Publicise the ICHC's role in providing financial support and expert assistance to disadvantaged States Parties.
	WH 2005 survey results: ***Threat sites*** *hurricane 11* *storm 9* *lightening 8* *SLRise 9* *erosion 8* *flooding 7* *rainfall incr. 4* *drought 3* *desertification 2* *temp rise 1*	Nomination form 4b (ii) addresses climate change threat. Decision 32COM 7A 32 to amend OGs (179,180, 181) to upgrade climate change threat. OGs 26, 32,	*Meeting of Experts, 2006, led to management response in the following <u>publications</u>:* ***1)Report on Predicting and Managing the Effects of Climate Change on World Heritage), 2)Strategy to Assist States Parties to Implement Appropriate Management Responses*** *and* ***3)Policy Document on the Impacts of Climate Change on***	**15.6** Review the Nairobi Work Programme's **'coping strategies' portal** to assess suitability of adopting a similar service for WH practitioners for incorporation into the ***WH:ICH:CC Database & Network.*** **15.7** Consider revising the six year Periodic Reporting regime to a four year simplified process including	*Article 19,20,21, 29, 30, Reporting - allows gathering of climate change data and assessing*	Periodic Reporting and OD 89 offer mechanism to implement new knowledge about safeguarding measures to	Gap in addressing physical and biological impacts on ICH.	**15.12** Consider establishing a Wikipedia style ***Global Database of ICH Events.*** Develop into a computer application (app). **15.13** Consider obliging States Parties to adopt, implement and report on climate change **management plans** requiring specific climate change information. **15.14** Consider amending *Operational Directives* **to include climate change threats** when nominating ICH. **15.15** Consider the **inclusion of climate change management for 3rd ICHC 'Practices' Listing.** **15.16** Be mindful of the challenges of **remaining true to the ICHC.**

	Articles 4, 5 and 29. Articles 5 and 6.	33, 74-66, 111,211-222, Annex 6, 7 are opportun-ities to address climate change through capacity building, awareness raising & education.	***World Heritage Properties*** and prioritising Africa and SIDS in all climate change actions. Book: *Case Studies on Climate Change and World Heritage*. 2007. Online 2010 manual: *Preparing World Heritage Nominations.*	specific climate change reporting. This may help facilitate the monitoring of threatened sites and give the plight of **such sites regular publicity.** **15.8** Encourage the WH Committee to review and support the specific **research priorities** for cultural heritage described in the *Policy* and insert 15.5.	*vulner-ability for global and local applications.*	address climate change impacts on ICH.		**15.17** Consider the following ICH **data opportunities** in WHC's *Operational Guidelines*, Annex 7, 11.5, the WH Periodic Reporting (section II). **15.18** Consider using the following **reporting and inventorying opportunities.** Insert 15.7. **15.19** Support the development of the proposed **World Heritage and Intangible Cultural Heritage Climate Change database and Network** (WH:ICH:CC Database & Network).
*To what extents are WH cultural properties protected, and ICH elements safeguarded, by the WHC and ICHC from the **'human impacts'** of climate change?* *They include: loss of land; decline in farming, fishing, forestry and tourism leading to livelihood insecurity; shortages of food and fresh water leading to famine, poverty exacerbation; migration; health deterioration; potential insecurity (conflict); social*	*Articles 4,5 and 29 offer*	**Gap in guidelines** about how to address the 'human impacts' on WH through OGs.	**Gap in addressing** climate change impacts on humanity and climate driven 'human impacts' on cultural properties. Decision 33COM 7c calls for	16.1, .2, 7 &.8 apply to the WHC. 16.3, 4, 5, 6 Apply to both the WHC & ICHC. **16.1** Urge rating climate change as a **top priority.** **16.2** Further to **15.2**, consider ensuring -that 'human impacts' on cultural heritage sites are addressed in the ***revised Policy Document on the Impacts of Climate Change on WH Properties***, that they are **socially inclusive** and that **community engagement** is included in policy planning and implementation, especially where indigenous communities/ minority groups have WH responsibilities. **16.7** Continue assistance for regional level **capacity building** and **awareness raising** workshops including the use of web based communication such as *Climate Frontlines* **16.8** Exploit UNESCO's WH status to **attract world attention** to: the 'silent crisis' of the climate change impacts on humanity; the disconnect of the developing communities at the frontline of climate change from those	*No specific references.* *Use Articles 1, 2.3,7, 11,12,13, 14,15, 16, 17,18 and Preamble paragraphs 4,7,and 13 to support actions to safeguard ICH from impacts of climate change.* *Use Article 8.3 to inform the*	**Gap in**	**Gap in**	16.1, .2, 7 &.8 apply to the WHC. 16.3,.4, 5,.6 Apply to both the WHC & ICHC **16.6** Consider addressing the following at the above joint meeting: current wellbeing of the traditional custodians; the implications of continuing climate change on cultural heritage 'owners'; the role of climate justice in reducing the human impacts of climate change; adaptive interventions to cope with 'human impacts' of climate change on WH sites e.g. capacity building, retention of traditional knowledge, knowledge of climate change science and resilience in WH protection. *Are there lessons to be learnt from the Nairobi Work Programme? Are there guidelines to be heeded about policy implementation from 'common pool resource management'? lessons to be learnt about **sustainability** from observations on the survival and collapse of past and present traditional cultures.* Consider enlightened and resilient present day traditional societies e.g. Ladakh, Bhutan, the Kwakiutl of the Pacific NW, other traditional societies with strategies **for managing local**

and cultural fragmentation; damage, deterioration and possible loss of culture; loss of social cohesion, dignity, sense of place and identity; and for some the ultimate loss – the loss of nation leading to - 'statelessness.'	*opportunities to introduce climate change protective measures*	OGs Annex 7 and Periodic Reporting Section II, re: minor opportun-ities for reporting 'human impacts'.	consistent approach using *Predicting and Managing the Effects of climate change on World Heritage* and *A Strategy to Assist States Parties to Implement Appropriate Management Responses.* and *Policy Document on the Impacts of Climate Change on World Heritage Properties.*	in industrialised communities who are at present relatively buffered and comfortable; and the greenwashing practices of many powerful western States Parties. **16.3** Consider having the WH and ICH Committees host a **joint meeting** of representatives and experts to address potential *'human impacts'* of climate change on both WH and ICH to contribute to the *Policy document on the impacts of climate change on WH properties*, the proposed *ICH climate change response and policy document and perhaps a joint climate change policy.* **16.4** Preliminary to the joint meeting consider **surveying SPs** about impacts of climate change and the consequential **'human impacts' on their cultural heritages**. The WH Committee has acknowledged that human impacts will overwhelm the physical and biological impacts of climate change on WH sites but has little knowledge on this. **16.5** Act on the results of the survey to **inform** the joint meeting. Provide invited experts with the latest climate change science. Aim to predict climate change impacts on humanity and the 'human impacts' on heritage, aware of the urgency of the situation.	***meeting of the ICH interested parties, practit-ioners and other 'experts** (Recomm-endation 15.18) to develop, write and **disseminate** proposed **ICH climate change response document.***	**directives** about how to address the 'human impacts' on ICH through ODs.	**addressing** climate change impacts on humanity and climate driven 'human impacts' on ICH.	**'commons'** and emerging stories of successful and unsuccessful strategies for addressing climate change impacts. *How can communities, mindful that some WH cultural sites are **community based**, meet the **health** challenges including mental health challenges of an unrelenting climate compromised world?* Consider possible resurgence of religiosity; fatalism; extremism and emergence of iconoclasts and zealots; **migration** and its impacts on WH sites, for instance, options for conserving cultural heritage (including cultural memory) prior to or following fragmentation or displacement of communities and even whole SPs. Consider lessons from exiled cultures such as the Tibetan culture .The question of cultural heritage protection and **loss**. *How to decide what to let go? At what cost?* Consider the contributions of **'holism'**, **Gaia theory and the Mother Earth** Movement. Consider innovations in organisational structures relating to possible improvement in implementation of climate change related policy/management plans e.g. **'nested' organisation.**
Do the WHC and ICHC address the effects of climate-induced human displacement? How do they address	No reference. Articles 4, 5	**Gap in guidelines** about how to address climate	**Gap in addressing** migration, and climate-induced-displace-	**17.1** Include all forms of 'displacement' **(voluntary, forced, internal & international) as a risk factor** in nomination, Periodic Reporting and monitoring processes. Consider the consequences of displacement on WH sites e.g. the gradual abandonment of the site by custodians, the severing of spiritual links, the mental health of the absent displaced custodians, the lack of	*No reference.*	**Gap in directives** about how to address climate	**Gap in addressing** migration, and climate-	**17.4** Raise awareness of the impact **climate change-induced migration** is and will have on intangible heritage. **17.5** Address **'gaps in knowledge'** by researching the effects of climate change induced displacement on ICH. Use the gained knowledge to inform the proposed ***Intangible Cultural Heritage Climate Change Response and Policy Document.*** **17.6** Make climate change-induced

climate change induced abandonment of cultural heritage through abandonment of homeland caused by e.g. SLR & subsequent potential loss of cultural memory?	and 6 address protection. Displacement is a factor in protection.	induced displace-ment and its impacts on cultural heritage through OGs.	ment and the consequent impacts on cultural World Heritage.	maintenance staff, and the potential occupation of site by non-related people. **17.2** Prioritise studies of potential impacts on WH sites from climate-induced displacement. Engage with traditional WH site custodians and indigenous 'owners'. **17.3** Support migration policies sympathetic to WH implications of displacement, driven by HRs needs rather than how new migrants may fit within SPs migration programmes.	*Use Preamble and Articles 2, 7, 13, 14 and 29.*	induced displace-ment and its impacts on ICH through ODs.	induced-displacement and the consequent impacts on ICH.	displacement a **priority** consideration in the safeguarding of ICH. **17.7** Support a **migration policy as** for WH. New Zealand has a settlement visa. While climate change was the impetus behind the programme, eligibility is based on age and language criteria. This limits effectiveness of that programme to address the HR and ICH needs of those impacted by climate change.
How do the WHC/ICHC address climate change-induced ***health*** *impacts on custodians & community?*	No reference. Articles 4, 5, and 6 address protection. **Health is a factor in protection.**	**Gap in guidelines** re: using WHC to protect WH from climate change induced health impacts through OGs and Periodic Reporting.	**Gap in knowledge** about addressing climate change induced health impacts on WH.	**18.1 – 18.2 APPLY TO WHC AND ICHC** **18.1** Frame 'health' in all WH and ICH policies, documents, guidelines, directives and reporting mechanisms as in the **Universal Declaration of Human Rights (UDHR)**: Article 25 "Everyone has the right to a standard of living adequate for the health and well-being of himself and of his family, including food, clothing, housing and medical care and necessary social services, and the right to security in the event of unemployment, sickness, disability, widowhood, old age or other lack of livelihood in circumstances beyond his control" and .(continued in far column).	*Use Preamble and Article 2 which includes commit-ment to UDHR.* *Address health as a HR in relation to WH and climate change impacts.*	**Gap in directives** about how to safeguard ICH from climate change induced health impacts through ODs.	**Gap in knowledge** about addressing climate change induced health impacts on ICH.	**18.1 – 18.2 APPLY TO WHC AND ICHC** Article 12(a) of the **International Covenant on Economic, Social and Cultural Rights (ICESCR)** which recognises the right of everyone to 'the enjoyment of the highest standard of physical and mental health'. **18.2** Encourage research into climate change induced health impacts, including mental health impacts such as solastalgia on WH and ICH to address gaps in knowledge.
How are the WHC & ICHC addressing climate change ***vulnerability and risk status of*** *cultural WH and ICH? Are developing SPs*		OGs paragraphs 118, 161, 162, 181, 241	Decision 28COM 10B, 2004 recommends SPs include risk preparedness in WH site management plans and training strategies. Decision 31 COM 7.1 requests specific	**19.1** Encourage SPs to consider their **climate change vulnerability** and to subsequently initiate major site recording programs targeting those places at risk of loss or major damage, and for which mitigation cannot be sufficient. **19.2** Acknowledge the **differential climate change impact** on WH related to **geographic location.** **19.3** Support policies that favour nominations from the **more vulnerable**		**Gap in directives** about how to address climate change		**19.8** Recognise Inscription rate success for smaller SPs. **19.9** Preferentially encourage ICH proposals from acute, severe or high vulnerability climate change impacted SPs. **19.10** Continue to preferentially assist those SPs with small populations and low areas which appear to be **disadvantaged** as far as the number of inscribed elements held.

marginalised with regard to climate change protection of their cultural heritage? *Do the WHC & ICHC address & influence* ***geographic balance & equity*** *in addressing climate change protection and safeguarding of cultural heritage?* *Is there a bias towards protecting the heritage of SPs in the* ***northern latitudes?*** *Is the heritage of SIDS* ***under-represented?*** *Do WHC & ICHC address* ***human development balance & equity*** *in addressing climate change protection and safeguarding of cultural heritage?* *Are developing SPs disadvantaged as regards having their heritage protected?*	*Article 11.4, 'List of WH in danger.'* *Article 11.5 offers: opportunity for climate change criteria definition for inclusion in 'in Danger' List.* *Articles 4, 5, 6, 11 commit to protection of WH and SP heritage.*	and Annex 2, 4B9(iii), Annex 7 re: disaster management. Expectation of OGs to have climate change vulnerability criteria in future. Decision 31 COM 7.1 urges DRM link with policy, OGs 4B (iii) include risk in nomination process. Periodic Reporting II.5 of risk.	climate change criteria relevant to vulnerability and risk status. Decision 31 COM 7.1 urges WH community 'to integrate actions pertaining to climate change risk preparedness into policies and action plans'. DRM is factored into WH climate change policy. DRM is factored into adaptation projects but offers part protection. Published *A Strategy to Assist States Parties to Implement Appropriate Management Responses for Reducing Risks and Disasters at WH Sites 2006.* *Online manual: Managing Disaster Risks for World Heritage. 2010.*	SPs that lack any inscriptions. **19.4 Reaffirm the preamble to the WHC, paragraph 8** 'Considering that, in view of the magnitude and gravity of the new dangers threatening them, it is incumbent on the international community as a whole to participate in the protection of the cultural and natural heritage of outstanding universal value, by the granting of collective assistance which, although not taking the place of action by the State concerned, will serve as an efficient complement thereto' **to raise awareness of the danger posed by climate change to WH and of our collective responsibility** to our 'WH commons' to protect it by working through the WHC for world climate change mitigation and a global acceleration to renewables. **19.5** Amend recommendation 15.7 to read: Develop 'standards, protocols, indicators and databases within the field of cultural heritage and climate change' **with particular emphasis on disadvantaged states parties that are especially vulnerable to climate change impacts** on built heritage and indigenous heritage. **These could be used to prioritise affirmative action when addressing the imbalance of geographic representation, development marginality and equity.** **19.6** Address marginality through further **capacity building,** practical assistance, expert advice and funding. Consider the resonance that the **Mother Earth movement** is having in the developing world. Are there opportunities for its strengths to be incorporated into addressing marginality? **19.7** Achieve **a positive inscription** status for all States Parties.	*No specific references.* *Use Articles 7, 13, 14, 29 and ODs.*	vulner-ability and risk. impacts on ICH through the ODs. Decision 4.COM 19; 4COM 1SUB/6 to amend ODs (Paragraph 20 amendment) to limit nominations per round. **Gap in directives** about how to further address imbalances between Lists through ODs.	**Gap in knowledge** about climate change vulnerability. vulnerability assessments and ICH. **Gap in knowledge** about climate change DRM and ICH. ICH Committee targeting SIDS and 'Focus on Africa.'	**19.11** Formulate intangible heritage climate change vulnerability assessment **tools** for inclusion in the proposed *Intangible Cultural Heritage Climate Change Response and Policy Document*. Vulnerability assessment research should invite full community **participation.** As this is a relatively **new area of inquiry** participants should be aware that current knowledge is limited and will benefit from interdisplinary contributions. A challenge will be, as with the nomination processes, to prioritise **community engagement** and make the vulnerability assessment tools **accessible** at village level. **19.12** Consider designing an online **joint WH and ICH climate change toolkit** that might incorporate the *Framework for addressing climate change impact'* (Chapter 28) and tables - *Assessing Intangible Cultural Heritage Vulnerability to Climate Change*, insert 30.1; *Intangible Cultural Heritage Risk Assessments* insert 30.2 and *Tangible Cultural Heritage Risk Assessments* at insert 30.3. Consider incorporating the 'toolkit' at WH nomination stage, in OGs, ODs and Periodic Reporting. Propose including Vulnerability Assessment Case studies of cultural heritage that have associated intangible heritage elements for other SPs to study and use, Such **case studies could be likened to List 3** of the ICHC-*Practices and process of safeguarding Intangible Cultural Heritage.* **19.14** Be aware of the **complexity** of vulnerability processes. Reviews of vulnerability 'tools' may be useful in determining their merits and demerits. **19.13** Support new manual: *Managing Disaster Risks for World Heritage*. Consider making additions to future editions to include ICH.

*Do the WHC and ICHC have the capacity for a multilevel **mitigation responsibility** and **mitigation mechanisms** (including **monitoring and reporting**) for addressing climate change impacts on cultural heritage?*	*Articles 5, 11.7, and 29.*	OGs 132, IV.A and V Annex 7. 'State of conser-vation' in Periodic Reporting. OGs 169-176, 199-210. Reactive Monitoring, Reinforced Monitoring, Periodic Reporting.	Decision 33COM7c calls for consistent approach to climate change management *using Predicting and Managing the Effects of climate change on World Heritage and A Strategy to Assist States Parties to Implement Appropriate Management Responses and Policy Document on the Impacts of Climate Change on World Heritage Properties.* UNESCO's 2009 'Climate Change Initiative' will use iconic WH sites to research mitigation strategies.	**20.1** Recognising that the WH Committee and its climate change publications have repeatedly called for **synergy with the UNFCCC** and taking into consideration the difficulties facilitating this, namely that the UNFCCC and the WHC share no single authority that can make Decisions about their work, and they have no cabinet through which to coordinate and acknowledging that **cooperative action is not occurring** propose that the WH Committee **take mitigation action** at the State Party and global levels by adopting **Resolution 16 GA 10 A** (15.1) and associated recommendations from Chapters 15, 20.1, 20.3, 21.6, 24.8, 25.2. **20.2** Support and publicise collaborative WH-related climate change mitigation programmes **between developed and developing** States Parties.	*No reference.* *Use Articles 7, 13, 29 and ODs Articles 12 and 29.*	**Gap In specific directives** about how to address mitigation and ICH: and monitoring and reporting on climate change impacted ICH through ODs. (Paragraphs 1-7, 79-99, 100-123 and 151-164).	**Gap in knowledge** relating to climate change impacts, mitigation, and ICH. **Gap in addressing** monitoring and reporting on climate change preventative measures to safeguard ICH.	**20.3** Propose that the ICH Comm. join with WH Comm. to develop ***a joint WH and ICH Climate Change Policy*** which would include a strong official stance on mitigation at the international and national levels, see 15.2, 15.21, 20.1, 20.5, 21.6, 24.8, 25.2 and 26.1. **20.4** Integrate **traditional knowledge** into an emerging **ICH body of knowledge** on climate change and its mitigation. **20.5 Monitor and report** climate change impacts to safeguard intangible cultural heritage. Presently reporting opportunities exist in the Reporting Forms under 'assessment of the element's viability and current risks'. Climate change impacts could be recorded here but **specific climate change information** seeking *ODs* are preferred as they enable better standardised data gathering. **20.6** Be prepared, when monitoring and reporting, **to address questions about how to record the 'changing' nature of intangible heritage.** **20.7** Be prepared also to suggest new **ways of documenting** a 'practice,' 'performance' or 'custom' to climate–protect it for survival. **20.8** Consider developing a self monitoring periodic **reporting** form for intangible cultural heritage (similar to that for WH) that factors in potential climate change impacts with input from community **custodians.**
Do the WHC and the ICHC	*No reference.* *Articles 4, 5, 6, 11 and 29*		Decision 33COM 7c calls for consistent approach to climate change management referring to: *Predicting and Managing the Effects of climate change on World Heritage and*	**21.1** Propose that climate change adaptation in WH management be **prioritised.** Incl. as a major consideration in WHC implementation. **21.2** Disseminate and **exchange** up-to-date **climate change knowledge** with stakeholders, including indigenous communities and minority groups prior to developing adaptive management policy and projects.	*No*	**Gap in directives** about how	**Gap in knowledge** relating to ICH and	**21.6** Suggest that the proposed *Intangible Cultural Heritage Climate Change Response and Policy document* (15.15) includes the safeguarding actions of **adaptation, management,** monitoring and reporting of climate change impacts and responses. **21.7** Be aware that the vulnerability assessment knowledge needed to make

have the capacity for a multilevel ***adaption*** *responsibility and mechanisms, such as maintenance, repair expertise, and financial resources to address climate change impacts on cultural heritage?*	*offer opportun-ities to introduce corrective measures through commit-ment to protection of WH and State Party heritage.*	OGs Nomination 132, Annex 5, Funding 239c, Annex 7 Periodic Reporting II.4.	*A Strategy to Assist States Parties to Implement Appropriate Management Responses* and *Policy Document on the Impacts of Climate Change on World Heritage Properties.* UNESCO's **Climate Change Initiative** recognises adaptative knowledge held by small islands, rural & indigenous communities. **Gap in knowledge** over adapting to potential loss of culture and choosing what to let go.	**21.3** Ensure the dissemination process is **culturally appropriate e.g.** climate change science can be meaningfully interpreted through a Buddhist perspective. **21.4** In consideration of 15.2 and as a consequence, **propose revisions** of the **report** *Predicting and Managing the Effects of Climate Change on World Heritage* and publication *A Strategy to Assist States Parties to Implement Appropriate Management Responses.* Add to the revisions a more detailed evaluation of climate change assessment options (incl. assessment of physical, biological and 'human impacts'), advice on technical assistance for assessments and illustrated examples of cultural WH site assessments. **21.5** Adopt the approach of the **Yamato Declaration on Integrated Approaches for Safeguarding Tangible and ICH.** Extend these to climate change impacts on WH properties and ICH. **Engage** traditional custodians and their communities in these processes.	*reference. Use Articles 1, 2.3, 13, and 29 and Reporting ODs 151-169.*	to address corrective measures including adaptation to impacts of climate change on ICH through ODs.	vulnerability assessment. **Gap in knowledge** about adaptation for addressing climate change impacts on ICH.	adaptation-related Decisions will itself raise **new areas of inquiry** for community leaders, heritage owners and practitioners. **21.8** Be prepared to consider **questions** such as *how do communities make judgements about their changing craft or performance in the light of a confronting and climate compromised future?* *How will communities predict the changing environment in which their ritual or ceremony will be performed?* *How can communities safeguard their intangible heritage when that heritage is vulnerable to change?* *Are there lessons to be learnt from the evolution or demise of past customary practices?* *Which approaches offer better means to assess ICH risk and vulnerability to climate change?*
How do the WHC and ICHC address climate change-induced threats to the ***Sustainable Development (SD)*** *of WH and ICH?* *E.G. how do the*	*No direct reference in WHC.*	OGs–Paragraph 6, 119 re: embracing concept of SD and acknow-ledge the contribution of heritage to sustainable development.	**Gap in addressing** need for WH management to protect SD programmes from climate change.	22.1 Propose making **'culture' the 4th pillar of SD. Support integration of** ***culturally sustainable development*****.** As UNESCO's DG Irina Bokova says 'We need a change of culture to tackle climate change" **22.2** Ensure that climate change is fully integrated into SD to prioritise scientifically established needs of mitigation and maintenance over economic goals by adopting **'climate change resilience' as the 5th pillar of SD.** This will necessitate **a strong definition of sustainability,** one that may be **operationalised** to plan WH development including 'sustainable tourism'. This should focus support on developing and using globally adopted	*Article 2.1 addresses the ICH Convention's*	ODs Paragraph 73 refers to ethical donations to SD.	**Gap in addressing** need for ICH safeguarding management to protect SD programmes from climate change.	**22.9** Include the current and anticipated impacts of climate change on intangible cultural heritage practices and practitioners associated with sustainable developments in the proposed *Intangible Cultural Heritage climate change response and policy document* and in the response requirements of the *Operational Directives.*

Conventions address breakdown in site/element maintenance caused by eco-tourism income loss that is itself caused by climatic extremes affecting both the associated tourist facilities and the traditional community of heritage custodians and/or practitioners and tourism staff?	*Articles 4, 5, and 6 address protection.* *SD is a factor in protection.*	2011 changes to OGs to incorporate **Sustainable Tourism.** **Gap in guidelines** re: climate change impacts on SD related to WH through OGs.	UNESCO supporting SD through **Intersectoral Platform on SIDS.** and contributions to **Mauritius Strategy** for 'the implementation of the programme of action for the SD of SIDS' and through 2009 UNESCO **Climate Change Initiative.**	'**boundary conditions for redesign**' for solutions (not fixes) to attain WH site sustainability and the adoption of scientific management strategies such as the ***Framework for Strategic Sustainable Development (FSSD).*** **22.3** Strongly support the recommendations in ***UNESCO's Input to the Rio+20 Compilation Document*** including WH's role in building 'green' societies & a determined approach and advocacy for breaking with BAU in development. It holds promise for holistic thinking and action leading to truly sustainable management of heritage. **22.4** Strongly support the concept for the proposed UNESCO Rio+20 **'new indicators' for development**, including 'development' in WH properties, especially sustainable tourism. **22.5** Propose that 'new indicators' have a strong environmental advocacy base and include a requirement for measuring the **carbon footprint** of sustainable tourism enterprises and SDs in WH areas. **22.6** Support the showcasing of WH sites that demonstrate appropriate SD, 'best practices' and innovation in conservation. **22.7** Exploit WH's 'green' credentials' to promote the protection of the 936 WH sites through the proposed campaign at 15.4. **22.8** Emphasis **'local' tourism** (involving only short journeys) and **online tourism** as the preferable and responsible forms of 'sustainable tourism'.	*commit-ment to ICH that is 'compatible with… the require-ments …. of SD.'*	**Gap in directives** about how to address climate change impacts on SD related to ICH through ODs.	UNESCO supporting SD through Intersectoral Platform on SIDS and contributions to Mauritius Strategy for 'the implementation of the programme of action for the SD of SIDS.'	**22.10** Support **research** into the safeguarding of ICH and its practitioners associated with sustainable tourism from the impacts of climate change. See **also Rio+20 postscript (Chapter 22)** with four suggested possibilities for the WHC and ICHC relating to the: power of vested interests, the benefits of multidisciplinary advice, local engagement and independently effected protective actions.
			Decision 33COM 7C welcomes the inclusion of illustrative material to educate online	**23.1 – 23.11 APPLY TO WHC AND ICHC** **23.1** Propose that the WH Committee request UNESCO to urge the **IPCC** to maintain an **inter-assessment updating service** that is active between issues of			General Assembly June, 2010, addressed	**23.1 – 23.11 APPLY TO WHC AND ICHC** **23.7** Refer to the ICH well developed knowledge sharing mechanisms and their potential role in climate change capacity building in the proposed ***Intangible***

Do the WHC and ICHC have ***knowledge sharing*** *measures (capacity building, training, awareness raising, education, cooperation and communication programmes, research dissemination, and networking) for addressing climate change impacts on cultural heritage?* *How are SPs informed of the latest science in relation to climate change impacts?*	*Articles 4, 5, 6, 7, 8 and 23.*	OGs paragraphs 212 – 222. **Gap in guidelines** about how to encourage specific knowledge sharing about culture and climate change through OGs.	about WH and calls **for consistent approach to addressing climate change** using books- *Predicting and Managing the Effects of climate change on World Heritage, A Strategy to Assist States Parties to Implement Appropriate Management Responses and Policy Document on the Impacts of Climate Change on World Heritage Properties.* UNESCO's 'Intersectoral Platform on Climate Change' supports 'UNESCO Strategy for Action on Climate Change and its Enhanced Plan of Action' to disseminate climate change knowledge. UNESCO's 2009'Climate Change Initiative' has 'education' as a core programme.	the seven yearly assessment reports. This might prevent the **inappropriate** referencing of IPCC Assessment Reports when they become outdated. **23.2** Include in the proposed *WH:ICH:CC Database& Network* a subsidiary portal to access studies, activities and events relevant to the impacts of **climate change on culture.** **23.3** Advocate **museums take on the** vital role of becoming places to learn about climate change generally and specifically through graphically illustrated local and regional case studies that the museums' natural & cultural history collections can tell. **23.4** Support the WH initiative to raise the 'overall profile of science' (UNESCO 2010l) and the introduction of a new category of WH inscription - **Heritage linked to Science and Technology.** **23.5** Provide authenticated and usable direct **feedback** to the source communities (especially indigenous communities) from which the climate data and information used in planning adaptive interventions is derived. **23.6** Appreciate that learning about climate change should be a **two way process involving specialists and traditional knowledge holders.** On this basis, combine the experience and observations of the traditional holders with the latest scientific information to arrive at the best possible shared knowledge base on which all can act.	Use Preamble and Articles 1, 2, 7, 13, 14, 19 and 29.	**Gap in directives** about how to use ODs- Chapter II 79-89. and Chapter III. 100-123. Chapter 7. 151-164 to address climate change impacts on ICH.	capacity building in developing SPs particularly Africa; training in legislative revision, inventorying, nominating and applying for assistance. Publication of attractive Fact Sheets. **Gap in addressing** how to bring together the findings of the 'new body of knowledge' about climate change and ICH with the full range of community engagement, knowledge sharing, capacity building and dissemination mechanisms of the ICHC.	***Cultural Heritage Climate Change Response and Policy document.*** **23.8** Propose that Category 2 Centres, e.g. the Intangible Cultural Heritage Centre for Asia and the Pacific act to safeguard **regional ICH** against climate change impacts by accessing regionally applicable climate change science knowledge bases; having access to and interpretative help with, climate and weather predictive data including climate maps, GIS information; climate change vulnerability and risk assessment tools, regionally tailored monitoring and reporting processes; and facilitation for cross disciplinary collaborative climate change programmes with other projects such as Local and Indigenous Knowledge Systems (LINKS). **23.9** Propose that **climate change vulnerable States Parties** including SIDS, PCDP and LDCs be assisted to nominate their ICH or **collectively nominate** their shared ICH such as for instance the knowledge of navigation in the Pacific. **23.10** Propose that the accreditations of NGOs (*Operational Directives* paragraph 88) include consideration of **the NGO's climate change qualifications.** **23.11** Endorse recommendation 23.3 in relation to the importance of **museums** in climate change education extending this role to include threats to ICH.
				24.1 - 24.11 APPLY TO WHC AND ICHC **24.1** Progress the development of a **Universal Declaration of Ethical Principles in Relation to Climate Change.** Inform heritage workers about the threats of climate change to tangible and ICH to gain their support in				**24.1 - 24.11 APPLY TO WHC AND ICHC** **24.6** Propose that the *"assistance"* referred to in 24.5 should recognise common but differentiated responsibilities and include **the transfer of relevant technology and adaptive projects** that "protect" from future impacts, recognising that the

Can the WHC and the ICHC address ***climate justice, human rights and cultural rights*** *issues relating to climate change impacts on cultural heritage?* *What roles do human rights law and cultural rights law have in relation to climate change- impacted cultural heritage in these Conventions?* *Consider references made to 'existing international human rights instruments' (ICHC Article 2) and*	*No specific reference.* *Use Preamble paragraph 7, Article 4 and Article 6.3a.* *Article 7 address ' "new dangers", "protect-Ing trans-mission to future generations' ' and "undertak-ing by States Parties not to take any deliberate measures which might damage... cultural*	**Gap in guidelines** about addressing the human contribution. deliberate and innocent. to climate change and consequent damage to WH and .... the UNFCCC equity and climate justice principle implicating a polluting State Party to pay a less polluting State Party to help this State Party adapt according to	The WH books- ***Predicting and Managing the Effects of climate change on World Heritage, Policy Document on the Impacts of Climate Change on World Heritage Properties*** and ***A Strategy to Assist States Parties to Implement Appropriate Management Responses*** advocate cooperation and knowledge sharing with UNFCCC, UDHR, IPCC, CBD, MAB, IOC, IHDP, COMEST about climate change issues. **Gap in addressing**	lobbying UNESCO leaders to progress the above Declaration **24.2** Support the development of a **HR approach** to the impact of climate change on heritage which would incl. the precautionary principle; the need to preserve culture in order to enable participation in such; the principle of sustainability; the principle of integrity as applied to ecosystems and the principle of safeguarding and promoting the interests of **future generations**. **24.3** Encourage the incorporation of the proposed **HR approach** into the revision of the current *Policy Document on the Impacts of Climate Change on World Heritage Properties* (15.2), and in the proposed *Intangible Cultural Heritage climate change Response and Policy document* (15.14) and the provision of **HR criteria** to be taken into account when SPs with inscribed WH or ICH seek resource allocations. **24.4** Ensure that, once reviewed, the current *Policy Document on the Impacts of Climate Change on World Heritage Properties* (15.2) and the proposed *Intangible Cultural Heritage climate change Response and Policy document* (15.14) **champion information regarding the need for climate change mitigation as a human right-to-know** as enshrined in UDHR, Art. 19. **24.5** Propose that WH and ICH Committees adopt stronger official positions **recognising that 'anthropogenic' climate change is threatening accelerating damage to WH sites & some ICH practices** incl. those in the least polluting & most poorly resourced SPs. Recognise that GHG pollution has proven and well known consequences that increasingly make it a "**deliberate measure** which	Use Preamble paragraph 11 and Article 2 and 13 which specifies UDHR and other 'rights' instruments. Use Article 8.3 to appoint a HR advisor to the Committee and to advise the proposed 'Meeting of	**Gap in directives** about how to address climate justice, human rights and cultural rights related to climate change impacts on ICH through ODs.	**Gap in addressing** climate change impacts on ICH including climate justice, and Human Rights and Cultural Rights.	"protection" in this context includes the "safeguarding" of the community's culture and cultural heritage. **24.7** Propose that States Parties be encouraged to develop **mutual obligations** mindful of the "development" and "human rights" needs of World Heritage custodians and intangible cultural heritage practitioners. **24.8** Adopt a human rights approach with respect to **disaster risk management** (DRM). This would address inequities by linking the WH / ICH risk policies to international HR law encompassing all relevant guarantees-civil and political as well as economic, social and cultural rights. **24.9** Integrate the **rights of Mother Earth** - living in harmony with nature - into the revised *Policy Document on the Impacts of Climate Change on World Heritage Properties* and into the proposed *Intangible Cultural Heritage Climate Change Response and Policy Document.* **24.10** Encourage more countries to ratify the two UNESCO Conventions and in so doing further implement appropriate just and ethical - legal, technical, administrative and financial measures and policies to ensure the safeguarding, the development and the promotion of WH and ICH at national and local level, with the participation of their communities, groups and individuals. **24.11** Build a **body of knowledge** by research and information sharing on human rights, ethics and culture in relation to the protection of cultural World Heritage and Intangible Cultural Heritage from climate change impacts.

associated Conventions, Recommend-ations, Covenants and Declarations.	*heritage" and "internat-ional co-operation and assistance" to conserve heritage.*	'common but different-iated responsib-ilities.'	human rights, cultural rights and climate justice in relation to climate change impacts on cultural WH.	might damage directly or indirectly the cultural and natural heritage... situated on the territory of other SPs" or "**threatens**" the purposes of the ICHC and the various measures of 'safeguarding' at national and international level and that it is beholden on the polluting and resourced SPs through "international co-operation and *assistance*" and "collective *assistance*" and "universal will" to support the inscription of WH and ICH in these least resourced SPs as a means to protect, "conserve, and identify that heritage" (WH) and "safeguard" ICH.	practit-ioners and other experts'.			
*How do the WHC and the ICHC address climate change-induced **threats to the security** of WH and ICH? E.g. How do WHC and ICHC address the threats to abuse of cultural heritage as a result of societal breakdown and conflict due to climate change exacerbated poverty?*	*Use Article 11 for immediate "serious & specific dangers" to WH leading to 'in danger.' Listing. Articles 4, 5 & 29 offer opportun-ities for addressing long term security threats from climate change.*	**Gap in guidelines** about how to specifically address long term security threats from climate change through OGs.	**Gap in addressing** long term security threats from climate change. Well managed, holistically planned adaptation projects offer longer term social cohesion and security. Increase in variety of engagement programmes as security insurance.	**25.1-26.4 APPLY TO WHC AND ICHC** **25.1 Appreciate** and learn from the WHC's 40 years of contributions to conservation and to awareness raising for the World's natural and cultural treasures, reflecting on its role in **social cohesion and stability**, and as a vehicle for raising national **identity and pride**. **25.2** Exploit the WHC's reputation of stability and security enhancement (see 25.1) by supporting **the campaign** proposed at 15.4 to draw attention to the vulnerability of WH to climate change.	Use Article 17 for grave threats & "threats to viability' against damage" leading to 'urgent' Listing. Articles 1, 7, 11,12,13,14 19, 29 offer opportun-ities for meeting long term security threats from climate change.	ODs 1.1u, 2a,b, 1.8, 160-164 for immediate threats. Opportun-ities to use reporting require-ments 157-159 for longer term threats. **Gap in Directives** over how to specify long term security threats from climate change.	**Gap in addressing** climate change induced threats to ICH and DRM and medium and long term security.	**25.1-26.4 APPLY TO WHC AND ICHC** **25.3** Appreciate that ICH contributes to **community strength and resilience** as well as international understanding and approbation making it a force for peace. **25.4** Recognise UNESCO, the WHC and the ICHC as global **forces for developing collective responsibility.**
				26.1-26.9 APPLY TO WHC AND ICHC **26.1** Assuming acceptance off **15.1, 15.2, 16.3, 20.1, 20.3, 21.6, 23.8, 24.2,**				**26.1-26.9 APPLY TO WHC AND ICHC** **26.7** Encourage a greater willingness by the WHC **to engage openly, inclusively and**

*How have the **legal powers** of the WHC protected WH from the impacts of climate change?*	*Insert 26.1 summarises over 71 Decisions relating to climate change.*	Climate change is included as a 'threat' in the Nomination Form.	See Insert 26.1: Decisions responding to climate change Petitions. Decisions relating to climate change strategies and policy. Decision relating to amending OGs in relation to climate change.	**24.8, 28.8 & 25.2** and appreciating the leading role the WHC plays in WH **site based protection** and informed by the **adaptability, capacity and integrity** of the WHC and the ICHC and reflecting on more than 60 **WH Committee Decisions related to climate change** tabled in Chapter 26 and noting with concern the **present limitations** to protecting WH properties from climate change impacts and recognising **the strong public interest** in WH, ICH and in their protection from future threats and witnessing the **progressive momentum** building in anticipation of the Rio+20 Conference, propose the development **of a joint** ***World Heritage and Intangible Cultural Heritage Climate Change Policy*** **with a strong position on mitigation.** **26.2** Urge the WH and ICH leaders to move beyond site specific mitigation actions **to national and international level action** by developing **a leading position on global climate change mitigation.** **26.3** Support the adoption of the **mitigation rationale** that: to ensure protection and safeguarding of heritage and the possibility of a sustainable, equitable future for all, concrete action is necessary to stabilise the climate by moving to renewables and reducing carbon emissions to 350 ppm. 350 ppm of CO2 is what many scientists, climate experts, and over 112 national governments say is close to the safe upper limit for CO2 in our atmosphere. **26.4** Urge the proposed joint *WH and ICH Climate Change Policy's* mitigation position to oblige SPs to make **clear commitments** to move to clean energy **by reducing GHG emissions according to the CBDR equity principle**, to make improvements in energy efficiency, and	See immed-iately above for ICHC and immediate grave threats and long term threats.	**Gap in directives** about how to address climate change threats.	**Gap in addressing** climate change induced threats to ICH and in legal protection. The ICHC has not been tested in relation to climate change impacts and safeguarding ICH.	**constructively** in the protection of climate change threatened heritage. **26.8** Assuming the acceptance of 15.2 (revision of WHC Policy document) re-examine the petitions that call for the listing of threatened WH properties to be inscribed on the 'in Danger' list due to the impacts of climate change **26.9** Encourage concerned citizens to **petition the Conventions** about the need to address climate change, the need for increased listing of WH sites on the ' in Danger' List and climate change threatened practices on the ' in Urgent Need' List. This would help combat political influence (especially from powerful polluting SPs) which has constrained the process to date. **26.10** Assuming acceptance of 15.6 & 20.1 & WHC Decisions 29COM 7Ba & 31COM 7.1 continue to **seek stronger ties with the UNFCCC.** It appears that the UNFCCC has made very little progress in negotiations relating to the Petition *State Parties Responsibilities to Reduce Emissions to Ensure the Protection, Conservation and Transmission of World Heritage to Future Generations* (September, 2009) and to the relevance it has for the WHC. **26.11** Noting 26.10, propose **mandatory consideration of UNFCCC advice** in deciding whether to list WH sites on the 'in Danger' List because of climate change. The advice would arise from a formal review of the science presented in petitions to the WH Committee as well as from WH State of Conservation reports. This process would further depoliticise decision making. It would also recognise that climate change judgements increasingly necessitate local, regional and global climate change science knowledge. As climate change impacts intensify, differentiating the part played by anthropogenic climate change becomes more important if compensation is linked to climate change threat or damage.

				to contribute to and/or benefit from universal access to low or no carbon energy services. **26.5** Advocate that the proposed joint *WH and ICH Climate Change Policy* co-ordinates with the proposed **heritage site vulnerability campaign** (using the Climate Change Vulnerability Scale with climate change crisis symbols for display at WH sites as at 15.4 and at ICH practices. **26.6** Uphold the objectives of the ICHC and the WHC that seek to prevent the devaluation of heritage. Failure of wealthy SPs to make ambitious cuts to their GHG constitute a deliberate avoidance of **responsibility that discounts the future** and runs counter to their undertakings as signatories to protect both their and other SPs' tangible and ICH from the impacts of climate change for the benefit of future generations..				**26.12** Invite a representative from the UNFCCC and an international environmental lawyer to advise on the latest climate change science and climate change litigation at all WH and ICH Committee, Conference and Assembly meetings. **26.13** Endorse action upon point 12 *"Further requests the SPs and the WH Centre to work with the IPCC, with the objective of including a specific chapter on WH in future IPCC assessment reports."* of the WHC decision 30COM7.1 re: *The Impacts of climate change on World Heritage properties 2007.*

Chapter 28

Framework for Addressing Climate Change

This framework brings together various factors discussed to date that have a direct bearing on addressing climate change and heritage. The factors are linked to form a process for investigating and assessing climate change impacts on tangible and intangible heritage. The framework is offered here as the template for the case study in the next chapter. Its development has drawn on the work of the Expert Meeting of the *World Heritage Convention on the Impacts of Climate Change on World Heritage 2006* (UNESCO 2006a,g) and the work of May Cassar and colleagues (Cassar 2005).

Integrated framework
For addressing climate change impacts on tangible and intangible cultural heritage.

Background information

Obtain a preliminary understanding of the likely regional physical and biological effects on heritage and effects on humans of climate change in the study area (consider e.g. temperature rise, changes in precipitation, extreme weather events, sea level rise and threats to livelihood, habitation and health possibly leading to displacement). Refer to climate models, climate and heritage experts and social researchers.

Determine likely climate change-related problems affecting heritage protection

Make a broad ranging list of likely climate change-related problems that might bear on the community's tangible and intangible heritage

Hold meetings with community groups, visit sites and survey by questionnaire

Draft a set of questions, each based on a climate change-related problem with each question introduced by background knowledge about the problem and how the problem may develop in the future. Survey heritage bearers, custodians, workers, managers, field officers, local councillors, and any other interested parties responsible for or concerned about heritage on and/or off site to gather informed views, comments, and recommended strategies. Also determine any possible conflicts of interest.

Workshop to consolidate views, recommendations and proposed actions

Hold a workshop of stakeholders to discuss and refine analysed survey responses. Determine main problems and formulate recommendations on further action. Invitees should be representative of the community population (including the disadvantaged) and should include artisans, representatives of community groups, interested individuals and representatives from community council.

Draft a management proposal

Integrate workshop views and recommendations on the protection of the community's tangible and intangible cultural heritage into a draft management proposal for discussion with local policy makers. The management proposal should be considerate of community issues such as population characteristics, socioeconomic factors, livelihood threats, health issues, displacement issues, local justice concerns and any relevant issues that are likely to be contested.

Develop local adaptive plans

Use the discussed outcomes of the above meeting to write local adaptive management plans for climate change sensitive tangible and intangible heritage. These should include climate change vulnerability maps, and advice on preservation strategies. The plans should encourage capacity building, information sharing, mitigation and continuing vulnerability and disaster risk assessments.

Implement plans after their approval by the appropriate authorities

Implement ongoing monitoring

Monitor developments and prepare progress reports for local use and for use by the WHC or ICHC.

Consider nomination for World Heritage and Intangible Cultural Heritage

Use adaptive and protective plans in nominations and if nominations are successful, use in application for funding.

Report to local, regional and national agencies

Respond to Feedback

Part IV

How can the Conventions 'Carry the Fire'?

Chapter 29

Climate Change and Cultural Heritage Case Study

Leh, Ladakh, India

We are still reaching for the sky. In the developed countries people are coming back down, saying, 'It's empty up there.' Gyelong Paldan, at a meeting in Sakti village, 1990. (Norberg-Hodge 2009 p157)

If Ladakh is ever going to be developed we have to figure out how to make these people more greedy. You just can't motivate them otherwise. Development commissioner in Ladakh, 1981 (Norberg-Hodge 2009 p141)

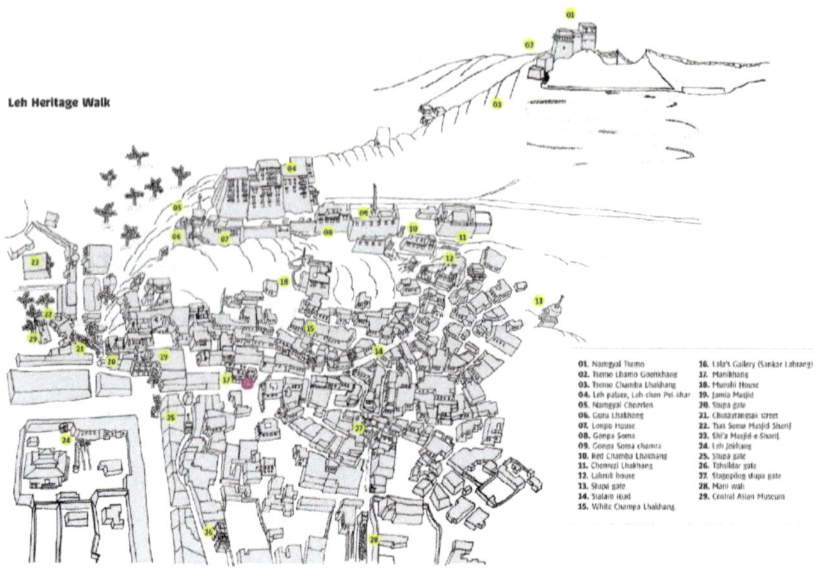

Insert 29.1 Diagrammatic representation of Leh Old Town, with permission from the artist P. de Azevedo; and A. Alexander, Tibet Heritage Fund (de Azevedo 2007; THF 2009)

The short theoretical case study described in this chapter had its genesis as an example in a post-graduate university assignment. It concerns itself with the impacts of climate change on two aspects of Ladakhi cultural heritage, one tangible the other intangible. This it does through the lens of the Integrated Framework described in Chapter 28. The story is developed as if the cultural aspects were contenders for nomination, one on the World Heritage List and the other on the Representative List of the Intangible Cultural Heritage Convention (ICHC). In every sense of the word this study offers an extraordinary, view of climate affected people living in and around Leh, the capital of Ladakh. Ladakh is a remote climate change threatened mountainous region within the State of Jammu and Kashmir of northern India. The town of Leh is home to 30,000 people; Ladakhis, Tibetans, migrants from southern India and seasonal traders from Kashmir.

The 16th Century Leh Palace with the 14th Century Namgyal Tsemo tower, chapel and Tsemo Lhamo Goenkhang (protector temple) perched further up the ridge. Leh Old Town, the world's best remaining example of Tibetan urban architecture, extends down the slope from the Palace to take in the near to level ground below.)

The major part of the study was written in mid 2010. Shortly afterwards, on the 6th August 2010, the city suffered an unprecedented downpour with flash flooding and landslides that took over 300 lives and left 5000 displaced. A postscript that describes the catastrophe has been included.

India is a State Party to both the World Heritage Convention (WHC) and the ICHC. As a consequence if nominations were made and accepted, India would become obligated to protect the accepted Ladakhi cultural heritage site and to safeguard the accepted intangible cultural element. Also, once listed, all other States Parties that are signatories to the WHC would become obligated to support the protection of the accepted Ladakhi cultural heritage site and all those that are signatories to the ICHC would support the safeguarding of the accepted intangible cultural element. In the absence of such listings no such expectations would apply to other States Parties. The Case Study hypothetically proposes that:

Leh Old Town (LOT), Ladakh – be nominated for inscription as a group of buildings on the World Heritage List as a cultural property

and that:

Ladakh traditional building design and construction practice (LTD&CP) - be nominated as an element for inscription on the Representative List of Intangible Cultural Heritage of Humanity.

It should be noted that Leh Old Town was placed on the **World Monuments Watch** in 2008 since which time support has been received from the **World Monuments Fund**.

The winter view, south across the Leh valley, to the Zanskar Range. In contrast to spring and summer the domestic gardens are bare of flowers and vegetables

Ladakh and Leh

Environment, History, People and Culture

Ladakh is a spectacular high, lunar landscaped, cold desert, about 87,000 km^2 in area, ranging in altitude between 2700metres to 4200 metres that is centred on the upper reaches of the Indus River. The region lies between the Karakorum Range in the north east and the Greater Himalayan Range in the south west and includes both the Ladakh and the Zanskar Ranges. The plateau, much of it over 3000 metres, is dramatically defined by the diagonally running parallel mountain ranges that rise to over 7,000 metres, see Insert 30.4. Ladakh 'the land of many high passes' is named such to indicate the significance of these high linkages between the populated valleys. It is a savage environment, a fragile mountain eco-system and a source of some of the subcontinent's major waterways. The climate is extreme characterised by intense sunlight, a high evaporation rate, strong winds and retreating glaciers. The unique cultural heritage is a blend of Tibetan culture, indigenous traditions and influences from the ancient Buddhist regions of Kashmir, and central Asia. It was an important centre at the crossroads of inner Asian trade. Today, most of Ladakh's population of 270,000 live a subsistence village or nomadic life, materially poor but culturally rich.

Nuns, Lamas, Mullahs, NGO workers, cows, cars, merchants, street sellers, local shoppers and students enliven the Main Bazar of Leh in winter. The 17th Century Jamia Masjid (central mosque) stands at the head of the Bazar with the Namgyal Chorten (Stupa), the Guru Lhakhang (Buddhist Shrine), Lonpo House and the Leh Palace (1616-1642) above

Prehistoric inscriptions and rock carvings show that early inhabitants were nomadic yak herders or Changpas from the Tibetan Plateau who in the 1st Century were part of the vast Kushan Empire which originated in south Central Asia - a branch of the Yuezhi confederation of nomadic peoples from Central Asia. Buddhism spread into western Ladakh from Kashmir in the 2nd century. Buddhist pilgrims known as the Mons travelled from India to Mt Kailash in Tibet migrating southeast alongside the Indus Valley, introducing irrigation, agriculture and forming settlements. In the 5th or 6th century, these groups were frequently accompanied by tribes of Indo-Aryan origin known as the Dards.

Insert 29.4 Map showing Leh situated between the Ladakh and Zanskar Ranges, close to the Indus River. Maps below show Ladakh as a region in the Indian State of Jammu and Kashmir. Much of the territory in the northern and western part of Jammu and Kashmir is in dispute and under Pakistani occupation.

The map above can be accessed at: http://en.wikipedia.org/wiki/File:Ladakh2.svg.

Although Buddhism became the dominant religion, Ladakh's religious heritage is part of the ancient shamanistic Bon-Po tradition with roots across the Tien Shen to southern Siberia. Today, a small Brokpa community live in the villages of Dha and Hanu (south west of Leh on the Pakistan border). Though nominally Buddhist they still perform animist and Bon rituals. The Brokpa people who are predominantly Caucasoid in appearance are thought by some to be descendants of the ancient Indo-European Dards.

A Brokpa elder from Dah- Hanu

Main Bazar, Leh, a centuries old trading venue and part of the ancient trading route between India and China. The border was closed in 1959

By the 9th century Ladakh had become a Tibetan kingdom extending all the way from Kashmir to Tibet. The Buddhist kings, descendents of the last Tibetan king, Langdarma, built imposing forts for protection and substantial Buddhist gompas (monasteries) in elevated and spectacular settings throughout the country. When the Islamic conquest swept through south Asia in the 13th century, Ladakh referred to Tibet for religious leadership, exchanging religious scholars. Buddhist sects vied for eminence and eventually the Gelukpa (Red Hat), introduced by the Tibetan pilgrim Tsongkhapa in the 14th century, became the most accepted belief system (Schettler 1981). For nearly two centuries till about 1600, Ladakh was subject to raids and invasions from neighbouring Muslim states, which led to the partial conversion of Ladakhis to Islam. The Baltistan Islamic leader Ali Mir gained control of the province in the 16th century, but Buddhism returned under Senge Namgyal (1570–1642). Famous for his work building monasteries, palaces and shrines King Senge Namgyal restored Ladakh's greatness annexing territories, and establishing Leh as the capital dominated by the Palace he built (1590-1620) (Rizvi 1996). Ladakh was finally annexed into the kingdom of the Dogra Rajas of Jammu in 1846. With the emergence of the East India Company, the whole region was taken over by the British who incorporated Ladakh and the neighbouring Baltistan into the newly created State of Jammu and Kashmir. After the partition of India in 1947 Baltistan became a part of Pakistan, while Ladakh remained in India as part of Jammu and Kashmir.

Partition put an end to trade and even personal contacts between regions to the north and west, effectively locking Ladakh into isolation and bringing its economy to a standstill. The border between Jammu and Kashmir and the Gilgit-Baltistan Area of Pakistan has remained in dispute ever since with conflict flaring in 1948, 1965, 1971 and 1999. On each of these occasions Kargil has been the scene of fighting with on the first occasion raiders coming within 30 km of Leh.

In 1949, China closed the border between Nubra (the northern region of Ladakh) and Sinkiang shutting the 1000-year old trade route from India to Central Asia exacerbating the stagnation of Ladakh's society, culture and economy.

In 1950, China invaded Tibet, and thousands of Tibetans, including the Dalai Lama sought refuge in India. Since the early 1960s the number of immigrants from Tibet (including Changpa nomads) has increased as they fled the occupation of their homeland. China closed the Ladakh-Tibet border in the 1960s ending the 700-year old Ladakh-Tibet relationship and further isolating Ladakh. In 1950, Chinese troops occupied Aksai Chin, which they claimed to be part of China. Their army marched deep into Ladakh building roads to link Tibet with Xinjiang. Though contested by India, the territorial dispute has never been settled. In 1962, China launched a full-scale war on India, with Ladakh as one of the fronts. In the aftermath of this war, Ladakh grew in importance as a military base. Due to its sensitive location, Ladakh remained closed to foreigners till 1974, when it was opened to tourism.

Ladakhis are renowned for their good humour

NGOs and a trickle of tourists started arriving via the Indian Army's road from the south. The road was built in the 1960s to defend borders with China and Pakistan. An air link with Delhi was also opened. Ladakh, and Leh in particular has seen enormous and sudden changes since the 60s with the development of schools, offices, shops and many small family run 'hotels'. In the summer months Ladakh is a popular international tourist destination for both trekking and for visiting ancient Buddhist monasteries. Since about 2000, Ladakh has also become popular with Indian domestic tourists. Leh, the largest town in Ladakh, is also home to a Tibetan community of 3,500. Ladakh became semi autonomous in the 1990s and is governed by separate **Ladakh Autonomous Hill Development Councils (LAHDCs)**, one in the predominantly Buddhist town of Leh the other in Ladakh's second largest and predominantly Shia Muslim town of Kargil. The Buddhist and Moslem traditions co-exist today as they have done over the centuries. Both religions are thriving. The number of Buddhist nuns has increased dramatically; the monasteries have been repainted with the small income from tourists, Mosques are also being built and the Buddhist festivals attract huge local enthusiasm and growing international interest. According to the Dalai Lama, communities with **long term secure religious bases** (Buddhism and Islam) and cultural bases that face demanding and extreme climates in fragile low carrying capacity ecosystems are particularly resilient (Dalai Lama 1990).

Before the Indian Government brought tourism and development to Ladakh in 1974, traditional practices, such as polyandrous marriage and having at least one family member become a nun or monk, maintained a balance between family size and the resource base, leading to low population growth. Subsequently the monetary economy introduced by the Indian Government undermined the Ladakhi relationship with the environment leading to population growth. Helena Norberg-Hodge attributes the population growth to competition between ethnic groups seeking political leverage to enhance their relative power bases (Norberg-Hodge 1994).

Padma Guesthouse, Leh with the vegetable garden in full production. All vegetables and fruit are grown over the short warm season

The Ladakhis are known for their kindness, generosity and their extensive community support systems. They look after and value their old people; they have a rich spiritual life, a relaxed lifestyle, and robust sustainable food producing systems (Trainer 2007). High altitude barley growing, fruit trees and vegetables are dependent on glacial melt for irrigation despite fiercely cold winters and a short growing season. At around 3,000 metres, subsistence farming is a precise practice. The growing season – sowing, harvesting, threshing, storing (mostly of barley) is short before the long cold winter. With increased weather unpredictability there is stress on fruit and crop growing. Humidity affects food storage as well as fabric preservation. The monks depend on the villagers for food. The monasteries and mosques form the spiritual and political hubs of villages. In Buddhist areas the monasteries provide careers that are both respected and help to stabilise population numbers.

> You mean everyone isn't as happy as we are?
> Tsering Dolma
> (Norberg-Hodge 2009 p83)

Agricultural production is labour-intensive, yet the pace of life in general is relaxed, with much time for ceremonies and religious observance. Few are isolated, lonely, or hungry. Ladakhis waste little, recycling almost everything, and most have little interest in power, domination or competition. They are very conscious of their dependence on nature, they are multi-skilled and practical, and they live simply with women enjoying high *status* and relative emancipation. Traditionally there has been very little crime, the people are frugal rather than poor with few drug problems and little social breakdown. Helena Norberg-Hodge, long time resident and founder of the **Ladakh Ecological Development Group (LEDeG)** relates that, in one village, people told her that "there has been no fighting in the village in living memory" (Norberg-Hodge 2009 p46). Above all Ladakhis are described as irrepressibly happy (Smith 2007; Trainer 2007; Norberg-Hodge 2009; Reynolds 1997). "Our vision for progress must be for human progress and not mere economic progress" claims Sonam Wangchuk (Wangchuk 2005b).

Land ownership across the State of Jammu and Kashmir has been restricted by law to remain in the hands of local people. This enlightened policy has protected Ladakhis from naively selling land to less beneficent outsiders and given Ladakhis the greater share of the tourist trade.

Nowshar Street Leh, showing women in traditional dress

Ladakh has had a roller coaster ride into the 21st century. The community led by its religious leaders, Councillors, small business men, and a group of educated professional Ladakhis together with NGOs have catalysed Ladakh's remarkably enlightened transition from a backwater where people were intimidated by 'western' influences, to a culturally confident society that is finding its own balance outside globalising forces. In Leh, traditional clothes, partially replaced in summer by western dress in the 80s and 90s are increasingly worn again. Traditional dress is now universal with women. The society, like many that exist in environmentally extreme areas, is marked by gender equality, lack of strict division of labour, has little hierarchy and near equal distribution of resources. In her classic work *Ancient Futures: Lessons from Ladakh for a Globalizing World*, Helena Norberg-Hodge, a Swedish linguist, traces a 34 year of history of these self-reliant, contented and isolated people who partially dismantled their culture when first challenged by the problems of 'progress'. In 1981, she initiated the establishment of the LEDeG with the aim of empowering Ladakhis to control the direction of the inevitable changes facing the region in a manner that would benefit the people. LEDeG promotes ecological and sustainable development, concentrating on renewable energy particularly solar energy. Some of those who came under the influence of LEDeG became leaders in Ladakh civil society and agitated for the establishment of the LAHDCs. Others formed groups such as The Women's Alliance. Now over 5,000 members strong, this Alliance has an esteemed reputation for working to preserve "the ecological and spiritual foundations of Ladakhi culture" (LEDeG p205) by setting up village womens' empowerment groups. Today there remain problems of pollution and some social dislocation, but overall there has been the re-emergence of Ladakhi self assurance and a cultural strength.

A women's empowerment group visiting SECMOL. Supported by the Women's Alliance, these villagers have come from Lingshed, a day's bus journey and 3 days of walking away

Annual Love and Friendship March for World Peace. This ebullient march is very much an expression of Ladakhis beliefs and is held for its own sake outside the tourist season. On each occasion there are probably more marchers than spectators. Ladakhis are committed to living by the tennants of their faiths and have a good time doing so

Ladakhis have a particularly strong intangible cultural heritage of festivals and religious observances. A strong culture of song pervades all aspects of life. For instance listen to Thukjay online {Thukjay 2004). Indicative of their predilections for idealistic public ceremony is the annual March for World Peace. This is a Ladakhi consciousness-raising and joyous celebration. Schools, social groups, monasteries and nunneries, are all represented. The streets are festooned with prayer flags and streamers and drinks are handed around for both participants and onlookers. The parade is held in the near freezing temperature of a winter day in the capital, Leh. At this time of the year snow has cut road access to the outside world. Although planes still fly several times a week, the city is almost devoid of tourists.

Ladakhis can be spontaneous in addressing matters of deep religious, cultural and social significance. An example comes from the remote Buddhist village of Lingshed. A village meeting was held to discuss the decline in numbers of young men joining the local monastery. In addition to being the spiritual centre of the village, monasteries own land which is worked by the village as a whole and by this means provide social security for the community. If any family is unable to care for their members, it is not unusual for younger sons to become monks and daughters to become nuns. In the monasteries and nunneries monks and nuns are looked after by the community in exchange for religious services (Norberg-Hodge 2009 p79). The Lingshed community, when faced with the prospect of their monastery becoming nonviable, immediately agreed to have 14 of their sons join as novice monks to trial their interest in the monastic life.

With similar pragmatism Ladakhis are managing the problems associated with the rising impact of tourists, tourists coming largely from the west, 527 arriving in 1974, the annual average now being about 30,000. A 1990s Tourist Information 'handout' produced by the state Government Tourism Department is frank in warning of unsafe drinking water. It also asks that unhygienic restaurants be reported. It outlines 'for cultural harmony and understanding' and 'for the sake of mutual respect and comprehension', please observe the following codes: a dress code; a behaviour code relating to public affection; a photography code; and a code for environmental protection. The Council has banned plastic bags, and when television first arrived, it was banned from being shown in shop windows. Also characteristically Ladakhi have been anti-corruption drives and the unique justice system under which, for example, thieves may be required to build a public Chorten (Buddhist monument) in keeping with the traditional punishment for theft from a monastery.

Students working at SECMOL's education campus at Phey, 18 kms from Leh, overlooking the Indus Valley. . Here students are given extra tuition to help them graduate from secondary school. Flat roofed buildings are made in the traditional style from compressed earth blocks

Some of the professionally skilled Ladakhi leaders, now in their 40s, were among the first of the tertiary-educated who chose to return to Ladakh as opinion leaders rather than take up lucrative positions elsewhere They have been rebuilding pride and self reliance in their communities through the embrace of democratic processes. Sonam Wangchuk, who founded SECMOL to reform the education system, is an outstanding example. For his work which involved cultural preservation and community and citizen participation he was elected to the Ashoka Fellowship in 2002 (social entrepreneurs) (Ashoka 2002) and named in The Week (India's version of Time magazine) as Man of the Year in 2001.Wangchuk's leadership has helped retain and build self-reliance, confidence and contentedness in Ladakhis. He has advocated traditional values encouraging students to live simply and sustainably. He advocates the teaching of climate change science in schools and by religious leaders including by monks, nuns and lamas. Listen to his 2007 interview at www.youtube.com/watch?v=tnwilxiuiow&feature=related. (Wangchuk 2007).

Community meetings and conferences are held to build solidarity and obtain consensus on matters such as how to improve primary and secondary education. **Village Education Committees (VECs)** have been formed to bring community pressure on parents to send children to school and to act as watchdogs to ensure teachers also attend school. An influential bilingual (Ladakhi and English) community magazine (Ladags Melong) (Wangchuk 2005a) was produced as a medium of change.

Ladags Melong published school matriculation results annually from 1997 when the failure rate was 95%, later falling to 45% in 2005 as a consequence of public action and better teacher preparation. Ladakhis did not have much in common with Indian life south of the Himalayas (Padmanabhan and Anantharajan 2005). Consequently these forward thinking leaders launched **Operation New Hope (ONH)**, to provide locally based culturally appropriate textbooks and reading materials to schools (Padmanabhan and Anantharajan 2005) and the **Students Education and Cultural Movement of Ladakh (SECMOL)** to provide in-service training for teachers and extra support for year 10 students in their matriculation year. SECMOL encourages International volunteers as 'paying' workers to help with language teaching, information technology and textbook writing.

In addition SECMOL established a graphics company, an eco-tourism company and a home stay industry (Wangchuk 2005b). News about contemporary issues including the work of SECMOL and ONH was reported through a regular TV slot created each week by students. Community feedback was welcomed and gathered through meetings and suggestion boxes in the main streets. Corruption was publically identified including that within the public service.

An 'outside' dry composting toilet. The fertiliser from the toilet sustains the flourishing summer garden

In 1993, **Sheyson Solar Earthworks**, a commercial, income-generating section of SECMOL began designing solar homes and buildings for the government, NGOs, and the Indian Army. The profits from Shesyon Solar Earthworks supported educational reform and environmental awareness. Buildings were traditionally designed and sustainable but with solar lighting and gas and solar cookers replacing polluting dung and wood burning black soot producing internal stoves. The traditional 'dry' composting toilets were retained. This system recycles waste by allowing it to accumulate over a season in the cold dry conditions. The decomposed waste is removed annually and sold, the owner being paid in cash or perhaps potatoes. Sheyson Earthwork's structures have adopted skylights as well as double glazed windows for interior lighting. The traditional low thermal conductivity rammed earth walls were further insulated with 'paper and plastic rubbish' incorporated into the gap between the double walls.

Dry composting elevated toilet. The 'throne' was added for the western volunteers. Office paper is used as toilet paper. SECMOL

Compressed mud block wall with traditional windows on 1st floor and introduced arches on the ground floor. Seyshon Earthworks introduced arches into Ladakhi architecture to avoid the need for expensive and scarce poplar beams

Arches were introduced and (mindful of the Muslim tradition) were accepted by all, as a stronger and cheaper doorway technique compared with wooden frames. Wood is a very precious building material in the region with no indigenous forests and few cultivated trees. Clear plastic sheeting was introduced as an experimental extra waterproofing layer for flat roofs. Ingeniously (and now routinely in public schools and army buildings) plastic sheeting was also used as a winter wrapping for the exterior walls of buildings to create a form of double glazing and 'greenhouse' for building side gardens and window boxes. **EcoDesign - Alternative architecture in trans-Himalayan cold deserts** has evolved as Sheyson Earthwork's successor (Rigzin Namgyal 2010)

It is probably the Inherent **engaging sociability** of Ladakhis that attracts and maintains an international following of supporters including UNESCO, NGOs, academics, volunteers, educators, photographers, and filmmakers. Ladakhis have benefitted from UNESCO eco-tourist workshops, (UNESCO 2002a; UNESCO 2003d), UNESCO Heritage mural and building restoration Projects(UNESCO 2004f; ACJP and Earthjustice 2009; UNESCO 2010hh) and UNESCO heritage awards for restoration (UNESCO 2004b, 2010ii). The THF celebrated the opening of the Central Asian Museum Leh (CAML) with two workshops 'Historical research and heritage conservation cooperation in Ladakh' in May and August, 2011 bringing together all interested parties (local and international) to enhance cooperation. Various initiatives are being taken to document and protect the cultural heritage of Ladakh. It was agreed that all participants benefited from greater communication and cooperation. The International Association of Ladakh Studies also held its 15th Conference in Leh in August, 2011 themed Responding to Climate, Biodiversity and Resource Changes in Ladakh and the Western Himalaya.

Unlike their Tibetan counterparts the Ladakhis have benefitted from the latest knowledge about climate change (Anja Byga 2009).

350.org. and Young Buddhist association climate change rally at Leh Polo Field on their 2008 international day of action to reduce carbon dioxide emissions. CO_2 levels need to be reduced to 350ppm or less to stabilise global warming.
Photo courtesy Conor Ashleigh

A Buddhist leader addresses the climate change rally at the Leh Polo Field.
Photo courtesy of Conor Ashleigh

Climate Change in Ladakh

The naivety with which the Ladakhis have embraced democratic processes has led to some disagreement with conservative religious leaders and world hardened bureaucrats, but overall has led to a remarkable capacity to face community problems including climate change threats.

97 year old Tashi Angchuck "charismatically shared with the attendees his experience of changing weather patterns" Photo courtesy of Conor Ashleigh

Ladakhis are experiencing the effects of **climate change first hand** and through their religious leaders, schools, NGO alliances, online sources and public media they are interested in knowing more about predicted impacts and relish the chance to show their concern through public action. In 2008, fifteen hundred Ladakhi gathered on the polo field in Leh to demonstrate their concern by participating in an event organised in conjunction with the US climate change action NGO 350.org. (Ashleigh 2008). 350.org is an international environmental organization, which specialises in annual co-ordinated global days of political action. It aims to build a global grassroots movement to raise awareness of man-made climate change, and to advocate the reduction of atmospheric CO_2 to the safer level of 350 ppm.

The **350 event** in Leh was organised by the Ladakh Young Buddhist Association. The crowd was addressed by religious leaders and 97 year old Tashi Angchuck from the village of Skurbuchn who "charismatically shared with the attendees his experience of changing weather patterns in the region such as unprecedented cold spells in the spring and autumn months in addition to the increasing rainfall in recent years". Ladakhis were urged to take action and address the impending climate disaster, with particular attention towards lowering the earth's carbon ratio to a sustainable level of 350 ppm. "Starting climate action groups, planting a tree on ones birthday and using less non-renewable energy sources were suggested as small steps to be taken in the direct future" (Ashleigh 2008). It can be confidently said that Ladakhis know more about their carbon footprint than most from industrialised societies. The revered Dalai Lama has publically claimed that climate change is of far greater importance than the freeing of Tibet. Community leaders and educators concerned about the climate threat talk to monks and lamas who in their teachings advocate care for the earth which sustains life, the danger of climate change and how it has arisen in industrialised societies through not caring for the earth. As found by Anja Byga when working with Tibetans, climate change is seen as a moral and spiritual issue (Anja Byga 2009).

An alleyway in Old Leh Town

According to the International Research Institute for the Himalayan Hindu Kush region in Kathmandu, Nepal's climate in the mountain region is warming more than the global average. In 2009, the French Development Agency GERES published a major study on climate change in the Indian Himalayas. GERES evaluated the past 35 years of data from the only weather station in Ladakh and interviewed hundreds of villagers. The survey found that the minimum temperature in winter months in Leh has increased by nearly one degree C. and the maximum temperature by nearly 0.5 degrees C. for summer months. The survey also indicated that less snow has been falling in winter (Hoerig and Dannenberg 2010).

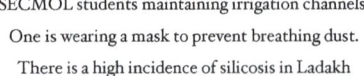

SECMOL students maintaining irrigation channels. One is wearing a mask to prevent breathing dust. There is a high incidence of silicosis in Ladakh

A home built over an alleyway in Leh Old Town

Wherever you have glaciers, you have a stream," said Nawang Rigzin Jora, a state parliamentarian from Leh. "Whenever you have a stream, you have habitation. Agriculture is spring fed, glacier fed. When the glaciers are receding, when there is less snowfall, obviously agriculture will become less sustainable. Things are going to be very difficult in the days to come." It is common knowledge that climate change is already having an effect on the amount and regularity of irrigation water available from melting glaciers. An ancient and time honoured irrigation network of channels covers over 50 square kilometres of agricultural land in Ladakh (Parvaiz 2010). From the Himalayan town of Dehra Dun where the Indian Government has, in 2009, set up a glacier research station, observer Dwarijka Dhobal reports, "The mouths of the glaciers through which the melt water flows out are receding by five to twenty meters every year. The volume is also reducing; the ice gets on average 30 cm thinner each year" (Hoerig and Dannenberg 2010). Water availability is becoming increasingly out of step with the growing season leading to poor barley, wheat and seasonal vegetable growing. 70% of Ladakh's water needs originate from glacial melt. Black carbon dust landing on the snow and ice surfaces have accelerated the rate of melting. This black carbon is from the polluted Asian air and had its genesis in domestic cooking, agricultural burning, industrial pollution and vehicle exhausts.

Sooty black carbon travels east along wind currents latched to dust – its agent of transport – and become trapped in the air against Himalayan foothills. The particles' dark colour absorbs solar radiation, creating a layer of warm air from the surface that rises to higher altitudes above the mountain ranges to become a major catalyst of glacier and snow melt (NASA 2009).

The more ground exposed as glaciers retreat, the warmer the earth becomes and the greater is the rate of glacial melt. Exposed mountain sides become unstable and more likely to slide away to fill valley floors and form dammed lakes which can later burst. Till recently what little precipitation there was in this shadow Himalayan area fell as snow. Now snow is becoming increasingly replaced by rain which can come as heavy precipitation events. Already there has been widespread flood damage leading to loss of life and collapse of homes which were not built to withstand floods and heavy rainfall. Higher humidity and warmer conditions are damaging agricultural production and health. Farmers now report seeing pests in high villages that used to be found only at lower altitudes. Some of Ladakh's limited pasture lands are drying up because of water scarcity (Parvaiz 2010). Locust plagues have devastated vegetation and crops in recent years. With more intense winds, more fine silica from the valley floors will be blown about and inhaled leading to an increased incidence of the serious respiratory disease, silicosis. Diesel truck exhaust fumes also contribute dangerous pollutants that add to the health burden.

Introducing WH Climate Change (Case Study 1): Leh Old Town (LOT)

The old town of Leh is a unique example of medieval Himalayan urban architecture and is the best preserved traditional Tibetan city in the world. It consists of 200 stone, mud and timber houses nestled together on level ground and the slope below the palace. The palace is a nine story stone structure erected around 1600 on a steeply rising spur behind the houses. A number of very lively Buddhist temples and monasteries and three mosques are also to be found in the old town. A few Buddhist stone carvings may be testimony to the beginning of Ladakh's recorded Buddhist history in the 10th century. LOT's living heritage goes back 11 centuries, being a former home to royalty, artisans, workers, migrants, families, Buddhist monks and lamas and Islamic mullahs. LOT meets four of the World Heritage criteria:

- to exhibit an important interchange of human values, over a span of time or within a cultural area of the world, on developments in architecture or technology, monumental arts, town-planning or landscape design;
- to bear a unique or at least exceptional testimony to a cultural tradition or to a civilization which is living or which has disappeared;

- to be an outstanding example of a traditional human settlement, land-use, or sea-use which is representative of a culture (or cultures), or human interaction with the environment especially when it has become vulnerable under the impact of irreversible change;
- to be directly or tangibly associated with events or living traditions, with ideas, or with beliefs, with artistic and literary works of outstanding universal significance.

Part of Leh Old Town. Old Leh Town is the best example of surviving early Tibetan urban planning and architecture in the world. Leh residents and staff from LOTI and THF are rehabilitating the area using traditional artisan skills and materials

An interview survey by the German NGO, **The Tibet Heritage Fund (THF)** (Alexander 2007), found that of the 178 residents of Old Town, one third were immigrant workers from southern India, one third low income Muslims and one third low income Buddhists. LOT had become a slum, with buildings in decay through neglect, poor drainage and poor access to water. Tensions existed along religious lines. Subsequently the Tibet Heritage Fund committed itself to driving the rehabilitation of the historic urban quarter with its private residences and public religious chapels, the oldest of which dates back 400 years. The THF assisted by other NGOs, the LAHDC and the community worked through a process of community inclusion, trust building, tension resolution and reconciliation towards the goal of rehabilitation. LOT was studied and documented, training positions were created, employment for artisans and labourers created, house owners who opted to have their homes rehabilitated were subsidised and overall advocacy for the project grew (Alexander and Catanese 2007). This integrated, holistic approach to urban conservation (UNESCO 2010ii) is incorporated into the case study.

Icy laneway bordered by stone walled gardens and homes. Winter, Leh

Introducing ICH Climate Change (Case Study 2): Ladakh Traditional Building Design and Construction Practice (LTBD&CP)

LTBD&CP is the 'element' of intangible cultural heritage relevant to the Convention domains of: (a) oral traditions and expressions, including language as a vehicle of the intangible cultural heritage; (b) social practices, rituals and festive events; (c) knowledge and practices concerning nature and the universe; (e) traditional craftsmanship. The LTBD&CP could be nominated for inscription once all of the following criteria are met:

Overlooking Leh from Lonpo House which was restored by the NGO LOTI and the THF. (#7 on map at insert 29.1). Lonpo House was the former residence of a minister to a past king

A resurfaced alleyway in Leh Old Town

R.1The element constitutes intangible cultural heritage as defined in Article 2 of the Convention. (Complies) R.2 Inscription of the element will contribute to ensuring visibility and awareness of the significance of the intangible cultural heritage and to encouraging dialogue, thus reflecting cultural diversity worldwide and testifying to human creativity. (Complies) R.3 Safeguarding measures are elaborated that may protect and promote the element. (Complies) R.4 The element has been nominated following the widest possible participation of the community, group or, if applicable, individuals concerned and with their free, prior and informed consent. R.5 The element is included in an inventory of the intangible cultural heritage present in the territory (ies) of the submitting State(s) Party (ies), defined in Articles 11 and 12.

Traditional building design and construction practice has begun to be safeguarded by the work of the THF and others (Naqvi 2005; THF 2010). The THF's long term and sustainable urban conservation and rehabilitation program is working with skilled builders, trainees, and residents to revitalise traditional building practice. The interactive web site appealingly animates the building stages (THF 2009a) describing each stage graphically and finishing with the local women consolidating the roof with traditional 'stampers' singing the song of completion. Every stage is, by tradition, sustainable and earthquake resistant. Climate change resilience measures will be proposed in the Case Study.

Case Studies 1, 2 and Postscript.

Addressing Climate Change Impacts on Tangible and Intangible Cultural Heritage in Ladakh

(A semi-theoretical study developed with reference to the framework process described in Chapter 28)

The tangible cultural heritage study	The intangible cultural heritage study element is
Leh Old Town (LOT)	Ladakh traditional building design and construction practice (LTBD&CP)

NOTE: This study draws on resources available to the authors who have visited Ladakh on five occasions, including once as volunteers at SECMOL. As such it is not a formal study and does not fit all the criteria of the integrated framework as summarised in Chapter 28. However it does demonstrate cultural heritage under dramatic climate change impact. Although some of the study is theoretical, much is based on actual observations, experiences, interventions and responses.

Background information

A preliminary understanding was obtained of the likely regional physical and biological effects on heritage and effects on humans of climate change in the study area. Ladakhi citizens, NGOs, the literature and personal experiences were referenced.

Regional physical and biological impacts

Ladakh's extreme and changing climate is described below.

Temperature: The region experiences a severe winter with minimum temperature as low as -40C. Over the last 35 years there has been an increase in the mean minimum temperature at Leh of nearly 1° C over the winter months which is almost twice the average global increase (Joe Thomas K and Rai 2005) and nearly 0.5°C over the summer months . This warming is causing glacial melt and contributing to a change in precipitation from snow to rain.

Snowfall: Precipitation in Ladakh is usually in the form of snowfall. 70% of the total precipitation over the entire year falls as snow in the winter months. Leh's records over the past 35 years show that there has been a definite decline in winter snowfall without a significant change in summer precipitation (rainfall). (GERES India 2009a)

Rainfall: Ladakh receives an average annual rainfall of 55 mm and has 250-300 cloud free days a year (LEDeG 2009). Precipitation as rain was uncommon till recently. Much of Ladakh is dry because it is in the rain shadow of the Himalayan and Zanskar Ranges. Rainfall varies in volume, intensity and frequency. In August, 2006, a series of intense rain downpours across the region resulted in flooding which destroyed over 20 houses near Kargil. Rainfall has led to drainage problems, increased erosion, instability of sloping unconsolidated land and increased incidence of mud and slush slides. It also damages the traditionally constructed flat mud-roofed Ladakhi buildings.

Glaciers: Over the past 35 years small regional glaciers have been retreating because of rising temperatures and diminishing snowfall. Melting of glaciers not only causes landslides and floods due to glacial lake wall collapse or overflow, but also increases annual variation in river water flows. Increased melting is causing water levels to rise in some lakes flooding roads and grazing land. As glaciers decrease in size summer glacial run off declines. (Glaciers retain water on mountains in high precipitation years since the snow cover accumulating on glaciers protects the ice from melting. In warmer and drier years, glaciers offset the lower precipitation amounts with higher melt-water outputs.) Because water from Himalayan glaciers comprise the principal dry-season water source of many of the major rivers, increased melting will initially cause greater flows followed by diminished flows as the source glaciers are depleted. The neighbouring Tibetan Plateau contains the world's third-largest store of ice. Temperatures there are rising four times faster than in the rest of China and glaciers are retreating at an accelerating rate. Up to 60% of glaciers may disappear by 2100. (ACJP, CANA, FOE Australia 2008). Glacier melt water is the sole source of fresh water for Ladakhi farmers who make up more than 70% of the Ladakh population of 270,000.

Water flow is unpredictable. There will be an increase in frequency and magnitude of floods (for a few decades) (Joe Thomas K and Rai 2005) followed by a reduction in water availability for irrigation of crops, leading to a reduction in crop production and shortage of fresh water supply for home use, domesticated animals, wild life, indeed for the whole ecosystem.

Humidity which is normally very low, is increasing in summer inducing favourable conditions for pest invasion (GERES India 2009a)

Pests: Migratory Locusts plagues have destroyed vast areas of vegetation. Such invasions will increase with rising temperatures.

Wind is Intensifying. Summers now bring more intense **dust storms.**

Black carbon aerosols, contributing to the 'atmospheric brown cloud' phenomenon are depositing on the Himalayan snows and causing temperature rise to accelerate even more (Shrager 2008). The bare earth exposed by loss of the reflective snow and ice, absorbs more heat providing a positive feedback (diminished albedo effect).**Impacts on humans**

Although there is a lack of formal weather and climate monitoring beyond Leh, similar trends have been noted by people in other Ladakhi towns and villages. For instance rural Ladakhis gave many examples of **glacial retreat.** Glacial retreat is leading to aesthetic, ecological and cultural degradation with the high altitude environment changing into one with more exposed rock and skree. Farmers only have to glance up from their fields to see the exposed swathes of dark rocky surface where once all was white. Ladakhis explain this from their observations that snowfall is less, thus glaciers are fed less, and **winters are** warmer thus melting is faster. In the long run, glacial retreat translates into drying up of natural springs, streams; depletion of underground water hence decreased water availability challenging the very existence of the mountain people (GERES India 2009a). **Water availability** is the gravest impact of climate change. Barley, wheat, potatoes, peas and fruit trees will suffer from lack of irrigation. Fruiting of trees is happening earlier. Crops suited to lower altitudes are being introduced. Wheat is now able to be grown at higher altitudes. Hybrid vegetable seeds of cauliflower, onion, cabbage, lettuce and Chinese cabbage have been distributed to Ladakhi villagers during the last 2 years by the Indian Agriculture Department. Climate change is causing and will continue to cause livelihood threats to builders, craftspeople, artisans and intangible heritage knowledge owners who rely on irrigation from glaciers for subsistence agriculture. Poverty and **loss of cultural heritage** will be a consequence. Extreme water shortages will also threaten the tourist industry, other sustainable development projects, agriculture and health. Migration of youth south for education and employment is likely to interrupt or prevent the transmission of design and construction knowledge, artisan experience and skills and reduce advocacy for traditional ways. This may be compounded by pressures from seasonal immigrant Kashmiri traders and tourist industry workers who do not promote or sell local Ladakhi handcrafts and other cultural products.

The Changpas are nomadic pastoralists who herd pashmina goats across the eastern plains. Their way of life is changing as conditions become more demanding under climate stress. Older herders are retiring to Leh; younger family members are leaving to attend school in Leh

Displacement: Some rural households have already moved due to failure of their glacier fed village springs and streams with an entire Zanskar Village having had to **relocate** due to the disappearance of the glacier that supplied its water (GERES India 2009a). The Changpas – the **nomadic pastoralists** of Changthang are selling their Pashmina goat herds and tents to settle down in Leh. They are **migrating** for better access to medical services and education, and because their families have been fragmented due to the older children leaving. Other reasons include pasture degradation, the harshness of nomadic life for the elderly, the threat of wild predators to their precious pashmina goats and a shortage of manpower. The decrease in pasture is due in part from competition from Tibetans, a lack of animals because many are being used in the tourist industry as pack animals for trekkers and because of the negative attitude of young Changthang to the nomadic way of life (Ahmed 2006). As climate change impacts intensify there will be more pressures on the Changpas to relocate compounding the loss of their way of life.

Spiritual concern: "Similarly, our planet is our house, and we must keep it in order and take care of it if we are genuinely concerned about happiness for ourselves, our children, our friends, and other sentient beings who share this great house with us. If we think of the planet as our house or as 'our mother - Mother Earth' - we automatically feel concern for our environment. Today we understand that the future of humanity very much depends on our planet and that the future of the planet very much depends on humanity. But this has not always been so clear to us. Until now, you see, Mother Earth has somehow tolerated sloppy house habits. But now human use, population and technology have reached that certain stage where Mother Earth no longer accepts our presence with silence. In many ways she is now telling us, 'My children are behaving badly,' she is warning us that there are limits to our actions" (Dalai Lama 1990).

Spiritual and secular hope: One of Ladakh's younger leaders, Sonam Wangchuk, who founded SECMOL to bring reform to the education system of Ladakh and who later introduced solar space-heating and energy efficient design into Ladakhi architecture, has taken his spiritually based message of sustainability to Buddhist leaders, Lamas and the public.

Hold meetings with community groups, visit sites and survey by questionnaire
Workshop to consolidate views, recommendations and proposed actions

The following possible problems were identified as a result of interviews by the authors, reference to the literature including the local magazine Ladags Melong and the work of NGOs including the rehabilitation study team from LOTI (Leh Old Town Initiative) and the Tibet Heritage Fund (THF). In all a cross section of people were contacted including LOT residents, community and religious leaders, professionals, NGOs and LAHDC officials.

The following problems were identified:

Problems related to (tangible) heritage: potential threats to OUV, integrity and authenticity	**Problems related to intangible heritage: potential threats to safeguarding**
Increased rainfall leading to **erosion** of centuries old mud brick and rammed earth buildings, especially those without traditional clay rain proofing treatment. LOT homes, heritage buildings such as the palace, monasteries, chapel, assembly hall, mosques, town gates, gompas and chortens need more traditional waterproofing. Supplementation with non-traditional materials e.g. plastic sheeting may need to be considered. • **Investigate Stability of the ridge and slopes** on which Leh Palace, a monastery, temples, tower with chapel and some homes are built as well as a mosque, and the LOT houses that are built beneath the steeper slopes. Consider close monitoring and engineering advice.	**Increased rainfall** leading to leaking roofs and stressed drainage of practitioners homes, and **flooding and erosion** of river flats (where willow and poplar trees are grown to provide raw materials for building construction and decoration), **Temperature rise** related glacial melt, inconsistent irrigation flow, and the unusual occurrence of rain, causing threats to traditional agriculture, introduction of weeds, pests, diseases of plants, animals and humans leading to impacts on livelihoods and less time and energy and resources for intangible pursuits and their transmission. **Water shortages** affecting domestic use, the growth of trees for building materials, the making of mud building blocks, clay whitewashing and consequently the livelihoods of practitioners and knowledge

<table>
<tr>
<td>

- Where **drainage** is poor, changes in water table will cause rising **damp,** damaging wall murals and increasing **humidity** within buildings.
- **Heavy downpours of rain** causing destructive **flooding.**

Temperature rise causing stresses, cracking and heaving of foundations, building walls, alley way walls, streets, roofs, wall murals and cracking of painted Buddhist banners and paintings (thangkas).
Water shortage affecting availability of water for traditional building methods and the small number of carefully cultivated trees used in roof, window and door rehabilitation.
</td>
<td>
holders. (Water has been trucked into Leh in recent years).
Health problems will become more common. Food shortages will contribute to malnutrition. Increasing wind strength and consequent dust storms may increase the prevalence of silicosis. Pollution from diesel exhausts may further contribute to lung and heart disease. Such ill-health may result in invalidism or emigration.
</td>
</tr>
<tr>
<td colspan="2">
Rundown and depressed condition of LOT due to scarce resources for the restoration, protection and maintenance of buildings and shared infrastructure. LOT residents are often tenants on low incomes and include seasonal workers from southern India.
Lack of community identity and empowerment in LOT due to poverty, lack of opportunity, racial and religious differences, lack of resources and the lack of leadership needed to foster social cohesion and trust.
Social tension. In the wider Leh community irrigation water is accessed through a network of channels which are strictly maintained and controlled by the community. There is potential for conflict and stress over access to this resource. The only open hostility seen in Leh (as observed by one NGO) was between women arguing over irrigation water.
Social divisions along ethnic and religious lines leading to suspicion between groups.
</td>
</tr>
</table>

Chutayrangtak Street Leh Old Town. Shopkeepers display their freshly baked breads and keep their shop fronts swept in time honoured fashion. Note the open drainage system and coppiced willow tree

A baker's shop in Chutayrangtak Street Leh Old Town

The view south down Chutayrangtak Street Leh Old Town

Draft a management proposal

The identified problems, views and recommendations on the protection of the community's tangible and intangible cultural heritage have been crafted into a theoretical draft management proposal for discussion with local policy makers. The management proposal should be considerate of community issues such as population characteristics, socioeconomic factors, livelihood threats, health issues, displacement issues, local justice concerns and any other relevant issues that are likely to be contested.

LOT Policy and Management Plan for climate change impacts

–Policy goals: Promote **inclusion** of LOT residents and those involved in addressing climate change interventions in LOT, especially women who, it has been shown, are likely to be impacted more by climate change (Sherpa 2009) .

Use as an exemplar for climate change projects, the THF's practice of employing Ladakh artisans, using local traditional materials and methods and of its ongoing **capacity building,** training **and employment program** which has gained community approval.

Encourage information **sharing** with traditional artisans, religious leaders, scientists, engineers, technicians, IT specialists and architects. Information sources include; climatologists and information systems experts working with remote sensing tools, reconnaissance by small camera over-flight, telecommunication and radio broadcasting system integrated with on-site installed hydrometerorological and geophysical instruments; NGOs, and welfare groups.

Support and propose extension of the THF's **documentation** and **monitoring** work to include climate change. This would involve complimentary reporting on the condition of LOT buildings, palace, monasteries, mosques, gates, streets, walled lanes, mansions, chapel and assembly hall.

Monitor immigration and **emigration,** population changes and social inclusion.

LTBD&CP Policy and Management Plan for climate change impacts

–Policy goals: Promote **collaboration** with a broad cross section of Ladakhis including those who own or are the custodians of heritage homes, monasteries, mosques and public buildings all being the work of traditional builders and designers.

Seek to address climate change impacts by safeguarding traditional practitioners from poverty, **livelihood threats** and possible **health and safety risks.** The plan should seek to include generational representatives of men, women, mullahs, monks, nuns and lamas to contribute to the welfare support net.

After consultation with intangible cultural heritage groups to seek free, prior and informed consent, plan to create an inventory of carpentry and woodworking practice and techniques to be managed by the new Central Asian Museum.

Carved support, Lonpo House, Leh Old Town

Raise awareness of proposed climate change related interventions in the LOT rehabilitation project as 'models' of climate change adaptation e.g. mural restoration; Seek to generate further **sustainable** restoration, rehabilitation and tourism projects with families e.g. selling local handicrafts; with NGOs e.g. revival of traineeships; with guided walks (Grover 2011); with cafe owners e.g. selling local food; with exhibition/community centre e.g. showing art and craft; and within the new Central Asian Museum, opened in Leh in 2011. Work to introduce **education** programme through schools, promoting tourist visitation and eco-tourism, (including home stays) and cultural programmes related to LOT.	 Tehsildar Gate, c. 1850, (# 26 on map, insert 29.1) at the entrance to a former mansion. It has now been restored by local craftsmen, at the request of the community, LOTI and the THF Ensure the transmission of **LTBD&CP** skills **by introducing them into the curricula**, and as recommended by carpenters, masons and carvers, through **traineeships.** Work to **integrate** the community based protective LOT project with the safeguarding of LTBD&CP. Seek to establish joint **sustainable development** and **awareness raising** programmes and a joint **Disaster Risk Management** Strategy and most importantly a **Mitigation Action Plan** with the LOT project. The latter could draw up guidelines for how to work towards a carbon neutral LOT society using renewable energy and the innovative Passive Solar Housing (PSH) design.

Develop local adaptation

The policy documents are used to guide the writing of proposed local adaptive management plans for climate change sensitive tangible and intangible heritage. These should include climate change vulnerability maps and advice on preservation/safeguarding strategies. The plans should encourage capacity building, information sharing, mitigation and continuing vulnerability and disaster risk assessments.

<table>
<tr><th>LOT Management Plan - Adaptation responses:</th><th>LOT Management Plan - Adaptation responses:</th></tr>
<tr><td colspan="2">Community engagement
The LOT Management Plan and the LTBD&CP project could prioritise community inclusion through regular formal and informal engagements. Of vital and central importance could be the collaborative approach taken by THF and the local NGO LOTI resulting in an integrated community programme which, by lifting the LOT community from slum conditions, has promoted social cohesion between its resident Moslems, Buddhists and Hindus. Informal social interaction opportunities could be created in the refurbished streetscape. The THF built a community centre, repaved and drained (with local input) an alleyway as a 'model' to demonstrate commitment to improving community life. The THF also developed sustainable income generating programmes and promoted women's roles in construction work (THF 2010). Such a joint heritage adaptation response could eventually work towards advocating for a LAHDC town plan that includes climate change protection and prioritises community engagement, traditional design, traditional building practice, (religious, retail and domestic) and employment of local artisans and knowledge holders.</td></tr>
<tr><td>Vulnerability assessment
The LOT Management Plan could add to the vulnerability assessment of LOT which has been partly done by the THF. The Plan could adopt THF's successful approach of inventorying buildings and occupant's socioeconomic status. To this it could add mapping, surveying and listing of further heritage buildings associated with LOT and the recording and monitoring of climate change impacts.</td><td>Vulnerability assessment
The Project could initiate a vulnerability assessment of LTBD&CP. This would involve inventorying the techniques, materials and designs; recording the number of practitioners, their tools of trade and design and their practice specialities; their socioeconomic status; and mapping, surveying and listing local examples of the practice plus the recording of climate change threats.</td></tr>
<tr><td>Physical and biological impacts
The responses to these impacts could include: installation of local flood early warning system, monitoring of slope stability and risk assessment. Consideration could be given to the expensive and controversial building of an upstream dam as a response to changing and potentially unreliable water supply (FOE 2009; Anja Byga 2009; Maneckscha 2010; GERES India 2009b). Ladakhi scientist Chewang Norphel began work on 'zings,' small tanks fed by run-off from melting glaciers</td><td>Physical and biological impacts
Adaptive responses could include safeguarding practitioners and their tools by monitoring and supporting their working, religious and home environments from rain, flooding downpours, rising damp, increased humidity and increased windy weather etc.</td></tr>
</table>

<table>
<tr>
<td>(Glacier is their forte. 1998). "I am the scientific data," said Norphel in 2009. "I have seen, for instance, the size of the Khardung La glacier since I was a child: it was solid ice then." Norphel, known popularly as India's 'glacier man', has more recently been building high-altitude water-conservation channels that freeze over as 'artificial glaciers' to prepare for the lack of water from the receding Himalayan glaciers (GERES India 2009b). In the longer term water resources will emerge as the climate-induced number one security concern.</td>
<td></td>
</tr>
<tr>
<td>Preservation strategies
Strategies include facilitation and support of the THF's scheme to subsidise half of the costs of house rehabilitation with owners and to revive identity by employing local designers and artisans. Rehabilitation interventions include roof raising, installation of skylights, equipping with bathrooms and toilets, making repairs and carrying out adaptations such as extensions to double glazed windows. Similarly, Ladakh Youth Coordination Committee's initiative to have a community suggestion post box in the main bazaar, to watch-dog for corruption, to erect notices to dissuade public toileting, to encourage rubbish recycling, to enforce, by peer pressure, the plastic bag ban (cattle eat them and die), to neuter pack dogs that roam town (Buddhists are loathe to harm animals) and to clean the streets leaving cows to roam could all be supported.

The Ladakh Youth Co-ordination Committee's Complaint Box stands in the Main Bazar, Leh</td>
<td>Preservation strategies
Encourage LTBD&CP by employing practitioners on the LOT rehabilitation with the LOT designers, NGOs and trainees. The integration would allow joint capacity building chances; joint publicity and awareness raising; increased opportunities for sustainable development programmes; information sharing meetings and collaborations with more NGOs, and more volunteer opportunities.
Record wall, brick, rammed earth and roof making techniques, their origins, availability and assessments of raw materials for different purposes such as traditional clay making, techniques of window and doorway carving and relationship to religious painting designs.
Safeguard the traditional proven sustainable growing of poplar trees for rafters and willow trees for roof joists with a reduction in fertilizer use which has a negative impact on soil quality. Special religious and secular construction practices could be documented and protected e.g. reinforcing structural beams with walnut timber.</td>
</tr>
</table>

By popular demand plastic bags are prohibited in and around Leh. Shopping bags are instead made from folded newspaper

Ban use of cement and steel building materials which, relative to earthen construction, are very poorly insulating and therefore cold. Such materials unlike traditional mud wall construction are likely to be lethal in an earthquake, are expensive and have a large carbon footprint.

Establish competent public drainage system adequate for increased rainfall, heavy downpours, local flooding and easy cleaning. Install convenient public water taps, and investigate how to maximise safe water storage. Water collection options for future such as tanks need to be considered Protect the clay mines and mining of various qualities of limestone clay for waterproofing layers from climate change impacts.

Collaborate with experts on new waterproofing techniques that could compliment traditional techniques.

Reusing and recycling should be enshrined in all aspects of the practice.

Sustainable development

Benefits of traditional building design and materials for climate change impacts

Star ceiling are built by traditional artisans. New Central Asian Museum, Leh.
With permission A. Alexander, THF

The usage of local stone for foundations and alleyway paving; superbly insulating sun-dried mud bricks for upper storey walls; wooden frames to safeguard from earthquake for doorways and windows and rammed earth walls which are thicker at bottom for lower floors (Manecksha 2010). Roofs are supported by poplar rafters and willow joists. The joists are made from branches pruned from sustainably grown willow trees once they reach the desired length (coppiced- pruned). Flat roofs are made with stones and clays. High quality local clay is used as a waterproofing roof layer and as a dust free plaster for interior surfaces.

Courtyard of a Leh Old Town Gonpa. (#9 on map, insert 29.1) Here the Ladakhi group LASOL is performing a Cham dance that is attracting the rapt attention of the local audience. The courtyard was restored by local masons and carpenters in collaboration with the THF

Retain and/or incorporate traditional roof parapets and skylights, interior dry composting pit toilets. These centuries' old methods and architectural styles are more resistant to earthquakes in this seismically unstable region (a condition which could become worse). Traditional houses may collapse but do not kill occupants unlike reinforced concrete buildings which injure and kill when ruined by earthquakes.

Traditional methods have the capacity to be modified for changing conditions. Increased water proofing measures include increased clay layering in roofs inserted between the 'mar-klag' layers. Better quality clay can be mined to improve these traditional clay mixes for roofing and for interiors for waterproofing. This is especially beneficial for protecting murals and painted and carved wooden interior decorative elements in all buildings, heritage restorations, rebuilding and repairs. Include strategies to prevent salty damp rise.

Sustainable tourism opportunities such as guided walks incorporating a museum visit, cafes celebrating traditional cuisine, traditional craft outlets, specialist educational tours on Tibetan architecture and Buddhist mural painting could be considered.

Information sharing	**Information sharing**
Under the LOT Management Plan the LADHC heritage manager could work with NGOs: THF, SECMOL (SECMOL 2010), LEDeG and Namgyal Institute for Research on Ladakhi Art and Culture (NIRLAC) to share information on Key Climate Factors. The Plan could exploit SECMOL's knowledge of technology e.g. the Trombe wall construction (a device for passive solar space heating); solar energy generation, water heating, and use of plastic sheeting in mud block ceilings, **Passive Solar Housing (PSH)** (Stauffer, Dawa, and Deen 2009; GERES India 2009c), improvement of traditional 'dry toilets' (SECMOL) and: LEDeG knowledge of solar ovens, food driers, **photovoltaic cells** (LEDeG 2009),and hydraulic pumps and windmills (LEDeG). The Ladakh company EcoDesign could be a sourced for design and architectural expertise in the use of rammed earth walls (gyapak) with insulation, double glazing of windows, and for further advice on **PSH** - the incorporation of ultraviolet-stabilised plastic sheets to encapsulate buildings to create insulation and greenhouses. The plastic sheets can act as solar heat traps for schools and accommodation buildings to create warmth and greenhouses for winter vegetable gardens. The plastic sheets are rolled up in the summer	Adaptive responses could include capacity building, training and information sharing with NGOs such as the THF, SECMOL, LEDeG and NIRLAC. These NGOs could employ traditional craftspeople and tradespeople to rehabilitate LOT for the local residents, for religious observances including Cham Dance Performances, for eco and cultural tourism programmes and contribute to recreating and revitalising Leh's identity as a former great caravan trading town renowned for its exceptionally cultured and friendly townspeople. Leh could continue into its sixth century as a time honoured religious city for Buddhists and Muslims and as a historic centre for royalty and craftspeople, traders and artisan (Alexander 2007; THF 2009a). Adaptive responses could include nominating the climate change threatened Changpas Nomadic Pashmina-herding traditions for inclusion on the Intangible Heritage in urgent need of safeguarding list, offering retraining in traditional building and construction in LOT and contributing to the new Central Asian Museum's exhibition development on nomadic life. This museum is already a living testament to the viability of **LTBD&CP** as it has been entirely traditionally designed and constructed.

to prevent overheating (SECMOL 2008); The sheets also discourage dogs and people from urinating on walls. Subsidised hybrid vegetable seeds have been distributed to Ladakhi villagers over the last 2 years by the Agriculture Department (Bukhari 2012). In winter Ladakhis have had to resort to stored vegetables or to purchasing expensive trucked in produce

Solar dishes power stoves through kitchen windows at SECMOL

Bringing down the plastic 'blankets' to insulate SECMOL for a minus 20 C winter

The freely available PSH designs for retrofitting and/or incorporating in new homes and work places could help intangible cultural heritage practices (e.g. carving, painting, mandala making) by looking after practitioners (including with a better diet) in greater comfort especially during the long frigid winter nights.

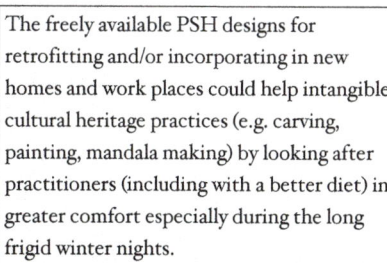

Ladakhi dancers, Leh festival

Health

Health initiatives could include the elimination of diesel fumes in Leh, wearing masks when quarrying stones to reduce silica inhalation from air borne stone particles; a continuation of women's' empowerment programmes for sustainable population and workplace health and safety (Sheridan 2004). The replacement of smoky unhealthy dung and wood burning bukharis by gas or solar ovens for cooking and heating should be supported. Improved water supply and drainage systems would also reduce health threats.

A stirring health message in the Tibetan refugee suburb Choglamsar, Leh

For a Buddhist and Islamic public - a notice promoting hygiene

Mitigation

The plan could develop **mitigation** strategies as part of the adaptation process. Strategies could include making LOT a no-car zone and revitalising bicycle riding in Leh. For both health and mitigation reasons it is imperative to ban diesel vehicles in Leh. Efficient energy vehicles could be deployed to relay commodities to and from a holding station out of town. Adapting buildings by insulating walls, double glazing windows, increasing solar heating, lighting and energy generation thus reducing emissions (LEDeG 2009) would be other important advance. The replacement of dung burning interior stoves would reduce black carbon pollution and assist health. Also to be encouraged would be the installation of water storage tanks and further afield 'small holding dams' as developed by the 'Ice Man,' (Chewang Norphel). Salvaged wood should be reused. (Ladakh is **treeless** except for carefully cultivated imported walnut, fruit, poplar and willow trees).

Education	**Education**
Public education and advocacy could be locally driven by the Leh LAHDC, NGOs, local radio and TV and internationally through academic and NGO web sites. The publication of this proposed study could be considered as there is an ongoing international interest in 'frontline' climate change cultural adaptation programmes.	Schools and monasteries could include culturally appropriate study of and practice of carving and woodwork design iconography, the religious origins of building and interior design, and the associated songs and dances that are connected to each stage. For example the building stages are specified as : Site, Digging, Foundations, Ground floor, First Stones, Walls, Lintels, More Stones, Pillars, Beams, Rafters, Branches, Pebbles, Pillarstones, 1st Floor, Pegur, Slate, Lintels, Pembe, Pillars and Beams, Rafters, Drainage, Karma, Slate, Window and door, Paint and Agra. The design is described as 'organic', rising from the earth with slightly inward-sloping walls. Buddhist and Moslem leaders should be included as design and construction corroborators and facilitators. Promotion would be given to religious practices that accompany many stages of building practice and design.

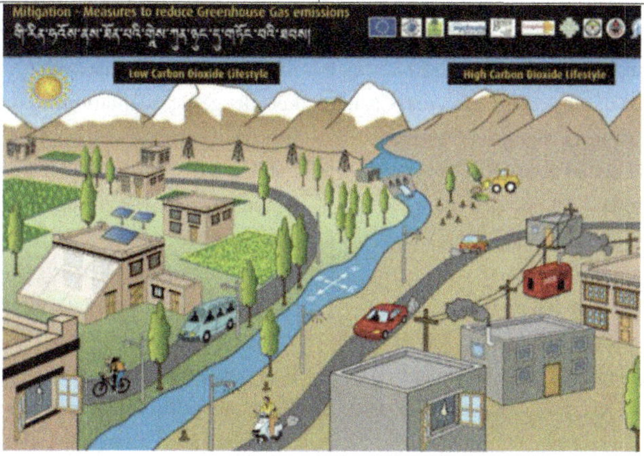

Poster educating about the carbon footprint of Ladakhi life styles

Climate Justice and resilience

Today the 'intangible' **LTBD&CP and** the 'tangible' **LOT** are fusing. What is emerging is a living, culturally rich, and identity-secure Leh community. Despite its very low material wealth and low human development index the community has a high happiness quotient. This augers well for resilience in the face of an 'acute' climate change vulnerability factor. It also places the LOT community which has contributed so little to GHG pollution but may suffer greatly from its effects, in a morally justifiable position to request **climate justice** – urgent technical and monetary support from wealthy States Parties - to adapt. The Dalai Lama is doing more than any other world leader for **climate justice** awareness raising when he advocates and teaches that addressing climate change should override national sovereignty issues (Dalai Lama 2010). "Both science and the teachings of the Buddha tell us of the fundamental unity of all things. This understanding is crucial if we are to take positive and decisive action on the pressing global concern with the environment" (Dalai Lama 1989).

Implement plans after their approval by the appropriate authorities

The LADHCs could approve plans using specific agreements such as the 2006, five-year **Memorandum of Understanding (MOU)** it has with the THF. The MOU regulates how the two sides will work together for the rehabilitation of the Old Town of Leh (THF 2010).

Implement ongoing monitoring

Monitor developments and prepare progress reports for local use and for use by the WH Committee or Intangible Cultural Heritage Committee.

Maintain and Monitor:	
• Factors related to buildings e.g. rising damp, cracking, erosion, collapse, water proofing efficacy, condition of mural painting. • Condition of thankas (traditional Buddhist hanging paintings on cloth sometimes called scroll paintings). • Condition of musical instruments, Cham Dance costumes, monastic clothing. • Monastic library books for deterioration, mould, pests. • Health of heritage custodians, LOT residents. • Ownership changes, migration, population changes. • Tensions between religious groups relating to recent building of religious centres.	• Status of Ladakhi traditional designers and craftspeople: builders, carvers, painters, mural artists. • Their well being, (including mental health). • Ability to partake in regular work (in all building projects), problems of access. • Recording and safeguarding of 'new knowledge e.g. adaptations to potential threats of extreme weather events. • Publication of current and historic techniques and designs together with iconography. • Tensions between religious groups relating to competition for jobs and globalising forces of a market. • Economy geared to growth. • Tensions within community relating to use of traditional Tibetan and Ladakhi languages.

- Tensions between political groups relating to how much autonomy from Indian central government should be claimed.

Cham dancers wear extraordinarily colourful and deeply symbolic costumes

Monastery door with mural and fabric surrounds

- Effectiveness at passing on traditional practices and interest of youth e.g. number of apprentices, workshops at schools.
- Problems of 'transmission loss' because of lack of support for education and health of Changpas' whose nomadic patterns of life are under threat.
- Problems of fall in educational standards, due to undermining of educated young leaders by conservative groups.

The ceiling of the ground floor of the new Central Asian Museum is built in early Ladakhi & Baltistani style. The capitals have been adapted from the Tsemo tower on the ridge above Old Leh Town, with the centre being a diamond ceiling as found in many early Ladakhi temples. Photo THF, with permission after THF

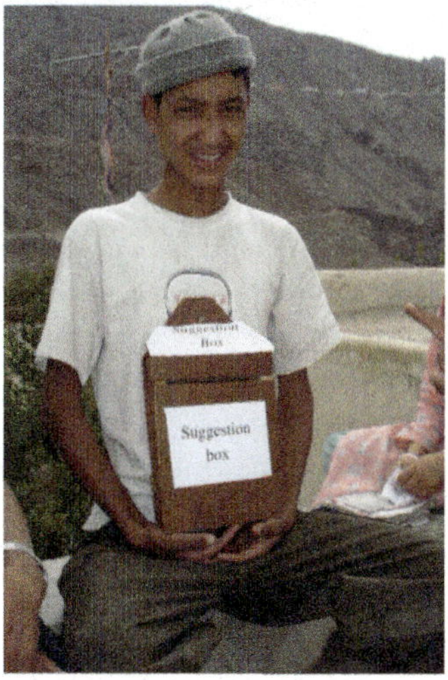

Young Ladakhis enjoy making suggestions as to how life may be improved on the SECMOL campus

Consider nomination for World Heritage and Intangible Cultural Heritage

Consider nomination of LOT for inscription on the WH List and the LTBD&CP for inscription on the Intangible Cultural Heritage Representative List using the adaptive plans in the nominations. Provided free prior and informed consent is obtained from the community submit the nominations. If nominations are successful, use in application for funding. Display nomination information in the new Central Asian Museum to further raise awareness Refer to WHC publications for guidance on nomination process: report on Predicting and Managing the Effects of Climate Change on World Heritage; Strategy to Assist States to Implement Appropriate Management Responses; Policy Document on the Impacts of Climate Change on WH Properties and; Strategy for Reducing Risks and Disasters at WH Sites Manual for Managing Disaster Risks for World Heritage; Nomination Manual, Sustainable Heritage manual, and WHC and Operational Guidelines and ICHC and Operational Directives including suggested format for gathering Intangible Cultural Heritage see Insert 15.7

Respond to Feedback

This case study identifies the areas of climate change protection offered by the Conventions such as 'information sharing' and recognition of the need for 'vulnerability assessments'. It also ventures into areas where 'gaps in knowledge' have been identified. The Ladakhi experience may offer fruitful lessons for Convention research. Ladakhis draw on their cultural and religious roots to 'adapt' to problems that may otherwise undermine their resilience, create a 'victim' mentality or damage their mental health. Ladakhis are 'people who know who they are' and are culturally, spiritually and ecologically equipped for a sustainable future. They also know 'who they are not' – greedy, individualistic, materially driven, growth oriented discounters of the future.

Post-Disaster Follow-up Study – Climate Change: Extreme Weather Event: Flooding of Leh, Ladakh, August, 2010

Physical and biological impacts

In the early morning of 6th August, 2010 Leh experienced its worst ever natural disaster. A downpour following cloudbursts dumped an estimated 11 centimetres (10 times average monthly rainfall) of torrential rain on Leh generating a flooding mudslide of rocks and debris. The mountain slopes surrounding Leh are covered in unconsolidated granitic debris. Large fans of loose deposits which are the product of several million years of erosion can be seen in many locations along the banks of streams and the Indus River. These sediments became dislodged and moved rapidly when the heavy rain saturated the loose sediment, setting off slides and flows charged with mud, boulders, trees, building debris and other objects which travelled down-slope towards the Indus River. Similar destruction occurred at locations throughout the Leh district. At Choglamsar (an outer neighbourhood of Leh and among the worst affected areas), the debris flow travelled approximately 10km from the epicenter of the cloudburst (near Saboo) spreading up to 2km. In Leh, the debris flow travelled about 3km, from an elevation of 3800m to 3410m. Indian meteorologists have linked this extreme weather event which was unprecedented in nature and scope (LREDA 2010) to global climate change. (Hoerig and Dannenberg 2010) Pakistan was affected by record floods at the time of the Leh disaster.

Impacts on humans

In all 9,000 people in Leh and 13 villages were affected. Over 300 died, buried under mud and slush or washed away by the flood. They were Ladakhis, Tibetans from the Tibetan refugee settlement 'Choglamsar' established in the 1960s, 26 from the Indian Army and 7 tourists. The Sonam Norboo Memorial Hospital, Leh Bus Stand, communications building, radio station, schools, roads and bridges were damaged and some destroyed. Over 600 homes and shops were washed away. Many more were filled with 1-2 metres of mud including the hospital. The 5,000 left homeless in and around Leh were placed in 15 tent camps set up by the Indian Army. Bamboo infrastructure was provided for educational institutions affected in the flash floods. Farmers in many villages downstream had all, or part of their fields, completely washed away by the floods, along with their crops. 1400 hectares of vital agricultural land, orchards; poplar and willow growing areas were swept away or submerged under mud. This layer of slush and mud was later analysed and found unsuitable for further agriculture (LREDA 2010). 269 large and 1090 small animals were lost.

The aftermath of the avalanche of mud and debris - the result of an unprecedented cloudburst - that swept through part of Leh killing over 300 people on 6th August, 2010. With permission of Andre Alexander, Tibet Heritage Fund

The lack of international media attention may have been due to attention being focused on the coincident flooding in Pakistan where 20% of the country and 13 million people were affected, 2000 dying.

Cultural heritage : Leh Old Town (LOT), Ladakh	**Cultural heritage :Ladakh traditional building design and construction practice (LTBD&CP)**
LOT and its potential OUV were largely unaffected. Monasteries, mosques, the Leh Palace and restored houses survived suggesting that these centuries old Leh building sites were chosen because they were well above water courses (Alexander 2010). The THF checked on the stability of walls of buildings and roof leakages in LOT. Some	The number of lost intangible cultural heritage practitioners is unknown. An unknown number of designers and builders were displaced to temporary camps. Surviving practitioners have been absorbed into the urgent rebuilding programmes. Further health threats posed by the severe cold of the

downtown newer reinforced concrete buildings suffered more than centuries old traditionally built compressed mud block buildings.

oncoming winter were a prime concern. The loss of poplars and willow trees has set back rehabilitation of Old Leh and rebuilding of homes.

Review of local biological and physical impacts, climate change vulnerability maps and advice on preservation strategies

Review could upgrade the magnitude of rainfall downpours in the proposed protection plan (see main study) and feature in the yet-to-be-developed climate change vulnerability maps. Lack of local weather stations meant that no reliable recording of the rainfall was made. That traditional Ladakhi buildings since the 12th century have been of mud block construction with flat roofs and not built to withstand anything more than the lightest rain indicates that heavy downpours are a new phenomenon. Protection and safeguarding actions could include emphasis on upgrading of weather forecasting and its communication and installation of an early warning system. Advice would be needed on the stability of unconsolidated slopes and the possible deployment of buffer zones, protective barriers and diversions as proposed in main the study and the consideration of triggering mechanisms such as is used in avalanche management by explosives.

In February, 2011, 5 Ladakhis died in an avalanche in remote Lingshed village. This extremely unusual event is probably the result of warmer weather and soft snow, conditions predicted under climate change.

The unconsolidated mountainsides were washed away forming a flood of debris filled mud which swept down Leh's main street swamping and destroying lives and buildings

Review of local physical and biological impacts on heritage

Review could include the need for additional protective measures such as further waterproofing of buildings and an enlarged drainage capacity to address the effects of torrential cloudbursts, flooding, rising damp, and increasing episodes of humidity on LOT buildings and their contents including fabric thankas, wall murals, ceremonial robes, costumes and

Review of local physical and biological impacts on heritage

The urgent need to clear mud and sediment from agricultural lands and from buildings will create a labour shortage that will have serious indirect repercussions on the rehabilitation of LOT. On top of this there is a livelihood threat to restorers through the need to engage elsewhere in unpaid reparative work. Design and construction knowledge holders might also be included amongst those affected.

In September, 2010, assessors gave the following advice for damage to agricultural zones:

Type A: Debris is less than 3 feet – cost of debris removal is Rs 300,000 per hectare (300 hectares of area in the district);

Type B: Debris is more than 3

instruments. The restoration of LOT will be impacted by the loss of traditional timber sources: the plantations of poplars and willows. Less flood prone locations for new plantations may need to be selected.	feet and less than 5 feet – cost of debris removal is Rs 400,000 per hectare (200 hectares of affected area in the district). Type C: Debris is more than 5 feet 920 hectares – ***not feasible to rehabilitate.*** As total cultivatable area for the whole of the Leh District is 10103 Hectares some 14% has been seriously affected (Deputy Commissioners Office Leh). The loss of subsistence farm land and the loss of income from tourism-based heritage projects could undermine efforts to transmit and teach design and building practice through LOT apprentice schemes. Farm and income loss could lead to loss of artisans and possible abandonment of traditional building work. Practitioners may have to adapt traditional practices to new challenges including new building standards (to accommodate seismic and flooding threats).
Review of Human Impacts **Management plans: information sharing,** The Leh experience has lessons for future protective planning – **short term.** The large resident Indian Army presence controlled emergency rescue operations after the flood. Army officers and staff worked with the local LADHC and a Relief Committee (formed a day after the flood) to deploy residents, tourists and 31 NGOs in a coordinated emergency response. 15 temporary camps were set up for 5,000 displaced people. Electricity was partly restored, the highway opened and emergency supplies flown in. Lack of earth moving equipment meant that mud dried quickly (in the high altitude arid atmosphere) and caused problems when it became windblown. The longer the mud was in situ the greater the damage to fields and homes.	**Review of Human Impacts** **Management plans: information sharing** The Leh experience has lessons for safeguarding planning - **short term** The Indian Army repaired infrastructure and provided food and shelter and bamboo structures for temporary schools. Lack of safety assessors and of weather predictions meant that residents were too frightened to return to their homes. Further immediate relief came in the form of bedding from Indian donors and rent relief to landlords provided by NGOs.

The Leh experience has lessons for future protective plans – **longer term** NGOs in Leh exploited online opportunities to **raise awareness** by **publicising** the news through NGOs, volunteer and tourism-related avenues. Many thousands of dollars were raised through appeals. NGOs were able to choose areas of need in which to assist from a list developed by the LADHC administration. All agreed that building basic shelter for some of those who lost their houses was a priority. WHC and ICHC protection may have led to greater international media exposure with consequentially greater international aid donations and help with respect to DRM, building safety issues and agricultural land rehabilitation.	The Leh experience provides lessons for the safeguarding of intangible cultural heritage over the **longer term.** A problem arose when the offer of aid to refurbish homes could not be rapidly implemented because of the delay in assessing the safety and long term security of sites. Convention protection could have led to more UN and UNESCO generated international media exposure of human plight and heritage threat on the world stage. The tragedy went largely unreported outside India. An international Conference/ Workshop: *Learning from Ladakhis* was held in August, 2011. This 2 week hands-on programme aimed to explore options for a safer and more environmentally sustainable future. The host - **Architecture Sans Frontières-UK** seeks to make community and international development issues integral to the practice and teaching of architecture. The workshop explored post-disaster rebuilding strategies for Ladakh, investigated the potential of local and traditional construction materials and techniques, and explored the role of new technologies in creating a safer and more socially and environmentally sustainable future.
Management plans: disaster risk assessments : The Indian Army was the lead agency in the disaster management response. A relief committee with representatives from all affected suburbs and villages was founded to assist the many residents who were totally unprepared and ill equipped to	**Management plans: disaster risk assessments** The proposed LAHDC's DRM plan would benefit from review by UNESCO experts in order to

deal with the disaster. Many traumatised Ladakhis rang friends from around the world to alert them to the disaster and request help. The disaster had no precedent. There was no **DRM plan** to address the scale and immediacy of the flooding. A DRM plan needs to be developed with UNESCO guidance and expert assistance and reference to the manual *Managing Disaster Risks for World Heritage (UNESCO et al. 2010).* The longer term response was uncoordinated. Knowledge about options for saving agricultural land should become a priority.  Precious irrigated fields rendered useless by the covering of hardened smothering mud which became a dust health hazard when dried and blown	safeguard intangible cultural heritage by attending to the well being of Ladakhi heritage workers including traditional designers and builders. Future post-disaster expert advice is needed on the rescue of agricultural land (for barley, wheat, potatoes and mustard and pea growing, grazing lands for Dzos, yaks, cows, sheep, camels and goats) from loss under mud, on the safety assessment of homes, on the safe placement of a new telecommunications station, on improvement of weather forecasts and on the assessment of options as how to access land clearing equipment in remote areas. To the majority of Ladakhis the rescue of their agricultural lands and the irrigation channels that feeds them would have been a top priority. No NGOs had the potential to help in this area. "Crops can only be cultivated on this land after the flood debris is cleared and the top soil is exposed," said Lobzang Tsultim, director of local NGO Leh Nutrition Project. The farmers can't clear this debris manually, they need JCB machines, which the government and NGOs need to provide to them." According to Tsultim, the government and NGOs are making no effort to restore the damaged land, on which the farmers' livelihoods depend, to its original state (Parvaiz 2010a).
Management plans: livelihood threat Future management plans need to consider the dangers of physical destruction from force of mud/rock/water slide and the difficulty of quickly removing hardening mud. Volunteers	**Management plans: livelihood threat** Of relevance to the caring for the 'safeguarders' of the LTD&CP

from all sectors of Ladakhi life, Buddhist, Muslims, soldiers, monks and tourists helped to dig out the mud filled homes after the disaster. The Army cleared roads and replaced washed away bridges. The THF helped prevent collapse of damaged historic buildings, mainly on the edge of the historic old town, and assessed buildings for their safety, advising whether families could return to their houses or not. An unknown number of families lost income earners. Acute shortages of labour and materials followed the emergency. The loss of agricultural lands undermines Ladakhi culture. The majority of families in Leh still have responsibilities to their relatives in villages. Many return for harvest time. Much Ladakhi cohesiveness is due to the traditional season cycle of work, relaxation, sociability, religious ceremony and celebration. Ladakhis are not used to being dependent on outside help.

New home being built in the traditional manner by local artisans using mud blocks. Photo LEDeG, with permission

was a home rebuilding program which was begun within a month of the tragedy funded by international donors. LEDeG became the onsite coordinator of an international online relief appeal which raised US$80,000 in 2 months. LEDeG teamed with SEEDS India which produced an ambitious and well planned reconstruction Strategy (LEDeG). SEEDS India is a post disaster construction group specialising in building small traditionally designed and constructed houses from mainly local materials using predominantly local skilled labour. The construction designs and materials minimise risk from future seismic events and floods, and keep homes warm in winter with reduced energy use. They are able to be built and extended with the participation of the 'recipient families' (SEEDS 2010). SEEDS claim it builds with a people-centric approach, being culturally relevant and being locally sustainable. Despite these aspirations SEEDS only built 14 homes before winter. A shortage of builders meant that workers had to be flown in. A shortage of materials was also a problem. Their work continued in the summer months. No retrofitting was done as residents were hesitant to return because of safety concerns over their buildings and sites. UNESCO Convention membership may have meant that there may have been appropriate DRM in place with quality post-disaster building and site assessment

<table>
<tr>
<td></td>
<td>available. In addition longer term protective and safeguarding plans may have been developed that recognised the threats to artisans and the community of LOT.</td>
</tr>
<tr>
<td>

Management plans: adaptation

Ladakhis managed their communal grief and 'adaptive' responses through their religious culture. A visitor reported that 5 days after the disaster "... the **Ladakh Buddhist Association (LBA)** organized a candle light march.... in prayerful support of the people affected. Vehicles were stopped to limit the dust, but the wind blasted everyone with it anyway" (Diwan 2010). The procession culminated at the top of Market Road, placing all the candles in a circle, with everyone's collective prayers for peace and harmony.

Candles accompany prayers to mark the end of March in support of those affected by the flood. Courtesy of Linkesh Diwan, Wise Earth Publishers

One NGO advised "Certainly, next spring the government and all concerned bodies will have to take a lot of precautions to prevent the catastrophe from recurring. Flood diversion channels can be built, drainage improved, and protective walls raised above settlements. Some building locations may have to be abandoned.

The THF offered a considered option. "The THF is proposing an alternative to building comparatively expensive shelters (between 2000-4000 Euros) that may not be suited to the local climate, and that may only be needed for six months. We found that it only takes two skilled masons and some helpers to make at least one room in each of the damaged buildings safe for the family to stay over winter. No expenses for new building materials are necessary, as everything from a traditional Ladakhi building can be recycled - including the mud bricks for the walls"(Alexander 2010). There is no record of this scheme being adopted.

An observer noted "there were two responses to the catastrophe. Many people in the affected villages opened

</td>
<td>

Management plans: adaptation

The government and Indian donors built one- or two-room shelters, from pre-cast concrete slabs that were trucked up across the Himalayas. These were not at all suitable to the climate of Ladakh and the life-style of its people. As one NGO commented "the coming winter will be very grim in a concrete box".

The force of the mud-flood was unprecedented

Some local Ladakhi NGOs advocated the building of shelters from concrete-enforced compressed bricks, but manufacture of such specialized materials is slow and many could not be built in time before winter.

The failure of some reinforced concrete frame buildings in Leh to withstand the mudslide may have been due to improper construction.

Future adaptive plans could assess these responses. Lobzang Tsultim asserts that adapting to the changing climate is the

</td>
</tr>
</table>

<table>
<tr><td>their homes to their devastated community members, welcoming them in, sharing, etc ... Others, more in Leh, which has suffered the 'development' agenda, looked for relief agencies to provide the housing, etc" (Diwan 2011).</td><td>better option. "You have to either adapt or become extinct," he says. "Though I am sure that we are paying for none of our faults, we have to think about adapting to the changes, which are taking place due to the actions of the developed world (in producing GHG)" (Parvaiz 2010b)</td></tr>
<tr><td colspan="2">Management plans: capacity building
This disaster demonstrated a lack of civilian DRM the response being dependent upon the military.
The SEEDS Strategy was admirable and would have been more achievable under a community-wide coordinated management programme.</td></tr>
<tr><td>Management plans: vulnerability assessments,
The disaster demonstrated the need for vulnerability studies of LOT in relation to future climate change impacts.</td><td>Management plans: vulnerability assessments,
Vulnerability studies in relation to climate change induced impacts on the practitioners and the practice of LTBD&CP are needed.</td></tr>
<tr><td>Management plans: health,

The new hospital which was nearing completion acted as the emergency hospital despite being flooded on the ground floor. Much equipment was lost.

The new hospital, nearing completion became the emergency hospital despite being flooded with mud</td><td>Management plans: health,
LOT's partly rebuilt drainage system survived. The Army effort ensured that immediate needs of food and drinking water were supplied. Future plans will need to address Leh's long term inadequate drainage, DRM preparedness for coping with the injured, protecting medical supplies, hospital access, sanitary conditions and psychiatric support for those affected by post-traumatic and longer term mental health instability. Windblown mud dust in addition to persistent high levels of diesel exhaust pollution added to health risks.</td></tr>
<tr><td colspan="2">Management plans: sustainable development
Small culturally sustainable businesses e.g. the cafe and gallery in LOT could have been impacted by the downturn of the tourist trade as a result of the flood disaster. However Leh has a defined</td></tr>
</table>

tourist season of March to September, so all businesses are familiar with very quiet trading periods. 5,000 Ladakhis were temporarily displaced by the disaster. It is not known how many were forced to, or chose to migrate from Leh, nor how many emigrated to urban Leh after losing their agricultural land.	
Management plans: mitigation Had there been WHC protection, **UNESCO** could have raised international awareness of this disaster as an example of the type of devastation that will become increasingly common with progressive climate change in particularly vulnerable regions. The community was already alerted by their religious leaders, some educationalists and NGOs (Ashleigh 2008; Wangchuk 2007) to an understanding of climate change as a global threat. Most Ladakhis blame climate change for this event. This is unlike in the west where climate change related disasters are dismissed as 'natural disasters' for which no one is responsible. Ladakhis are quite open in talking about climate change, saying historically it has never rained much in Ladakh. 12th century wall-paintings in simple buildings with flat mud roofs seem to bear witness to this (Alexander 2010). Many farmers have observed that their fruit trees are flowering earlier, that migratory birds are staying longer, and that wheat can be harvested at elevations previously only suitable for barley. "The snowfall which we do get, melts quickly," said Tashi Namgiyal, a farmer." We are now seeing pests in upper villages that used to be found only in villages lying lower. We are also witnessing shifts in sowing and harvesting of barley. Our region is arid and we have small glaciers which we draw water from. But over the last several years, many of these glaciers have receded. Not only this, we have seen some of our limited pasture lands drying up because of water scarcity" (Parvaiz 2010a).  A Leh resident rebuilding in traditional style with assistance from NGOs funded by Indian and international donors. Photo LEDeG, with permission. after LEDeG	**Management plans: mitigation** ICHC membership could allow such 'hotspots' as Ladakh to be included in the **UNESCO Climate Change Initiative**, so adding to the body of knowledge about iconic sites, climate change and possible mitigation and certainly 'adaptive' strategies The Leh flooding episode would under WHC and ICHC shielding become perhaps a 'case study' and have added weight to UN, UNESCO and World Heritage advocacy to stop atmospheric pollution. (Note the Dalai Lama has agitated for the US to pressure China to allow traditional Tibetan 'herd animal grazing' life style, which is sustainable, but which is being disallowed in favour of dam building, fenced farms and a population influx of sedentary Chinese farmers) (Wikileaks). Indeed, when speaking to ambassador Roemer the Dalai Lama remarked that the political agenda "should be sidelined for five to ten years" while the U.S. seeks to engage China on climate change and the environmental peril on the Tibetan plateau. Specifically, the Dalai Lama was concerned about melting glaciers, deforestation, and increasingly polluted water from mining projects. These problems, he explained, "cannot wait," adding that China's dams had displaced thousands of

<table>
<tr><td></td><td>Tibetans while leaving temples and monasteries underwater. In addition, the Dalai Lama recommended that China compensate nomadic peoples by providing vocational training which would emphasize alternative skills like weaving (Dalai Lama 2010).</td></tr>
<tr><td colspan="2">Management plans: population, justice and education security
As has been addressed in Chapter 22, the WHC's and the ICHC's protective capacities could be increased if a human rights approach was embraced. The Leh disaster presents as a classic case for potential benefaction from Convention membership. It also offers 'lessons' for heritage sites and practices. The majority of residents in Leh are Buddhists. They follow The Dalai Lama's teachings with enthusiasm. The Dalai Lama teaches environmental responsibility and has been successful in leading millions to an understanding of climate change "In the first place we must strive to overcome these states of mind by developing an awareness of the interdependent nature of all phenomena, an attitude of wishing not to harm other living creatures and an understanding of the need for compassion. Because of the interdependent nature of everything we cannot hope to solve the multifarious problems with a one-sided or self-centred attitude. History shows us how often in the past people have failed to cooperate. Our failures in the past are the result of ignorance of our own interdependent nature. What we need now is a holistic approach towards problems combined with a genuine sense of universal responsibility based on love and compassion" (His Holiness the 14th Dalai Lama of Tibet 1991). "Destruction of nature resources results from ignorance, lack of respect for the Earth's living things, and greed". (The Office of His Holiness the Dalai Lama 2011)</td></tr>
</table>

Chapter 30
Conclusion
The Future is with us

As a rule, indeed, grown-up people are fairly correct on matters of fact; it is in the higher gift of imagination that they are so sadly to seek. The Golden Age. The Finding of the Princess (1895) Kenneth Graham

The Question is 'What is the Right Thing to do Now'?

This is how Professors Lyndel Prott and Patrick O'Keefe, veteran convention negotiators and international heritage law specialists, open their students' minds to the study of UNESCO's Conventions. They explain that this is the starting point for Convention creators. It is also the stance when interpreting these global standard setting instruments. It serves well in orienting this examination of the World Heritage Convention (WHC) and the Intangible Cultural Heritage Convention (ICHC) towards the future.

The genesis of this book was a dissertation that one of the authors (Rae) wrote for a Master's degree in Museum Studies at the University of Queensland. The lead-up course on international heritage protection was taught by Professors Prott and O'Keefe. On learning more about cultural heritage under the guidance of Professor Amareswar Galla, and about the Conventions' efficacy (despite the lack of specific consideration of climate change) from my learned teachers, I became increasingly impressed, at times gripped, by the Conventions' embodied ambition and potential value in the global care and protection of humanity's most special natural and cultural places and practices. Further, I was awed by the prospect of universal ratification. The inquiry I undertook has deepened my respect for how difficult and strenuous were the making of these international agreements. It is my belief that the Conventions retain their status as grand instruments with the capacity to contribute even more greatly in future. The WHC has stood tests of interpretation and found not to be wanting for far sightedness. Perceptive wording has injected checks and balances such that global interests are always the priority. The ICHC is in rapid formative mode but shown to be open-minded to the climate change challenge, to which, so far, it has not been formally challenged from within or without.

Ladakhi women spinning their prayer wheels at a climate change rally in Leh organised by the Young Buddhist Association and the international group 350.org.
Photo courtesy of Conor Ashleigh

The Means by which the WHC and the ICHC can be Updated

The study set out to recommend 'the means by which' the WHC and the ICHC can be updated to address the impacts of climate change on cultural heritage.

Thirteen lines of inquiry in the form of questions were asked of the Conventions, these answered and recommendations proposed. The recommendations are anchored within the terms of reference of the WHC and the ICHC and their unequivocal aims of protection. With respect to the WHC each State Party recognises their duty " of ensuring the identification, protection, conservation, presentation and transmission to future generations of the cultural and natural heritage" on its territory from village to state level and internationally and it "will do all it can to this end, to the utmost of its own resources" (WHC Art 4) and "that such heritage constitutes a world heritage for whose protection it is the duty of the international community as a whole to co-operate" (WHC Art 6). Similarly with respect to the ICHC each State Party agrees "to safeguard" ..."to ensure respect for" ..."to raise awareness" ... "ensuring mutual appreciation" of intangible cultural heritage (ICHC Art 1) and will work "ensuring the viability of the intangible cultural heritage, including ...protection ... enhancement, transmission ...as well as the revitalization of the various aspects of such heritage" (ICHC Art 2.3) and that each State Party will "keep the public informed of the dangers threatening such heritage" and to "promote education for the protection of natural spaces and places of memory whose existence is necessary for expressing the intangible cultural heritage" (ICHC Art14).

It does well to reacknowledge the 'greatness' of these two cultural Conventions; how much heritage has been brought under international agreement, how much cultural creativity has been substantiated in practice and in records, how many iconic nationally protected natural and cultural areas and performances, customs and expressions are monitored, to help one appreciate the potential contribution, taking into account how much is at stake, that the two Conventions could make to protecting and safeguarding heritage against climate change impacts.

The recommendations arising from the 13 questions reinforce and elaborate the theoretical approach to addressing climate change taken by the World Heritage Committee as described in their suite of dedicated but uneven climate change publications and propose that a similar process (see Chapter 15) could be taken by the Intangible Cultural Heritage Committee. The recommendations include recognising and incorporating climate change impacts on humanity particularly livelihood insecurity, displacement and health and the consequential costs to tangible and intangible heritage of these 'human impacts' into the policy documents of the respective conventions. The discussions and recommendations arising from the examination of the WHC guide this book into an inquiry exploring the capacity of the ICHC to address climate change impacts. In so doing areas of overlap, similarities and differences emerge in the potential responsiveness of the two Conventions to the protection/safeguarding of heritage. The summary table ***Comparing the Conventions,*** in Chapter 27, allows a quick visual perusal of these and in addition assists the identification of strengths, of needs and of gaps in our knowledge that may be ripe for research.

It is hoped that the questioning approach adopted in this book has helped in the exposure of climate change problem areas which should be central to heritage management. Important examples are mitigation and migration that either intentionally, or through paucity of knowledge are marginal to the WHC's focus. The avoidance by the Conventions of areas that they consider to be beyond their 'competencies' and under the ambit of the UNFCCC is challenged and re-examined. The discussions clearly differentiate the capacity of the World Heritage Committee and of UNESCO to review, redirect and reform from within and the ability of the World Heritage Committee, UNESCO and the UNFCCC to review and reform in response to 'petition' pressure. There are identified capacities within the Conventions, some partly recognised and some that need to be tested further, for improving protection.

The statistical history and climate change vulnerability/Human Development Index analysis of the WHC reveals not only how successful has been the ratification uptake but how equitable has been the inscription attainment over 40 years, an admirable but unrecognised achievement. These analyses, including of the ICHC, and with reference to the Data Set in Appendix 1, clearly identify to which States Parties and non-States Parties future advocacy and effort should be directed. With every new member of the WHC and ICHC community and with every new inscription the security enhancing roles of the Conventions are strengthened.

The ethical dilemma of climate change for UNESCO is the nub of all challenges and the one that perhaps calls upon the skills of convention professionals the most. Presently it is like an iceberg of potential action with the glistening exposed ice surface being the aspirations for a **Universal Declaration of Ethical Principles in relation to Climate Change** and the many times larger but inapparent submerged section being international political opposition and inertia. The aspirations of the Universal Declaration of Ethical Principles in relation to Climate Change desperately need support from the World Heritage and Intangible Cultural Heritage communities. The central moral question of climate change and intergenerational justice, although enshrined in both Conventions, has not been acted upon in relation to climate change impacted heritage.

The question of climate change's health impacts, like that of climate change's impacts on displacement and migration, has been hidden within other causation categories such as poverty and natural disaster. Health professionals and humanitarian authorities are attempting to have these impact areas recognised and Convention climate change policies need to do likewise. Matters such as the mental health burden of climate change, that ranges from solastalgia to suicide, need to be recognised in future Convention policy.

Whilst the physical and biological impacts of climate change are being monitored with increasing accuracy at global level by the UNFCCC and environmental information from some World Heritage sites is being supplied to the UNFCCC there appears to be no systematic mandated collection of possible or probable climate change related World Heritage site damage. Whilst prospective global climate change consequences may be comparable in scale to an 'asteroid' impact, (Gilding 2012; Hansen 2012b), and although climate change is being referred to seriously in the WHC's publications, the

WHC and ICHC are lagging in translating the gravity of the scientific predictions into appropriate World Heritage and ICH global positions on climate change mitigation and into adequate protective and safeguarding actions at site and element level.

The role of culture and of cultural heritage in addressing climate change, even within the time frame of the writing of this study, has been drawn more decisively into the debate... now to be seen as central to 'tackling climate change' (UNESCO 2012a). The time has come for "Culture" and "Climate Change Resilience" to be added as the 4th and 5th pillars of sustainable development.

The legal history of the WHC, Chapter 26, provides some insight into the contrast between the manoeuvrings employed by some States Parties' representatives in order to favour their nation's short term economic interests over the longer term interests of humanity and global heritage, and the more immediate interests of other Convention participants from higher minded more objective States Parties.

It is evident from pre-decision background meeting reports that the World Heritage Committee has struggled with the problem of protecting World Heritage from the impacts of climate change. The eventual Decisions reflect diplomatic compromise and not infrequently loss to the well articulated agendas of wealthy States Parties. The decisions can sometimes show a sidelining of scientific advice, a reticence to engage with the latest science, and an inability to facilitate a closer working relationship with the UNFCCC and the IPCC. This may in part be due to a paucity of scientists on the advisory bodies, a lack of scientific briefing of the World Heritage Committee and a lack of scientifically qualified World Heritage advocates. Whilst political pressures prevented the IPCC from naming climate impacted World Heritage sites within its fourth Assessment Report, the World Heritage Committee has requested the IPCC to dedicate a chapter to World Heritage in their future assessment reports. To date there is no indication of progress.

The *Policy Document on the Impacts of Climate Change on World Heritage Properties* has stimulated strong legal criticism and generated petitions on deeply concerning matters such as for example, the petition on *The Role of Black Carbon in Endangering World Heritage Sites Threatened by Glacial Melt and Sea Level Rise* (ACJP and Earthjustice 2009). The petitioners in the foregoing example explain that the UNFCCC does not collect data and therefore does not address this critical short lived climate warming agent that results from the inefficient burning of fossil fuels, biofuels and biomass. Black carbon is considered to be the second most powerful contributor to global warming after carbon dioxide, and because of its darkening of snow, warms the planet three times more efficiently than CO_2. As the petitioners explain World Heritage sites are unprotected from this threat. "The World Heritage Committee has the opportunity to advance critical early action on this issue and take important steps to preserve World Heritage until the effects of the UNFCCC (Kyoto) process can be realised. Because the causes of climate change are inherently *transboundary*, and because the impacts of climate change on such World Heritage sites as the Ilulissat Glacier in Greenland could trigger catastrophic tipping points that would accelerate damage to other diverse sites around the world, protection of World Heritage requires new thinking and approaches" (ACJP and Earthjustice 2009 p2). The World Heritage Committee has responded to this petition as tabled in Chapter 26, but has done little to address the global dimension of the problem.

The legal history of the WHC also reveals both the flexibility and potential of the WHC. That the increasingly popular yet ever less environmentally prioritising term, sustainable development, is now embraced by the WHC to encompass tourism suggests that the World Heritage decision makers are compliant to persuasion by States Parties to accept what may primarily be the States Parties' economic aspirations. The potential lies in the prospect that the World Heritage Committee's stance on addressing climate change could, in a readily justifiable display of flexibility, make environmental sustainability its highest priority.

Further Means to Protection and Safeguarding

Overall these 'further means' centre on:

- improving the dissemination of up to date accessible and authoritative information about climate change threats, climate change mitigation and heritage protection,
- the inclusion of Human Rights principals especially the right to know and the right to intra- and inter-generational climate justice and
- assistance with adaptation (often meaning 'repairing' or 'coping with') to the inevitable impacts of climate change upon cultural heritage.

The means by which the Conventions can be updated to address the impacts of climate change partly involve the further development of the Conventions' capacity to:

- use and extend their world wide reach,
- capitalise on UNESCO's knowledge base pertinent to climate change (latest science, latest strategies; knowledge accessible from networks and links with the UNFCCC, UNDP, DHR; historical and archaeological knowledge about past communities that succeeded and failed, knowledge about current thriving materially poor yet culturally rich traditional communities such as Ladakh and knowledge about communities that successfully negotiate the management of 'commons') and on the feedback from communities concerning their climate change observations and experience,

- exploit their status of high repute and authority to cut across the barriers of partisan interests to maintain a higher order level of understanding and objectivity towards climate change from which authoritative climate change information and advice can be disseminated,
- further hone their already high order World Heritage and Intangible Cultural Heritage (ICH) institutional intervention skills to effect climate change programmes. Convention staff work at all levels from village, through regional hubs, area offices, to the halls of power. They have, over four decades capacity built, trained, produced education materials and developed web sites. They are masterfully equipped to encourage, urge, use moral persuasion, set good examples, use emotional and culturally diverse leverages, and yet always protect their reputation. These skills maximise co-operation. The text of the WHC contains fifteen references to 'international assistance', emphasising the centrality of global co-operation and partnerships to the protection of world heritage (Baxter 2006 p70),

to help implement community-based climate change management strategies in communities with tangible and ICH requiring protection.

Priority 1

The following recommended undertakings are additions to the currently practised actions of the WHC and the ICHC:

A proposal to develop a joint WHC and ICHC **'Cultural Heritage Vulnerability Campaign' (including proposed 'heritage site vulnerability campaign' 15.4)** with a *'decade to make a difference'* style message and comprising of two levels of action:

- A site/element based sub-campaign exploiting the proposed **climate change vulnerability scale** and the proposed climate change crisis **symbols** to raise awareness, educate and cause concern about the impacts of climate change on cultural heritage.

This proposal derives from recommendations 14.4, 15.2, 15.3, 15.4, 22.7, 25.2, 26.5.

- An internationally based sub-campaign to openly address climate change mitigation as the transboundary problem it is. This **global mitigation movement** would aim to protect cultural heritage by making greenhouse gas (GHG) reduction, under CBDR, an obligatory part of Convention implementation.

This proposal derives from recommendations 15.1, 15.2, 15.10, 19.4, 20.1, 20.3, 20.4, 24.4, 25.2, 26.1, 26.2, 26.3, 26.4

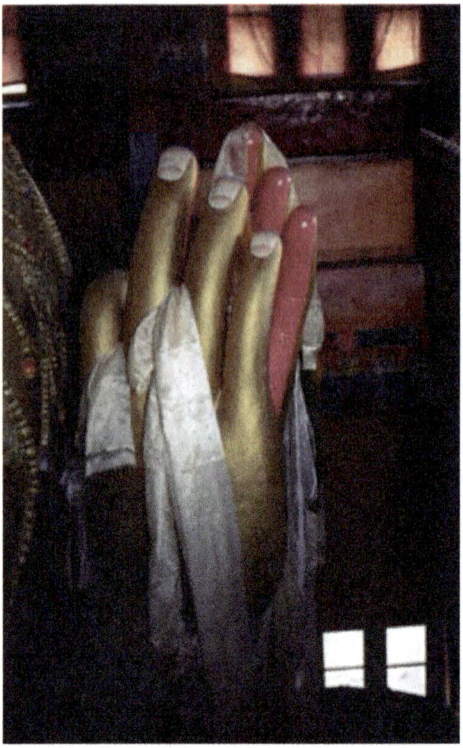

Shey Gompa. The hands of compassion

The above two proposed movements are within the Conventions' legal jurisdiction. Authoritative legal opinion argues that States Parties with large carbon footprints are breaking their WHC agreement by not making deep cuts to GHG emissions (Baxter 2006; Rothwell 2004). World Heritage advisers have themselves said "The threats to World Heritage properties from the effects of climate change must be confronted head-on and with specific legal obligations spelled out very clearly if the States Parties are to take these issues seriously (UNESCO 2007e p38)."

The WHC and the ICHC are based on *international cooperation* and *international assistance*. Altogether these phrases occur *33 times* in the WHC and the ICHC. Under these treaties, States Parties agree to do all they can to protect and preserve World Heritage sites and ICH, including agreeing not to take "deliberate measures which might directly or indirectly damage" World Heritage (WHC Article 16.3) and to "ensuring respect for" and "mutual appreciation of" ICH (ICHC Article 1) while keeping "the public informed of the dangers threatening such heritage" (ICHC Article 14b). From these commitments arise obligations to stabilise and reduce GHG emissions for it has been established that these are causing climate change and that climate change is, and will increasingly damage tangible and ICH. Given these obligations and mindful of the reluctance of the United States to sign the Kyoto Protocol, the WHC and ICHC could provide an international forum beyond the UNFCCC and the Kyoto Protocol that could meaningfully address both climate change mitigation and adaptation. The WHC provides the opportunity to bring the USA and similarly inclined States Parties into dialogue and for the global community to discuss both the effects and solutions related to climate change (Thorson 2008 p123).

Some observers are forecasting that the transformation of society away from a carbon and growth driven economy to one reliant on 100% renewables will be driven from the bottom up (Guilding 2011). To have the WHC and the ICHC advocating mitigation from site level up, would be in keeping with what the 3rd Nobel Laureate Symposium called "the Mind-shift for a Great Transformation" (3rd Nobel Laureate Symposium 2011a). The WHC and the ICHC are ideal intermediary bodies through which to bring about such planetary change as they allow direct citizen participation through a petition process (Thorson 2008 p38). Sympathetic with the Nobel Laureates is UNESCO, which this year released its *Input to the Rio+20 Compilation Document* that reveals UNESCO's well considered advocacy of a greener and more equitable form of sustainable development with the adoption of a radical new agenda to create green societies with green economies that will necessitate a break with BAU (see pre-Rio+20 discussion and post Rio+20 postscript in Chapter 22).

The site-based sub-campaign at priority 1 could be expressed through the emblem regulations such that every site based World Heritage and ICH emblem and related publication could be accompanied by a climate change crisis symbol, a climate change prognosis sign. The internationally based sub-campaign at priority 1 that addresses GHG gas mitigation could assist protection and safeguarding of world tangible and intangible heritage and contribute to *the possibility of a sustainable*, equitable future for all, by obligating States Parties to move to renewable energy sources thus reducing carbon emissions.

This may well be the last decade of opportunity to promote the message of universal responsibility for the protection/safeguarding of our world cultural heritage. This could be assisted by publicising at every World Heritage Site, with every ICH practice and in every communication the priority of GHG pollution reduction. Such a Cultural Heritage Vulnerability Campaign offers opportunities for interested individuals and groups, through to the World Heritage and ICH Committees and the UNESCO General Assemblies and Conferences to generate greater concern. Such a programme could assist the World Heritage and Intangible Cultural Heritage Committees in their international endeavours to promote climate change mitigation.

As described in Chapter 22, the World Heritage Committee sees no ambiguity in embracing profit motivated sustainable development in the form of tourism despite the environmental cost of its transport, accommodation and ancillary carbon- intensive activities. In the sustainable development methodology example, consideration of climate change was ranked 14th out of a total of 16 considerations. Sustainable development is now mainstream

World Heritage management, in fact a condition of site management. **Sustainable tourism** has been enthusiastically embraced as applicable to *almost every* WH site. *Operational Guidelines* have been significantly amended, reports and research studies have been published and a manual is planned. 'Sustainable development' has however made no specific accommodation for its most serious longer term threat, climate change. With the call for 'a break with BAU' to survive into the next century, 'sustainable tourism' may have to be reconsidered in terms of its total associated carbon footprint. For instance what is the carbon footprint of visitations to Easter Island and how truly sustainable are such visits?

Many World Heritage sites are national economic power houses. ICH festivals, performances and rituals could develop likewise. Some have already. Climate change puts this income stream in jeopardy. The WHC has been a party to this neglect by deflecting its international and national mitigation role by not obligating States Parties to legislate specifically to reduce GHG emissions and in not setting prescriptions for including climate change considerations in the management plans of all World Heritage sites and ICH practices threatened by or potentially threatened by climate change, recognizing, of course, that World Heritage protection will be only one, but a highly significant benefactor if the global need to drastically curb GHG emissions is met (UNESCO 2007e p36-8).

Envision the Sydney Opera House behind a dyke or as a harbour island. How will the OUV survive? Will the world's 20th century architectural masterpiece qualify for 'in Danger' listing? No longer can mitigation be 'one of the mixes' of World Heritage actions in response to climate change. Mitigation should be *the* first concern because unless the right decisions are implemented within the ten year frame for effective action, all other actions will be for naught. The case in point – the contested issue of not listing many climate change affected World Heritage sites on the 'in Danger' list except on a case-by-case judgement using no clear quantitative criteria, and with the emphasis on adaptation rather than mitigation, is becoming increasingly untenable. To accept the emphasis on climate change 'adaptation' as against mitigation is to accept a palliative care stance on climate change threatened heritage rather than a preventive or rehabilitative stance. The history of this problem is detailed in Chapter 26.

Priority 2

Reconsider the means by which objective information (new knowledge and indigenous knowledge) may be generated and disseminated paying special regard to the mediating role of culture. Recommendations regarding this are described in Chapter 23 but it is appropriate here to refer again to the role of culture in knowledge sharing. The physical and biological impacts of climate change influence, and are in turn perceived through the lens of, our cultural framework e.g. the Mother Earth movement and the Buddhist belief in responsibility for the Earth. New knowledge generation in the context of this examination of the Conventions would include that arising from research undertaken in response to the identified 'gaps in knowledge' – gaps that relate to the role of culture in communicating the understanding of climate change as well as gaps that relate to predictions about how our cultural heritage will be impacted, see Chapter 27. As recommended in *UNESCO's Input to the Rio+20 Compilation Document* (UNESCO 2012a p17) the integration of local and traditional knowledge and management practices of both women and men provide valuable insight and tools for mitigating the effects of climate change. It is a reflection of our 'individualistic' consumption driven life paths with our illusion of exceptionalism that determines our level of objectivity and steers our behaviour. Our narcissism and subjectivity combined with our diminishing scientific literacy make us easy prey for powerful information distorting vested interests and a high proportion of politicians. Matters are worsened by a predominantly ill-informed and sometimes manipulative media the focus of which is influence and sales, not truth. All this leads us to a greater separation from objectivity and to the consequent alienation of science. Unquestionably the status of science has been undermined in the west.

The devaluation of science with the consequent rejection of climate science has caused some scientists to leave their labs to become more effective advocates of the climate change message. Scientists are not yet leading the powerful world think tanks, but they are becoming, some reluctantly, public intellectuals, journalists, philosophers, psychology researchers, writers of popular books, proposers of economic policy, activists and legal advocates. They are setting the way for they know 'the future is with us'.

Priority 3

Make the **listing, documenting and inventorying of further WH properties and ICH elements an urgent priority.**

The benefits from these actions will give States Parties an improved chance to protect heritage and safeguard against the impacts of climate change.

Priority 4

Develop a 'holistically thought through' **WH and ICH climate change strategy framed in a human rights** perspective that prioritises cultural heritage and takes into consideration

- socioeconomic impacts
- migration and displacement
- health
- sustainable development
- security

Enthusiastic young performers in traditional dress at the Leh Festival

Priority 5

Implement climate change **adaptation projects** that are well based on the methodologies

- vulnerability and risk assessment
- information sharing
- carbon neutrality

to protect and safeguard cultural heritage.

Knowledge Bases

The sources that provide information to supplement and update the Conventions knowledge bases vary in nature and quality. Leading environmental lawyers said as far back as 2007 'Taking into account the latest scientific research and data concerning climate change and its effect on biological diversity, protected areas and the cultural heritage, the urgency of conducting law and policy research in this area cannot be overstated. The threats to World Heritage areas from changes in the climate are already very significant, as demonstrated by prominent examples such as the melting of World Heritage-listed glaciers, as well as by the less obvious but equally important threats and changes to floral and faunal habitat which will increasingly require a reconsideration of the World Heritage criteria for which many areas are listed (UNESCO 2007e p38).

The knowledge of physical and biological threats is substantial. The knowledge of their impact on immoveable world and national heritage has a defined research pathway and young and small body of knowledge. The human impacts of climate change on World Heritage have been partly recognised, but the knowledge base is nascent. The awareness raising of physical, biological and human impacts of climate change on ICH is coming first hand and online from the voices of those working at the desperate climate frontlines in the warming, drying and starving regions of Africa, the melting Arctic, the Pacific islands undergoing relentless inundation and the mountainous regions of the world which are losing more than their glaciers and regular water supply. These experiences have yet to attract significant research.

With the recommended upgrading of the right to scientific knowledge for heritage carers, the Convention Committees themselves, through lobbying from heritage workers, may themselves be better informed and so anticipate further petitions and submissions that will arise from concerned conservation groups and other engaged groups. The Convention Committees may act proactively in developing better protective mechanisms, especially in relation to initiating action through the courts of the States Parties and through the International Court.

There are inescapable knowledge realms that are both central to heritage survival and as such should be openly discussed, yet are treated as 'classified', as if climate change is a political and economic player in a 'war footing' diplomacy, and thus outside the public arena. Aubrey Meyer has alerted us to disturbing top level thinking on the costs of climate change mitigation, and the abandonment of poorer nations for the sake of the wealthy (Flannery 2008 p208), see Chapter 10.

Another area of great concern where **the right to know** seems irrefutable, relates to the demonstrable obfuscation and distortion of information by global corporations allied to the fossil fuel industry. We would argue that any inquiry into the protection of global heritage has a duty to recommend confronting interests whose actions will undermine the value of cultural heritage. Leading and authoritatively briefed governments are being thwarted in their actions to reduce emissions by pressure from the fossil fuel lobby which is supported by quasi-scientific but 'authoritative looking' think tanks, which themselves are funded by fossil fuel interests. This, and their preoccupation to deliver political messages to sure up the elections of supporting political parties, are angering climate scientists who are urging mitigation. The interests of the young and unborn are being dismissed – discounted without any informed public conversation. The relationship between government and vested interests has become even more sinister with the emergence of *geoengineering strategies* many of which might entail known and unknown risks and might be used to prolong the use of fossil fuels and delay mitigation. In mid 2011 the IPCC discussed, without wide public debate, the options for funding research into geoengineering our future to alleviate the symptoms of climate change (IPCC 2011). UNESCO also called a first-ever expert meeting on geoengineering to explore its potential for mitigating climate change. A policy brief for government decision-makers and the publication *Meeting Report of the Intergovernmental Panel on Climate Change Expert Meeting on Geoengineering* (IPCC 2012) resulted. As far back as 2005 Tim Flannery expressed concern. "One of the most disturbing things about this issue (geoengineering solutions) is that governments in the US, Australia and elsewhere are debating with industry right now, behind closed doors, on how much risk they will accept on behalf of their constituencies, and how much industry will bear" (Flannery 2008 p253). Heritage workers are among the first to be appalled at this irresponsibility, at the unfathomable dangers such actions may pose for heritage, knowing bitterly, that many of those who are traditional heritage holders have originated from or are at the heart of cultures where responsibility to *future* generations is an integral concept.

The intricate woodworking and colour design of Hemis Gompa (monastery)

Methodological knowledge gaps are also apparent. Climate change vulnerability assessment and to a lesser extent risk assessment are demanding and complex disciplines. When they are applied to immoveable cultural heritage an across-the-board range of knowledge is required. When they are applied to ICH there are no precedents. How do practitioners assess the vulnerability of their ICH? How do artisans, performers, elders match the safeguarding of their valued ICH with the challenges of transmission in the ever changing climate impacted world? Perhaps their ICH has a relevance and resilience that the immoveable cultural World Heritage does not? How do ICH holders control, and transmit their ICH under threat of societal fragmentation following displacement, a pandemic, or more likely the insidious erosion of livelihood? Is ICH amenable to monitoring in relation to climate change impacts? How do ICH holders deal with loss of their intangible heritage? And how do they decide what to let go? This area of ICH assessment, whether it is being worked out within communities or universities, informs and influences and invites new and innovative methodology.

Over the course of this study the authors have had the chance to compile three tables (Assessing Intangible Cultural Heritage Vulnerability to Climate Change, Insert 30.1; Intangible Cultural Heritage Risk Assessments, Insert 30.2 and Tangible Cultural Heritage Risk Assessments, Insert 30.3). The tables may best serve as mind expanders – opening up the plethora of climate change impacts and their many ramifications. Assessors and their communities could select their lines of inquiry for data gathering across the tables, being confident that they had 'covered' the most important parameters.

Insert 30.1 **Assessing Intangible Cultural Heritage Vulnerability to Climate Change**

Climate Change-induced Culture indicator	Culture Risk	Impact on Intangible Cultural Heritage
Migration or birth/death rate related changes in community size may cause changes in the numbers and relative proportions of people from differing constituent groups.	Changes in community population size, age structure, constituent group proportions and marginalisation of minorities may adversely affect cultural practices, even causing discontinuance then loss.	ICH practise, transmission, and awareness, modified, reduced or lost with creation of a different balance. Exceptions exist: A fall in numbers of ageing monks at a remote Ladakhi monastery was brought to the attention of the small community who immediately volunteered 14 of their children to train as monks.
'First' language less in use.	Language loss causes reduction in key cultural identity domain.	ICH practise, transmission, and awareness, modified, reduced or lost.
Loss of land.	Local, regional or international displacement, leading to competition for space and resources, displacement of cultural by economic needs, loss of special places and special resources that are integral to cultural practices and, loss of identity, loss of relevance of culture under new circumstances.	ICH practise, transmission, and awareness, modified, reduced or lost. ICH that is tied to no longer accessible sites or specific resources that are no longer available will need to be either modified or no longer practised.
Population pressure.	Competition for resources spread of disease.	ICH practise, transmission, and awareness, modified, reduced or lost.
Water stress.	Water shortages lead to crop failure, death of domestic animals, restricted water for cooking, bathing and drinking causing poverty and malnutrition. These lead to increased susceptibility to disease. Water contamination causes water-borne diseases, debility and death. These will interfere with the pursuit of cultural practices to which a lack of raw materials for cultural practices may contribute.	ICH practise, transmission, and awareness, reduced or lost. e.g. loss of faith in water rituals affected a Mexican ICH inscription, loss of plants that supplied raw materials for dyes for paper making or textile weaving.

Less resources from oceans.	Lack of ocean resources may lead to changes in dietary sources and food shortages. If fishing practices or seafood diet were linked to ICH then changed circumstances will impact on ICH.	ICH practise, transmission, and awareness, modified, reduced or lost.
Biodiversity change loss of species, change in distribution, flora & fauna extinctions.	Shortage of ritual food, ritual animals, and raw materials for cultural attributes; loss of symbolic native plants & animals; pestilence.	ICH practise, transmission, and awareness modified, reduced or lost.
Loss of income.	Fewer resources for festivals, parades, performances and other cultural practices.	Reduction in ICH practise.
Unemployment.	Relocation, separation, fragmentation, poverty	Cancellation of festivals e.g. annual cultural festivals of dance & song, drama and games. Causing stress and loss of ICH.
Reduction in tourist numbers.	Loss of income	Selective reduction in ICH then and reduced community awareness.
Resources being redirected away from ICH to disaster management and medium term climate change projects.	Less time & resources to practise ICH	Reduction in ICH practice.
Infrastructure change.	Less/more movement of people, education changes, standard of living drop/rise.	ICH Practise, transmission, and awareness, modified, reduced or lost with the creation of a different ICH balance.
Community fragmenting due to migration.	Too few people to maintain culture.	ICH loss, transmission difficulties, not recreated, loss of cultural diversity and human creativity.
Remembering the lessons of *Collapse* by Jared Diamond and 'sharing the commons' by Elinor Ostrom, see Chapter 9. ICH has to be sustainable and has to change, otherwise facing change would be handicapped and could be disastrous (Diamond 2005). Climate change is a multiplier of threats.		

This table presents generalised information applicable to Schroter's 8 Step Approach to assessing vulnerabilities to the effects of global change. (Schroter, Polsky, Patt 2005).Other approaches can be found in the following references (Adger 2006, Eakin 2006, Fussel 2007, Fussel and Klein 2006, Liverman 2008, Parry 2007).

Young Monk who when asked whether he thought he might become a Head Lama replied "not in this life"

The terminology and knowledge of climate change science has yet to penetrate our everyday language. We speak of the Southern Oscillation Index or of the futures market with greater authority than 'thermal inertia', 'acidification of the ocean 'and 'positive feedbacks'. Because the threat is not being communicated, our universal instruments - the Conventions - are not being stretched to perform to protect and safeguard. Unfortunately the 2012 *UNESCO's Input to the Rio+20 Compilation Document* also avoids direct terms such as greenhouse gas emissions that are at the very base of climate change. UNESCO's superlative diplomatic expression can soften the message (see Chapter 22). From a future vantage point, it is well to be unambiguous and clear. James Hansen, a pre-eminent climatologist explains why he speaks out. He has a number of grandchildren. He realized he did not want them to say, "Opah understood what was happening, but he didn't make it clear." (Hansen 2012b).

The Cultural Conventions

By ratification standards both of the Conventions are successful. They have brought unparalleled organisation, merit and esteem to humankind's heritage. As short term protectors of heritage from the threat of climate change the Conventions are the Davids who will be increasingly challenged by the

Goliath of climate change. Evidence has shown that the WHC is able to deliver short term climate change responses at the local level and has been able to generate and share experiences of site specific, mainly adaptive, climate change initiatives. There is evidence, though, that the Conventions have the potential to be more effective. The recommendations and priorities 1 to 5 have been proposed to address this. The Conventions have a mainstream role as part of the momentum to awaken and prepare mankind for the reality of the difficult climate change adjustments ahead. No matter what the past perceptions of the 'competencies' and limitations of the Conventions, the question to ask is "What is the right thing to do now? The Conventions are well equipped to be strategic, measured, powerful and persuasive. In one to three generations time under business as usual (BAU) policies the Outstanding Universal Values that the WHC strives to preserve and the intangible cultural heritage the ICHC with such commitment strives to safeguard, may be very severely affected and even more grievously threatened by what by then will be 'in the (climate change inertia) pipeline'. Continuation of present carbon intensive economic activity will by 2100 see the Conventions as palliative care hospices for an accelerating succession of ailing sites and dying practices. It is concerning that under the current protracted World Heritage case-by-case decision determination process the inclusion of climate change threatened sites on the 'in Danger list' (Decision 30COM 7.1) is subject to the proviso clause that climate change threats "must be those which are amenable to correction by human action" (most often meaning by the actions of the State Party) (Decision 32COM7A.32). It needs to be recognised that most of the world's presently climate impacted listed cultural heritage may not qualify for 'in Danger' listing because the States Parties cannot on their own address the global causation of climate change. However this is a false argument as the climate change impact can be alleviated by concerted human action to withdraw from the carbon economy.

Insert 30.2 **Intangible Cultural Heritage Risk Assessments**

Climate Indicator	Climate Change Risk/Risk Multiplier	Physical and Biological Impacts on the Environment and their consequential effects on the ICH's Custodians and their Artefacts	Consequential Human Impacts on Intangible Cultural Heritage
Temperature. Change. (often in association with changes in precipitation)	Warmer weather. Seasonal changes in growing season. Warmer, rising oceans. Glacial retreat and loss. (Polar ice shelf and ice sheet loss – see under sea level rise.)	Crop failure, domestic animal deaths, livelihood impacts, food shortages and poverty. Influxes of plant and animal pests further impact on food production causing famine and malnutrition. Changes in agricultural practices with need for crops and grazing animals that are heat tolerant. Impacts of food shortages on morbidity and mortality especially of infants and the elderly (who are the principal bearers of traditional knowledge). Changes in disease distribution and prevalence. Warmer wetter climates favour arthropod-borne diseases such as dengue and malaria. Fall in biodiversity of ecosystems including sea-life. Species extinctions, loss of symbolic 'clan' animal and plant species. Reduction or loss of connection with 'nature'. Erosion of responsibilities and obligations to home environment. Loss of raw materials for economic production, religious use, and for tool, instrument, craft, artefact and traditional medicine making.	ICH practise, transmission, and awareness, modified. Hunting, gathering, fishing practices and knowledge may become unsustainable. Loss of land, loss of community cohesion, conflict with neighbours, mental health issues, lack of food and poverty may all lead to neglect of ICH. Loss of 'Critical Space' or appropriate venue may threaten ICH performance or practice. Reduction in tourist numbers may cause a reduction in cultural practice. Traditional practices may also be enhanced by hardship. ICH of disadvantaged minorities tends to be most at risk. People may be too busy, too few, or too weak to practise ICH. There may be too few ICH stakeholders: performers, singers, medicine makers,

		Undermining of protective rituals, belief systems, seasonal customary practices, with social impacts being greater on traditional communities. Loss of traditional knowledge. Changes in family structures may occur as sources of livelihoods may become more distantly dispersed. Abandonment of traditional lands with regional or international displacement. Major dislocation may lead to societal stresses that result in abandonment of religion, fatalism, emergence of religious cults, profound distrust of government with anti-government riots, social disintegration, civil violence and even terrorism.	craftspeople to practise ICH. Differential susceptibility to disease and mortality amongst the elderly custodians of ICH diminishes transmission to younger people. If ICH is no longer relevant it may be discontinued: e.g. seasonal celebrations, performance times, feasts, and practices related to a crop that is no longer sustainable, or livestock that are no longer available. Fewer community members to embody and practise ICH leads to cultural loss. Loss of income and societal fragmentation threatens ICH. Relocation threatens loss of ICH. Social disintegration, civil war and displacement would be very damaging to ICH. Climate refugee camps will be unlikely to foster the ICH of the detainees.
Changes in and lack of predictability of rainfall and snowfall.	Drought. Fall in water table. Drying of rivers, lakes etc. Changes in seasonal precipitation	Desertification. Water shortages. Water stress. Flood water damage. Soil erosion. Subsoil instability: ground heave and subsidence affecting buildings and cultural spaces. Increased humidity causes increase in moulds that may affect foods, buildings and artefacts.	ICH practise, transmission, and awareness, modified, reduced or lost. Hunting, gathering, fishing practices and knowledge may become unsustainable.

(In association with rising temperatures)	e.g. timing of the monsoon. Flooding.	Contamination of fresh water by runoff. Changes in agriculture with need for crops and grazing animals which are tolerant to drier or wetter conditions. Changes in the distribution of pests and weeds. Plant and animal pests cause increasing crop and livestock losses that impact on food production causing famine and malnutrition. Impacts of food shortages on morbidity and mortality especially of infants and the elderly (who are the principal bearers of traditional knowledge). Changes to the nature of the seasons. Lack of seasonally available irrigation water. Economic stress through lack of water and unpredictability of water for gardens, domestic and wild animals, for human consumption and washing. Crop failure, domestic animal deaths, livelihood impacts, poverty and starvation. Collapse in biodiversity of ecosystems. Species extinction, loss of symbolic 'clan' animal and plant species. Loss of connection with 'nature' leading to erosion of responsibilities and obligations to home environment. Loss of raw materials for economic production, religious use, and for tool, instrument, craft, artefact and traditional medicine making. Undermining of protective ritual, belief systems, seasonal customary practices. Social impact especially on traditional communities. Impacts on health of the community:	Loss of land, loss of community cohesion, conflict with neighbours, mental health issues, lack of food and poverty may all lead to neglect of ICH. Loss of 'Critical Space' or appropriate venue may threaten ICH performance or practice. Reduction in tourist numbers may cause a reduction in cultural practice. Traditional practices may also be enhanced by hardship. ICH of disadvantaged minorities most at risk. People may be too busy, too few, or too debilitated to practise ICH. Fewer ICH stakeholders: performers, singers, medicine makers, craftspeople to practise ICH. Differential susceptibility to disease and mortality amongst the elderly custodians of ICH diminishes transmission to younger people. ICH may no longer be relevant so is discontinued: e.g. seasonal cultural events such as festivals abandoned as no longer relevant due to changed climate.

		malnutrition, water borne diseases, arthropod borne diseases etc. Changes in family structures may occur as sources of livelihoods become more distantly dispersed. Abandonment of traditional lands. Major dislocation may lead to societal stresses that result in abandonment of religion, fatalism, emergence of religious cults, profound distrust of government with anti-government riots, social disintegration, civil violence and even terrorism.	Loss of income and societal fragmentation also threaten ICH. Relocation threatens loss of ICH. Less opportunity to transmit cultural knowledge. Loss of cultural memory. Climate refugee camps will be unlikely to foster the ICH of the detainees.
Sea level rise. (As caused by warming seas and melting of glaciers and polar ice sheets)	Increasing gradual submersion of low lying islands, coastal areas and river deltas. (See under extreme events for storm surges)	Less area to grow crops and graze animals. Conflict over land availability. Loss of crops through salination of gardens, livelihood impacts, poverty and food shortages. Salination of ground-water which, if the source of potable water, may have adverse effects on the health of both humans and other animals. Collapse in biodiversity of ecosystems. Species extinction, loss of symbolic 'clan' animal and plant species. Reduction or loss of connection with 'nature'. Erosion of responsibilities and obligations to home environment. Loss of raw materials for economic production, religious use, and for tool, instrument, craft, artefact and traditional medicine making. Changes in family structures may occur as sources of livelihoods become more distantly dispersed. Abandonment of traditional lands. Displacement may be either regional or international. Major dislocation may lead to societal stresses that result in abandonment of religion, fatalism,	ICH practice, transmission, and awareness, modified, reduced or loss. ICH of disadvantaged minorities most at risk. Loss of land, loss of community cohesion, conflict with neighbours, mental health issues, lack of food and poverty may all lead to neglect of ICH. Loss of 'Critical Space' or appropriate venue may threaten ICH performance or practice. Reduction in tourist numbers may cause a reduction in cultural practice.Traditional practices may be enhanced by hardship.People may be too busy, too few, or too debilitated to practise ICH. Fewer ICH stakeholders: performers, singers, medicine makers, craftspeople to practise ICH. Lack of raw materials needed for ceremonies, rituals etc lead to abandonment of ICH.

		emergence of religious cults, profound distrust of government with anti-government riots, social disintegration, civil violence and even terrorism.	e.g. (Leahy 2009b). If ICH no longer relevant it may be discontinued. Loss of income and societal fragmentation also threatens ICH. Relocation, overseas education and being married outside traditional group all threaten loss of ICH. Less opportunity to transmit cultural knowledge. Loss of cultural memory. Climate refugee camps will be unlikely to foster the ICH of the detainees.
Extreme weather events. (For flooding see under changes in and lack of predictability of rainfall and snowfall)	Cyclones and severe storms. Storm surges. Damaging wind. Heat waves. Wildfires.	Erosion and flood damage to ICH artefacts in heritage keeping places, schools, homes and museums. CO_2 from increasingly intense and frequent fires add to GHG burden thus adding to a vicious cycle that causes further climate change with impacts on the environment and people. Damage and loss of forests, agricultural land, housing and community infrastructure. Increasingly frequent storm surges causing flooding of low lying islands, coastal flooding as well as erosion. Disasters cause immediate injury and death which may be followed by further morbidity and mortality from diseases related to lack of clean water, food and shelter. Damage and loss of buildings, stages – places for ICH performance.	ICH practise, transmission, and awareness, modified, reduced or lost. Resources may be redirected away from culture to disaster management and medium term climate change impacts. Relocation of displaced people break societal bonds and weaken ICH. Undermining of traditional knowledge and ICH customary practices e.g. knowledge of winds in Samoa (Tiatia 2008).

Ocean acidification. (in association with increasing warmth)	Coral bleaching. Changes in ocean life. Loss of shell making species.	Disturbance of marine ecosystems affecting the availability of seafood economically valued sea- life and tourist attractiveness.	ICH practise, transmission, and awareness, modified, reduced or lost. Changes in livelihood of traditional settlements. Undermining of relevance of traditional knowledge and customary practices such as sea-life dependency. Lack of tourists may accelerate loss of saleable ICH with resort cultural performances ceasing.

For use in integrated risk assessments of ICH in relation to the ICHC and the WHC. Modified and expanded from table 1 in the report *Predicting and Managing the effects of climate change on World Heritage* (UNESCO 2006a p25).

The Convention Community – 'The People by which'

The means by which the Conventions can be improved with respect to offering climate change protection can be explored by considering the Convention communities and those with whom they interact. Those with influence have included Committee members, UNESCO advisors, special experts, policy working group members and petitioners. Regrettably as has been detailed in Chapter 26 several determined World Heritage Committee members who represented some of the wealthier energy intensive States Parties were successful in limiting the WHC's capacity to address climate change impacts to mainly restricted localised interventions. This outcome raises questions about the undue influence of powerful interests.

A further disappointment was the failure of UNESCO and the World Heritage Committee to spontaneously recognise the need to initiate action on climate change. It took environmental petitioners to precipitate such action. That such a fundamental threat to world cultural heritage as climate change, which has been widely recognised since the UNFCCC was established in 1992, was not addressed spontaneously *from within* the organisation is a matter of concern. Although the early petitioners failed in their goals of having four climate change threatened World Heritage Sites protected through placement on the 'in Danger' List, their petitions did have the effect of rousing the World Heritage Committee into starting processes that have led to its current stance on climate change. The acceptance of the petitions, the subsequent 'draft' Decision that specifically disagreed with the petitioners' requests and the following series of meetings and Decisions that developed the climate change documents and implementation guidelines

(Chapter 15 and 26) indicate the Committee's willingness to respond in good faith although to a limited degree. As was noted in Chapter 26, the people who attended the Meeting of Experts that was concerned with updating the Convention included heritage workers, climate change experts, biologists, environmental advocates and delegates from numerous countries, but regrettably no environmental lawyers, this resulting in some limitation to the Convention' effectiveness. Also unfortunately, at that time, knowledge relating to the protection of cultural heritage sites from climate change impacts had not been extensively researched. The later *Policy* 'Working Group' lacked scientific expertise and although it invited submissions from States Parties', it was guided heavy-handedly by Australia and Canada, both wealthy carbon intensive States Parties. The Committee drew upon 'expert' people and adopted 'best' participatory practice to achieve this. The records of the working group meeting suggest that the influence of 'realpolitic' and national self interest overrode that of environmental law – an outcome that has hampered the WHC's capacity to become a 'talons and teeth' agreement.

The WHC has been criticised for its lack of legal clout, for being overly sensitive to national pride when considering placing Listed sites on or withdrawing them from the 'in Danger' List. It has also been criticised for not having a truly protective function in failing to prevent popular listed sites being damaged by tourism and by under-resourced management. Some think there are too many listed sites to look after adequately given the UNESCO resources.

The World Heritage Committee's meeting reports are well written. Their clarity is welcome as it allows outsiders to read the well reasoned rationales for the Decisions. However in December 2010 an *Expert meeting on the decision-making procedures of the statutory organs of the World Heritage Convention* (UNESCO 2010j) found that participants were noticing an increased politicisation of Decisions, "Similarly, experts were concerned about the conflict of interest in being a member of the Committee and presenting new nominations to the World Heritage List. There is some statistical evidence that Committee members achieve a significantly higher level of inscriptions, this being remarked upon by external stakeholders, thus damaging the credibility of the Convention" (UNESCO 2011f p5). The outcome was a recommendation to increase the transparency of decision-making procedures. Also of concern to the World Heritage critics was the lack of media access to meetings, lack of regular detailed reporting of meetings, the issue of confidentiality and use of secret ballots plus the lack of time to address the big-picture strategic questions and the need to resolve these longstanding issues. Unsurprisingly the participants at the above meeting were impressed by *the Nairobi 2010 meeting of the Intangible Cultural Heritage Committee* which was *live-streamed* over the web to enhance transparency of its operations.

In addition to the above concerns difficulties relate to the practicalities of attendance by some of the smaller developing States Parties caused for instance by issues of affordability and political unrest. World Heritage Committee meetings attract around 700 participants, mainly governmental and non-governmental heritage experts. Membership numbers 21 for the World Heritage Committee and 24 for the Intangible Cultural Heritage Committee, with representation based upon the 'equitable representation of the different regions and cultures of the world' for the WHC Committee and the 'principles of equitable geographic representation and rotation' for the ICHC Committee. The democratic processes of the ICHC have been highly praised, "the 2003 UNESCO ICHC is significant for locating individuals, communities, research units and non-governmental organisations in the most democratic operational framework within the entire suite of UNESCO's cultural conventions" (Galla 2009).

Thus the stage has been set for the WHC and the ICHC to work together in developing and implementing policy, including a mitigation policy as recommended in 'priority 1' above. Tangible and intangible heritage under climate change are inextricably intertwined and interdependent. The prioritised recommendations 1 to 5 would, if implemented, reduce the chance of calamitous damage to world cultural heritage by obligating States Parties to drastically curb GHG emissions.

The next question is whether the WHC's review process, WHC's Art 37 and Resolution 16 GA 10 (UNESCO 2008b) (see recommendation 15.1) and the responsiveness of the Intangible Cultural Heritage Committee to ICHC's preamble and Articles 1,13 and 14b are adequate to stimulate meaningful climate change policy. To date they have not been. The trigger to initiate the present climate change policy has come from the outside, from the nine citizens' petitions. However the petitioners' legacy has not been strong enough to pressure for direct international mitigation action. This may mean that 'the means by which' the Conventions will advance to the next stage of climate change action will again need to come from outside, stirred by a cascade of reports of deteriorating World Heritage properties and threatened ICH holders. To this, add the prospect of confusion as to why the former are not on the 'in Danger' List and growing concern as to the inability of the WHC to protect what is becoming increasingly threatened. Also, as yet, the ICHC lacks a stance on climate change.

Community Engagement

Perhaps cleverly the WHC knows its fortes and limitations in the realpolitik world. The World Heritage Committee has to be diplomatic. It has woven a delicate path within 'its competencies' to influence each State Party by couching directives with terms such as 'to the extent possible' and 'within the available resources' and 'feasible'. Could these qualified provisions taken together with the overall thrust of the WHC's cooperative arrangements lead to new strategies? Is it feasible that the culture of cooperation created by the diplomacy of the WHC and the ICHC may generate independent coherent regional strategies (involving a number of State Parties) to address GHG emissions because of climate change impacts on heritage. Is it also possible that developing States Parties might cooperate to demand the development of a world heritage global climate change mitigation policy? If this were to occur a basis might be established for going beyond the tenor and intent of the *'Policy'* Decision 31COM 7.1.

Some lawyers are of the opinion that as the WHC has no enforcement mechanism, States Parties will not take Convention-driven climate change mitigation measures seriously. Provided there is no abrogation of responsibility under climate change then as properties deteriorate under climate impacts, the World Heritage Committee's expertise for delivering tangible heritage risk preparedness and adaptive measures will struggle to cope with ever reducing States Parties' budgets and the World Heritage community will need to build much greater expertise in disaster risk management as the future becomes more painfully predictably unpredictable. With increasing responsibilities and demands in relation to sites and elements the failings of the WHC and the ICHC will inevitably stimulate more criticism. The 'means by which' the Conventions may be made more effective will amount to engaging with the community more fully and effectively.

The recent disappointment of the UN Rio+20 Conference on Sustainable Development (see Rio+20 postscript in Chapter 22) and the resultant disillusionment in multilateral processes are reasons for suggesting that the Convention Committees should expand their capacities to connect directly with local communities to initiate local actions to protect and safeguard heritage by encouraging and facilitating independent relationships between and among outside groups.

The Conventions' publicity machines celebrate the diversity of our cultural heritage, publicising new inscriptions, drawing attention to colourful events and telling success stories. To reflect greater inclusion more stories need be heard about the 'silent crisis' and how heritage is being impacted globally. Affected community members need to be given a voice at all levels of Convention organisation thus having the opportunity to directly influence the World Heritage Committee to review its position on global mitigation, and possibly influence regional economic blocks such as the European Union into forming mitigation agreements to protect the profitability of their heritage industry. The World Heritage Committee has recently decided to make some of their meetings more accessible. However further inclusive opportunities need to be created for dialogue and two way learning. The aware world is expecting petitions from low lying Pacific States Parties such as Vanuatu, the Solomon Islands and Papua New Guinea, that will mount cases for in 'urgent need of safeguarding' listing of ICH inscriptions and 'in Danger' listing of World Heritage sites such as Roi Mata's Domain site, the latter perhaps citing loss of OUV from 'deliberate damage' by polluting States Parties. The Convention Committees should encourage and welcome such petitions and use their legal powers to responding far more fully (Recommendation 26.9).

Petitioning is a form of communication that is free of government influence. The WHC (Thorson 2008 p5) and the ICHC are unusual in that they are amongst the few international multiparty instruments allowing direct citizen participation through the petition process. Petition power in the World Heritage context has already demonstrated its potential as an agent of change. It is proposed in Recommendation 26.8, that in the light of current climate science, there is a case for re-examining past petitions that presented well reasoned and substantially evidence-based cases for climate change impacted World Heritage sites to be moved to the 'in Danger' List. Reflecting on repeated requests from the WH Committee for closer ties with the UNFCCC, Recommendation 26.10 proposes that, under a formal agreement with the WHC, the UNFCCC review the science presented in petitions and advise the WH Committee as to its validity.

To be true to the Conventions' protective intent, heritage workers may have to step into discomfort zones, develop 'new 'competencies and fill roles never anticipated by the Convention writers. These roles may range from being world spokespeople on the global climate change stage to being educators standing alongside religious leaders on a tiny Pacific atoll nation. These roles may range from defending heritage, sustaining cultural heritage, raising awareness of the shifting paradigm to enduring sustainability to taking a leading role by advocating through human rights perspective for climate justice.

Insert 30.3 **Tangible Cultural Heritage Risk Assessments**

Climate Indicator	Climate Change Risk/Risk Multiplier	Physical and Biological Impacts on Tangible Cultural Heritage	Effects on Humans and Consequential Human Impacts on Tangible Cultural Heritage
Atmospheric moisture change (rain, snow, sleet, hail).	Flooding, (sea, river). Intense rainfall. Changes in water table levels. Changes in soil chemistry. Ground water changes. Changes in humidity cycles. Increase in time of wetness. Sea salt chlorides.	Erosion of inorganic and organic materials due to flood waters. pH changes to buried archaeological evidence. Loss of stratigraphic integrity due to cracking and heaving from changes in sediment moisture. Erosion of clay, rammed earth structures, buildings, that are not waterproofed. Data loss preserved in waterlogged / anaerobic / anoxic conditions. Eutrophication accelerating microbial decomposition of organics. Physical changes to porous building materials and finishes due to rising damp. Damage due to faulty or inadequate water disposal systems; historic rainwater drainage not capable of handling heavy rain and often difficult to access, maintain, and adjust. Crystallisation and dissolution of salts caused by wetting and drying affecting standing structures, archaeology, wall paintings, frescos and other decorated surfaces. Biological attack of organic materials by insects, moulds, fungi, invasive species such as termites. Subsoil instability, ground heave & subsidence. Relative humidity cycles/shock causing splitting, cracking, flaking and dusting of materials and surfaces. Corrosion of metals. Other combined effects e.g. increase in moisture combined with fertilisers and pesticides.	Food shortages threaten customary practices of looking after heritage properties and cultural artefacts.

Temperature change.	Diurnal, seasonal, extreme events (heat waves, snow loading). Changes in freeze-thaw & ice storms, & increase in wet frost, fires, Melting glaciers, Melting ice caps Reduction of sea ice.	Deterioration of facades due to thermal stress, Freeze-thaw/frost damage. Damage to brick, stone, ceramics that have become wet then frozen before drying. Biochemical deterioration. Changes in 'fitness for purpose' of some structures. For example overheating of the interior of buildings can lead to inappropriate alterations to the historic fabric due to the introduction of engineered solutions Inappropriate adaptation to allow structures to remain in use. Subsoil instability, ground heave subsidence. Drying of wood, splitting. Exposure of graves, Arctic archaeology.	Changes in growing season > livelihood threat. Monsoon time changes affecting agriculture timing. Disease distribution changes. Older custodians, teachers more prone to heat stress. Changes to agriculture, grazing. Lack of seasonally available water. Crop failure. Erosion threat to Arctic fauna traditionally hunted by indigenous Inuit. Inability or less time to care for heritage properties and cultural artefacts.
Sea level rise.	Coastal flooding. Sea water incursion.	Coastal erosion/loss of archaeological properties. Intermittent introduction of large masses of 'strange' water to the site which may disturb the meta stable equilibrium between artefacts and soil. Permanent submersion of low lying areas. Loss of land.	Loss of land. Contamination of fresh drinking water. Relocation. Threat to culture tied to land. Loss of ICH 'context space'. Submergence of culture (Chief Roi Mata's Grave site, Vanuatu). Less area to grow crops> livelihood threat. Pressure on land resources > migration. Population migration/ statelessness. Disruption of communities. Loss of rituals and breakdown of social interactions. Nations inundated , "Statelessness". Inability or less time to care for heritage properties and cultural artefacts.

Increase in Wind Changes in seasonal wind directions.	Wind-driven rain. Wind-transported salt. Wind-driven sand. Winds, gusts & changes in direction. Severe weather events e.g. cyclones.	Penetrative moisture into porous cultural heritage materials. Static and dynamic loading of historic or archaeological structures. Structural damage and collapse. Deterioration of surfaces due to erosion. Destruction, damage to properties.	Threat to traditional fishing equipment, boats. Homelessness, disruption to custodianship, stewardship. Inability or less time to care for heritage properties and cultural artefacts.
Desertification.	Drought. Heat waves. Fires.	Erosion, Salt weathering, sand inundation and collapse of sites.	Impact on health of population. Heat stroke. Loss of forests, damage to agricultural land, abandonment of heritage properties. Loss of cultural memory. Threat to sustainable development e.g. eco-tourist industry.
Climate and pollution acting together.	pH, precipitation. Changes in deposition of pollutants > positive feedbacks> accentuating loss/ change.	Stone recession by dissolution of carbonates. Blackening of materials. Corrosion of metals. Influence of bio-colonialisation.	Health impacts. Disease spread. Vector increase and distribution. Inability or less time to care for heritage properties and cultural artefacts.
Climate and biological effects.	Proliferation of invasive species. Spread of existing and new species of insects (e.g. termites). Increase in mould growth. Changes to lichen colonies on buildings.	Collapse of structural timber and timber finishes. Reduction in availability of native species for repair and maintenance of buildings.	Transformation of communities, health, education, transmission of knowledge about cultural heritage, dislocation, social unrest. Changes to the livelihood of traditional settlements. Changes in family structures as sources of livelihoods become more dispersed and distant.

	Decline of original plant materials. Threats to agriculture.	Changes in the natural heritage values of cultural heritage sites. Changes in appearance of landscapes.	Unhealthy cultural custodians, malnutrition, starvation, famine, disease increase due to water borne and insect borne diseases, less cultural involvement, Migration, loss of status, identity, sense of place, intergenerational cultural loss, Distrust of leaders. Erosion of social cohesion, competition for resources > cultural mutation, extreme elements, impacts on religious beliefs, conflict> inability or less time to care for heritage properties and cultural artefacts.
Plate tectonic stress.	Possible increase in earthquakes, volcanoes, tsunamis.	destruction of monuments, buildings, tunnels, canals.	Loss of life. Physical and mental invalidism amongst survivors. Inability or less time to care for heritage properties and cultural artefacts.
New pests.	...eat wood, vegetation.	destruction of artefacts and wooden structures.	Locust plagues destroy crops, undermining livelihood of cultural custodians. Inability or less time to care for heritage properties and cultural artefacts.
Species extinction.	Ecological impoverishment e.g. bees, bats pollinating.	Collapse of ecosystems. Loss of raw materials for crafts.	Undermine customary hunting, gathering and fishing. Cultural loss. Religious dilemma; loss of/ undermining of faith. Inability or less time to care for heritage properties and cultural artefacts.

Acid oceans.	Coral bleaching. Loss of shell making species.	Destruction of coastal properties.	Loss of fishing livelihood > migration> abandonment of cultural properties. Inability or less time to care for heritage properties and cultural artefacts.

For use in integrated risk assessments of tangible cultural in relation to the ICHC and the WHC. Modified and expanded from table 1 in the report *Predicting and Managing the effects of climate change on World Heritage* (UNESCO 2006a p25).

What is the next step? ...The Imperative

Climate change is a transformative issue which has life threatening implications. It is diabolical (Garnaut 2008; 2011) because it is uncertain, insidious, long term, international and potentially dangerous. It is a planetary emergency in the making (Flannery 2008, 2009, 2010; Spratt 2008) but is too impersonal to engender public fear. But there is no doubt, according to the Australian Climate Change Commission that "This is the critical decade. Decisions we make from now to 2020 will determine the severity of climate change our children and grandchildren experience" (Steffen 2011). 'And so it goes' that decisions made or not made soon will determine how diminished will be the world's cultural heritage.

As the climate science becomes more sophisticated, the message becomes more complex and more urgent. The time when the Earth's resources can no longer sustain life as we know it is within a generation. Mitigation actions can ameliorate and stabilise the impact but not stop it. The dynamics of how the impacts of climate change will eventuate is an open question. Many think the impacts will follow a 'drift and shift' pattern or visualising it another way – a 'ratchetting' pattern. In other words, there will be impacts such as continuing droughts which are incremental - 'drifting'- possibly leading to migration, increased poverty, and border insecurities. There will also be momentous sudden changes or 'shifts' such as the impacts from the failure of the monsoon, or the collapse of a marine ecosystem.

Our heritage will deteriorate. There will probably be a multispeed decline which will be destabilising. It is unconscionable that the World Heritage Committee should urge adaptive responses by developing States Parties to their climate changed sites yet not support the obligating of polluting States Parties to drastically reduce their carbon emissions. Our heritage is in unprecedented peril, the problem and the material solutions are understood by most if not all world leaders, yet there is no effective model for change being followed. Others must act and the WHC can do just that. It is well to reflect that "only States have the power to enforce governmental obligations and only those embodied in international treaties are enforceable against them" (O'Keefe and Prott 2011 p5).

What is happening to I-Kiribati (people of Kiribati) and Tuvaluans is that high tides, storm surges and rising sea levels are eroding cultural heritage that has not as yet been inscribed. These people are in the front lines of climate change confrontation, a confrontation that will affect us all. Those who speak for the Conventions – its Committees, Conferences and Assemblies - are speaking as the symbolic guardians of the World's finest cultural heritage. May the confidence of those who speak, be justified in the time of climate change by the future actions of the Conventions.

The respected *Global Footprint Network* has calculated, considering *only living renewable resources*, that humanity has overshot the earth's carrying capacity by a factor of 1.5, with Westernised countries using 5 to 9 times more than nature can regenerate. Taking into consideration non-renewable resources such as minerals, fossil fuels, artesian (fossil) water, pollution, soil erosion, etc., humanity's overshoot may be as much as 400 times (GFN 2012). We are living far beyond our means and GHG emission rates continue to increase. The 'sustainability' movement is growing, the sophistication of the emerging 'sustainable tourism' business is testimony to changing attitudes, but none of this is reflected in the trajectory of CO_2 emissions. There may be a decade to make the decisive changes that can blunten the inevitable impacts of our profligy. Humankind may still have a chance to redefine the future and avoid overwhelming dehumanising impoverishment. It all depends on our willingness and ability to make the necessary changes with the rapidity needed to deflect the worst consequences of a complex and seemingly distant threat. These changes depend on our willingness to embrace the science and foster the wisdom to salvage what opportunity we can for a stable future for ourselves, our planetary companions and those who follow us. We must recognise global limitations and demand and fully participate in the massive collective readjustment needed to bring us to a truly sustainable carbon constrained economy in which there is the possibility of a healthy culturally rich future.

And finally two World Heritage tales to ponder: Chief Roi Mata's Domain, Vanuatu and Mapungubwe Cultural Landscape, South Africa.

It is incumbent upon UNESCO to ensure that WHC and the ICHC contribute to their full potential in facilitating this change.

Chief Roi Mata's Domain, Vanuatu

The essential question remains. **Chief Roi Mata's Domain (CRMD), Vanuatu** is World Heritage Listed as an outstanding example of a 400 year old continuing cultural landscape representative of Pacific chiefly systems (UNESCO 2008g). Chief Roi Mata was the last paramount chief. For many contemporary people in Vanuatu his Domain which consists of three 17th century sites on the islands of Efate, Lelepa and Artok including the Chief's residence, the site of his death and his mass burial site, still lives as a source of power evident through the landscape and as an inspiration for people working out their lives. As has been recognised by Meredith Wilson, Chris

Ballard and Douglas Kalotiti, "Global warming and sea-level rise pose an obvious and perhaps inevitable threat to the Chief Roi Mata's Domain, and especially to the low-lying sites of Roi Mata's grave on Artok Island and his residential home at Mangaas. There are few mitigation measures available that might realistically address this problem at a local level, while retaining the integrity of the location of the sites within their cultural landscape. The cultural heritage of the Pacific region is confronted with this form of threat to an exceptional degree, but the region makes a negligible contribution to global warming and Pacific states can exert little or no influence over the major polluters" (Wilson, Ballard, and Kalotiti 2007 p.5-6).

The inundation will be tracked through the WHC processes of monitoring and reporting, despite there being no foreseeable 'adaptation' or intervention that can be made to protect this developing State Party's immoveable heritage. Transferring it to the 'in Danger' list would at least be a 'means by which' the WHC could respond to bring world attention to the need for collective action to stabilise climate change. Additionally, labelling it with a climate change crisis symbol would emphasise the actuality of the potential loss.

Chief Roi Mata would have reason to wonder why the climate change knowledge-rich 'Chiefs' of today (our leaders) who have known what needs to be done, have done little or nothing leaving his traditional knowledge-rich descendants as the ones to suffer.

Mapungubwe Cultural Landscape, South Africa

The inscription on this World Heritage site confronts the WHC and the ICHC with a foretaste of their most important challenge (UNESCO 2003c). The site was inscribed on the basis of four cultural criteria:

Criterion (ii): The Mapungubwe Cultural Landscape contains evidence of an important interchange of human values that led to far-reaching cultural and social changes in southern Africa between AD 900 and 1300.

Criterion (iii): The remains in the Mapungubwe cultural landscape are a remarkably complete testimony to the growth and subsequent decline of the Mapungubwe State which at its height was the largest kingdom in the African sub-continent.

Criterion (iv): The establishment of Mapungubwe as a powerful state, trading through the East African ports with Arabia and India was a significant stage in the history of the African sub-continent.

Criterion (v): The remains in the Mapungubwe cultural landscape graphically illustrate the impact of climate change and record the growth and then decline of the Kingdom of Mapungubwe as a clear record of a culture that became vulnerable to irreversible change.

(The climate change referred to above was regional.)

Present Global climate change whilst virtually irreversible when measured in terms of human life spans might, if faced squarely with the resolve 'to do what is necessary' be stabilised sufficiently to permit civilisation to endure. The WHC and ICHC give the global *heritage* community a chance to lead in climate change awareness raising – a chance that might help trigger a universal 'awakening' to the climate change emergency, a step in the transformation of global society. These, our most inspirational cultural Conventions should demand that we, their custodians, do what is necessary to protect cultural heritage and humanity.

"Wait a moment", you may think, "this is beginning to seem pretty difficult!" Well, I never said it was going to be easy, but we can and will win this. Just remember: we have a remarkable planet that is worth fighting for and never give in to naysayers. James Hansen, Butterfly Report + Jeremiah, the Frog, 2012 http://www.columbia.edu/~jeh1/

Appendices

Appendix 1
Data set

Table 1 Alphabetical List of States Parties and other nations

- **the total number of their inscriptions to the WHC, 2012**
- **the number of cultural inscriptions to the WHC, 2012**
- **the number of inscriptions to the ICHC, 2012**
- **Human Development ranking, 2011**
- **Human Development Index 2011**
- **Latitude**
- **Climate change vulnerability ranking**
- **Climate change vulnerability factor, 2010**
- **Area ranking**
- **Area**
- **Population ranking**
- **Population, 2011**

Data from UNDP(UNDP 2010), Climate Vulnerability Monitor (DARA and Forum 2010)and UNESCO web sites.
The graphs in this book drew on this data.
Gaps indicate that the nation is not a State Party.

0 indicates that the State Party has no inscriptions listed.
Small Island Developing States (SIDS) are marked with an '*'
Least Developed Countries (LDC) are marked with a '+'

A monastery library, Ladakh. Books are loose leaved, long and rectangular in shape. The leaves are held in place by wooden covers and the whole is wrapped in fabric. Monks and Lamas carry these religious books on the public march that marks Buddha's birthday. The marchers wave the books over the heads of onlookers as a blessing

States Parties and other soverign States	World Heritage sites 2012	World Heritage cultural sites 2012	ICH 2012 inscriptions	Human development 2011 ranked	Human development 2011	Ranked north to south by latitude	Climate change vulnerability	States Parties area ranked	Area sq kilometres	States Parties 2 population ranked	Population 2011
Afghanistan+	2	2	0	172	low	62	acute	41	625090	42	29117000
Albania	2	2	1	70	high	51	high	143	28748	137	3195000
Algeria	7	7	1	96	medium	67	severe	11	2381741	35	35423000
Andorra	1	1		32	very high	50		184	468	189	84082
Angola+	0	0		148	low	170	acute	22	1246700	60	18993000
Antigua & Baruda*	0	0		60	high	107	severe	187	442	188	89000
Argentina	8	4	1	45	very high	191	moderate	8	2780400	32	40518951
Armenia	3	3	2	86	high	58	high	141	29743	136	3238000
Australia	19	7		2	very high	175	moderate	6	7692024	54	22421417
Austria	9	9	0	19	very high	27	low	115	83871	93	8372930
Azerbaijan	2	2	4	91	high	56	high	114	86600	91	8997400
Bahamas*				53	high	84	severe	159	13878	174	330000
Bahrain*	1	1		42	very high	89	moderate	177	758	161	807000
Bangladesh+	3	2	1	146	low	85	acute	95	143100	7	164425000
Barbados*	1	1	0	47	very high	125	moderate	188	430	178	257000
Belarus	4	3	1	65	high	13	high	85	199951	87	9471900
Belgium	10	10	9	18	very high	22	low	139	30528	77	10827519
Belize*	1	0	1	93	high	105	acute	151	22966	175	322100
Benin+	1	1	1	167	low	132	severe	102	112492	90	9212000
Bhutan+	0	0	1	141	medium	82	acute	134	38394	164	708000
Bolivia	6	5	3	108	medium	177	severe	27	1098581	85	10031000
Bosnia & Herzegovina	2	2	0	74	high	37	high	127	51197	128	3760000
Botswana	1	1	0	118	medium	189	severe	47	580367	148	1978000
Brazil	18	11	5	84	high	154	moderate	5	8514877	5	193364000
Brunei	0	0	0	33	very high	155	moderate	168	5765	173	401890

Bulgaria	9	7	2	55	high	40	high	105	109886	98	7576751
Burkina Faso+	1	1	0	181	low	115	acute	73	270467	63	16287000
Burundi+	0	0	0	185	low	168	severe	145	27834	92	8519000
Cambodia+	2	2	2	139	medium	126	severe	89	176215	70	13395682
Cameroon	1	0		150	low	122	severe	53	462840	59	19958000
Canada	15	6		6	very high	1	moderate	2	9984670	36	34207000
Cape Verde*	1	1		133	medium	110	severe	170	4033	169	513000
Cent African Rep+	1	0	1	179	low	138	severe	44	603500	121	4506000
Chad+	0	0	0	183	low	94	acute	20	1284000	75	11274106
Chile	5	5	1	44	very high	188	moderate	37	752612	61	17114000
China	41	34	36	101	medium	18	high	3	9629091	1	1339190000
Colombia	7	5	7	87	high	123	moderate	25	1141748	29	45569000
Comoros*+	0	0		163	low	180	high	173	2235	166	691000
Congo	0	0		137	medium	158	severe	63	342000	129	3759000
Congo, Dem Rep+	0	0	0	187	low	153	severe	12	2344858	19	67827000
Cook Is*	0	0						195	236	198	19569
Costa Rica	3	0	1	69	high	136	moderate	128	51100	120	4640000
Cote d'Ivoire	3	0	1	170	low	139	severe	68	312685	55	21571000
Croatia	7	6	12	46	very high	35	high	126	56594	124	4435056
Cuba*	9	7	1	51	high	96	high	106	108889	76	11204000
Cyprus	3	3	2	31	very high	70	moderate	167	9251	162	801851
Czech Republic	12	12	4	27	very high	24	moderate	117	78865	79	10512397
Denmark	4	3	0	16	very high	11	low	132	43094	112	5540241
Djibouti+	0	0	0	165	low	128	acute	150	23200	159	879000
Dominica*	1	0	0	81	high	114	high	178	751	191	67000

States Parties and other soverign States	World Heritage sites 2012	World Heritage cultural sites 2012	ICH 2012 inscriptions	Human development 2011 ranked	Human development 2011	Ranked north to south by latitude	Climate change vulnerabil-ity	States Parties area ranked	Area sq kilometres	States Parties population ranked	Population 2011
Dominican Rep*	1	1	2	98	medium	102	high	130	48671	83	10225000
Ecuador, Equator	4	2	1	83	high	162	high	77	245857	69	14228000
Egypt	7	6	1	113	medium	78	high	29	945087	16	78848000
El Salvador	1	1		105	medium	118	high	153	21041	108	6194000
Equatorial Guinea+	0	0	0	136	medium	157	acute	144	28051	165	693000
Eritrea+	0	0	0	177	low	106	acute	101	112622	115	5224000
Estonia	2	2	3	34	very high	9	high	131	45227	155	1340021
Ethiopia+	9	8	0	174	low	120	acute	26	1104300	15	79221000
Fiji*	0	0	0	100	medium	182	high	155	18272	160	854000
Finland	7	6		22	very high	5	low	64	331212	114	5366100
France	37	34	10	20	very high	23	low	42	622984	20	65447374
Gabon	1	1	0	106	medium	159	high	76	256369	153	1501000
Gambia+	2	2	1	168	low	121	acute	163	11295	149	1751000
Georgia	3	3	1	75	high	44	high	121	69700	123	4436000
Germany	36	33		9	very high	16	low	62	342000	14	81757600
Ghana	2	2		135	medium	134	high	81	238391	48	24333000
Greece	17	17	1	29	very high	57	low	97	130373	74	11306183
Grenada*	0	0	0	67	high	131	severe	190	344	185	104000
Guatemala	3	3	2	131	medium	109	high	107	103000	68	14377000
Guinea+	1	0	1	178	low	129	severe	78	242900	81	10324000
Guinea-Bissau*+	0	0		176	low	130	acute	136	36125	152	1647000
Guyana*	0	0		117	medium	145	acute	84	207600	163	761000
Haiti*+	1	1	0	158	low	101	acute	146	27750	84	10188000
Holy See	2	2						201	0.44	201	829
Honduras	2	1	1	121	medium	112	acute	103	111369	96	7616000
Hong Kong, China				13	very high			175	1104	101	7067800

Hungary	8	7	2	38	very high	28	high	110	93028	86	10013628
Iceland	2	1	0	14	very high	7	moderate	108	100200	176	317900
India	28	23	8	134	medium	71	acute	7	3287263	2	1184639000
Indonesia	7	3	6	124	medium	150	high	15	1910931	4	234181400
Iran	13	13	9	88	high	60	high	17	1628750	17	75078000
Iraq	3	3	1	132	medium	63	severe	58	406752	39	31467000
Ireland	2	2		7	very high	14	low	120	70273	122	4459300
Israel	6	6		17	very high	74	moderate	152	22072	97	7602400
Italy	47	44	3	24	very high	33	low	71	300000	23	60340328
Jamaica*	0	0	1	79	high	104	moderate	164	10991	141	2730000
Japan	16	12	20	12	very high	39	low	61	357114	10	127380000
Jordan	4	4	1	95	medium	73	high	112	89342	105	6472000
Kazakhstan	3	2	0	68	high	15	acute	9	2724900	64	16197000
Kenya	6	3	1	143	low	152	acute	48	527968	31	40863000
Kiribati*+	1	0		122	medium		acute	180	726	187	100000
Kosovo						46	high	165	10908	150	1733872
Kuwait	0	0		63	high	80	moderate	156	17818	138	3051000
Kyrgyzstan	1	1	2	126	medium	47	high	86	196722	111	5550000
Laos+	2	2	0	138	medium	97	high	83	214969	107	6436000
Latvia	2	2	2	43	very high	10	moderate	124	64589	143	2237800
Lebanon	5	5	0	71	high	72	moderate	166	10452	127	4255000
Lesotho+	0	0	0	160	low	194	high	140	30355	145	2084000
Liberia+	0	0		182	low	143	acute	104	110879	131	3476608
Libya	5	5		64	high	75	acute	16	1759540	104	6546000

States Parties and other soverign States	World Heritage sites 2012	World Heritage cultural sites 2012	ICH 2012 inscriptions	Human development 2011 ranked	Human development 2011	Ranked north to south by latitude	Climate change vulnerability	States Parties area ranked	Area sq kilometres	States Parties population ranked	Population 2011
Liechtenstein				8	very high	32		92	160475	194	35789
Lithuania	4	4	3	40	very high	12	high	123	65300	134	3329227
Luxembourg	1	1	1	25	very high	25	low	172	2586	171	502207
Madagascar+	3	1	1	151	low	181	acute	46	582000	57	21146000
Malawi+	2	1	2	171	low	176	acute	100	117600	66	15692000
Malaysia	3	1	1	61	high	147	moderate	66	323782	44	28306700
Maldives *	0	0		109	medium	148	acute	192	300	177	314000
Mali+	4	4	6	175	low	92	acute	23	1240192	67	14517176
Malta	3	3		36	very high	68	low	191	316	172	416333
Marshall Is *	1	1			lack of data	117	acute	196	181	192	63000
Mauritania+	2	1	1	159	low	88	acute	28	1030700	133	3366000
Mauritius*	2	2	0	77	high	178	high	174	1969	156	1297000
Mexico	31	27	7	57	high	76	moderate	14	1964375	11	108396211
Micronesia, Fed St*	0	0		116	medium	140	acute	181	702	183	111000
Moldova	1	1	0	111	medium	29	high	138	33846	130	3563800
Monaco	0	0	0		lack of data	43		200	1.9	195	33000
Mongolia	3	2	9	110	medium	21	severe	18	1564100	140	2768800
Montenegro	2	1	0	54	high	45	lack of data	160	13812	142	2536288
Morocco	8	8	4	130	medium	69	acute	57	435244	38	31892000
Mozambique+	1	1	2	184	low	179	acute	34	800000	51	23406000
Myanmar+	0	0		149	low	81	acute	40	632759	24	50496000
Namibia	1	1	0	120	medium	187	acute	33	801590	144	2212000
Nauru*					lack of data	164		199	21	220	9322
Nepal+	4	2	0	157	low	79	severe	94	143998	40	29853000
Netherlands	9	8		3	very high	19	low	135	37354	62	16609518
New Zealand	3	1		5	very high	174	low	74	267668	126	4383600
Nicaragua	2	2	2	129	medium	116	acute	98	120538	110	5822000

Niger+	2	0	0	186	low	93	acute	21	1267000	65	15891000
Nigeria	2	2	3	156	low	127	acute	31	912050	8	158259000
North Korea	1	1	0		lack of data	48	acute	194	260	118	4896700
Norway	7	6	0	1	very high	3	low	99	118484	47	24346229
Niue*	0	0						67	322463	50	23991000
Oman	4	4	1	89	high	86	moderate	70	301336	139	2905000
Pakistan	6	6	1	145	low	66	acute	35	783562	6	170260000
Palau*	0	0	0	49	high	146	high	185	459	197	20000
Palestine			1	114	medium			147	27000	125	4400000
Panama	5	2	0	58	high	142	high	118	75417	135	3322576
Papua New Guinea*	1	1	0	153	low	167	acute	54	450295	102	6888000
Paraguay	1	1	0	107	medium	190	moderate	59	390757	106	6460000
Peru	11	9	7	80	high	163	high	19	1285216	41	29461933
Philippines	5	3	2	112	medium	99	high	72	272967	12	94013200
Poland	13	12	0	39	very high	17	moderate	69	309500	34	38167329
Portugal	13	12	1	41	very high	53	low	111	92090	78	10636888
Qatar	0	0	1	37	very high	90	moderate	162	11586	151	1696563
Republic of Macedonia	1	1	0	78	high	52	high	149	25713	147	2048620
Romania	7	6	2	50	high	30	high	82	236800	56	21466174
Russia	24	16	2	66	high	2	high	1	17098242	9	141927297
Rwanda+	0	0		166	low	166	severe	148	26338	82	10277000
Samoa*+	0	0		99	medium	184	acute	171	2842	180	179000
San Marino	1	1			lack of data	41	lack of data	197	61	196	32386
Sao Tome & Principe*+	0	0	0	144	low	160	acute	176	964	182	165000

States Parties and other sovereign States	World Heritage sites 2012	World Heritage cultural sites 2012	ICH 2012 inscriptions	Human development 2011 ranked	Human development 2011	Ranked north to south by latitude	Climate change vulnerability	States Parties area ranked	Area sq kilometres	States Parties population ranked	Population 2011
Saudi Arabia	2	2	1	56	high	77	moderate	13	2149690	46	26246000
Senegal+	6	4	2	155	low	111	acute	87	185180	73	12534228
Serbia	4	4	0	59	high	36	lack of data	113	88361	99	7319712
Seychelles*	2	0	0	52	high	169	severe	186	451	190	84000
Sierra Leone+	0	0		180	low	141	acute	119	71740	109	5836000
Singapore*				26	very high	161	moderate	182	694	117	5076700
Slovakia	7	5	1	35	very high	26	high	129	49037	113	5426645
Slovenia	2	1	0	21	very high	34	high	154	20273	146	2062700
Solomon Is*+	1	0		142	low	171	acute	142	28896	167	536000
Somalia+					lack of data	133	acute	39	637657	89	9359000
South Africa	8	5		123	medium	192	severe	24	1221037	25	49991300
South Korea	10	9	14	15	very high	61	moderate	109	99828	26	49773145
South Sudan								43	619745	94	8260490
Spain	43	40	12	23	very high	42	high	51	488100	27	46951532
Sri Lanka	8	6	0	97	medium	144	moderate	122	65610	58	20410000
St Kitts & Nevis*	1	1		72	high	108	high	193	261	193	38960
St Lucia*	1	0	0	82	high	119	moderate	183	616	181	173765
St Vincent & Grenadines*	0	0	0	85	high	124	high	189	389	184	109000
Sudan+	2	2	0	169	low	98	acute	10	2505813	33	39154490
Suriname*	2	1		104	medium	149	acute	91	163610	168	524000
Swaziland	0	0		140	medium	193	severe	157	17364	157	1202000
Sweden	14	13	0	10	very high	6	low	55	447400	88	9366092
Switzerland	11	8	0	11	very high	31	low	133	41284	95	7782900
Syria	6	6	1	119	medium	65	high	88	181035	53	22505000
Taiwan						87		137	36008	52	23061689
Tajikistan	1	1	1	127	medium	59	severe	96	131957	100	7075000
Tanzania+	7	4	0	152	low	165	acute	30	923768	30	45040000

Thailand	5	3		103	medium	100	high	50	505992	21	63525062
Timor-Leste+				147	low		acute	158	14874	158	1066582
Togo+	1	1	1	162	low	137	severe	125	56785	103	6780000
Tonga*	0	0	1	90	high	185	high	179	747	186	104000
Trinidad & Tobago*	0	0	0	62	high	135	moderate	169	5130	154	1344000
Tunisia	8	1	0	94	high	64	severe	93	147181	80	10432500
Turkey	10	10	9	92	high	54	moderate	36	756102	18	72561312
Turkmenistan	3	3	0	102	medium	49	high	52	475442	116	5177000
Tuvala*+					lack of data	172	severe	198	26	199	10544
Uganda+	3	1	1	161	low	156	acute	80	238533	37	33796000
Ukraine	5	4	0	76	high	20	high	45	587041	28	45871738
United Arab Emerites	1	1	2	30	very high	91	moderate	116	83600	119	4707000
United Kingdom	28	24		28	very high	8	low	79	241550	22	62041708
United States	21	9		4	very high	4	high	4	9600000	3	309975000
Uruguay	1	1	2	48	high	195	moderate	90	163820	132	3372000
Uzbekistan	4	4	4	115	medium	38	high	56	446550	45	27794000
Vanuatu*+	1	1	1	125	medium	183	acute	161	12189	179	246000
Venezuela	3	2	0	73	high	113	high	32	824268	43	28888000
Vietnam	7	5	6	128	medium	95	acute	65	330803	13	85789573
Western Sarah						83		75	266000	170	513000
Yemen+	4	3	1	154	low	103	acute	49	513120	49	24256000
Zambia+	1	0	1	164	low	173	severe	38	676578	71	13257000
Zimbabwe	5	3	1	173	low	186	acute	60	377930	72	12644000

Appendix 2
Glossary

Adaptation: individual or governmental actions to reduce adverse effects or future risks associated with climate change (The Climate Institute 2010). The IPCC defines adaptation as the "adjustment in natural or human systems in response to actual or expected climatic stimuli or their effects, which moderates harm or exploits beneficial opportunities." Adaptation means to change behaviour: - strategies, programs, investments or individual behaviour - based on our knowledge about climate change risks (Smit and Wandel 2006). Adaptation will involve changing lifestyles, infrastructure, and businesses operations to deal with climate change. According to the IPCC, however, "adaptation alone is not expected to cope with all the projected effects of climate change" (The Climate Institute 2010). See also Chapter 3.

Adaptive capacity: The ability of a system to adjust to climate change (including climate variability and extremes) to moderate potential damages, to take advantage of opportunities, or to cope with the consequences (Garnaut 2008). The terms *resilience* and *adaptive capacity* are often used synonymously despite some subtle differences. We prefer the term *adaptive capacity* to *resilience* because it suggests the possibility of change. Resilience, as defined by the Third Assessment Report of the IPCC (McCarthy et al., 2001), is "the amount of change a system can undergo *without changing state.*" In contrast, the capacity to adapt can be determined by the system's ability to *change into a state* that is less vulnerable than before (Schröter 2005). ***Resilience*** *– Amount of change a system can undergo without changing state. (IPCC 2001).*

Aerosols: A collection of airborne solid or liquid particles, with a typical size between 0.01 and 10 micrometres (a millionth of a metre) that remain in the atmosphere for a relatively short time.

Albedo effect: The amount of solar radiation reflected by a surface or object, often expressed as a percentage.

Anthropogenic climate impacts: Climate change impacts caused by humans and our activities, including industry and agriculture, rather than by nature.

AOSIS: The Alliance of Small Island States, a group of 43 low-lying and small island countries. The group represents the voice of these small island nations that will be deeply affected by climate change due to rising sea levels and increased storm activity (The Climate Institute 2010).

Binary nature of natural and cultural heritage: Lahey explains that in indigenous societies cultural heritage includes the 'natural' as well as the 'cultural' (Lahey, 2010). Both are inextricable to understanding the impacts of climate change on culture. The land and the sea, the mountains and the rivers, the biodiversity are melded to cultural (diversity) as indigenous societies have always understood. The WHC deals with 'natural', 'cultural' and 'mixed' heritage properties.

Biochar: A form of charcoal which comes from heating biomass (organic materials such as waste or manure). The process of creating biochar can serve as a form of carbon capture and storage by locking carbon into the charcoal formed, and then ultimately putting it back into the soil. Once in the ground, the structure of biochar makes it difficult to break down, meaning that the captured carbon can remain in the ground for a long period of time.

Business as usual: A scenario of future greenhouse gas emissions that assumes that there would be no major changes in policies on mitigation.

Carbon cycle: The carbon cycle is the complex cycle by which carbon is exchanged between living matter, air, water and the underlying earth. It is one of the most important cycles of the earth and allows for carbon to be recycled and reused throughout the biosphere and all of its organisms.

Carbon sink or reservoir: Parts of the carbon cycle that store carbon in various forms.

Climate change (presently occurring): Refers to the change over time of climate that is attributed directly or indirectly to *human activity* that alters the composition of the global atmosphere and that is in addition to natural climate variability observed over comparable time periods (ACJP and Earthjustice 2009).

Climate displaced person: Person displaced temporarily or permanently due to environmental causes, notably land desertification, sea level rise and weather-related disasters (GHF 2009). Climate displaced people is used in this dissertation to describe people who predominantly involuntarily are forced to move or are displaced, either permanently or temporarily, because of climate change, through its impacts and shocks. It is possible to estimate the numbers of Climate Displaced People in global terms, since, for instance, a correlation can be made between the great increases in the number of severe weather events – much of which can be attributed to climate change – over the last decades, and the number of additional people that these events displace. It is, however, virtually impossible to single out individual people or even scenario specific situations as being attributable to climate change. This means that the definition of Climate Displaced People carries almost no practical application today. It is however, a useful estimative indicator of the additional burden that climate change is placing on the international community, on existing legal frameworks of protection and assistance, and on local communities in areas where climate impacts are most acute. (GHF 2009)

Climate sensitivity: A measure of the climate system's response to sustained radiative forcing. Climate sensitivity is defined as the global average surface warming that will occur when the climate reaches equilibrium following a doubling of carbon dioxide concentrations.

Climate system: A highly complex system consisting of the atmosphere, the water cycle, ice, snow and frozen ground, the land surface and plants and animals, and the interactions between them.

Conservation: This is the management of change. It includes measures to extend the life of cultural heritage while strengthening transmission of its significant heritage messages and values (ICCROM, 1998). In the domain of cultural property, the aim of conservation is to maintain the physical and cultural characteristics of the object to ensure that its value is not diminished and that it will outlive our limited time span (UNESCO, 1988). Climate change is the most significant global challenge facing humanity and the environment today.

Convention: This is an agreement under international law entered into by States and that establishes rights and obligations between each party and every other party.

Cultural diversity: This refers to the many ways in which the different cultures of groups and societies find expression. These cultural expressions are passed on within and among groups and societies, and from generation to generation. Cultural diversity, however, is evident not only in the varied ways in which cultural heritage is expressed, augmented and transmitted but also in the different modes of artistic creation, production, dissemination, distribution and enjoyment, whatever the means and technologies that are used.

Cultural heritage: As defined in the WHC, article 1, includes artefacts, monuments, a (or a group of) building(s), site(s) and landscape(s) that have a diversity of values including symbolic, historic, artistic, aesthetic, ethnological or anthropological, and scientific. When not referring to a WH property, it is used in the wider sense which includes traditions or living expressions inherited from our ancestors and passed on to our descendents, such as oral traditions, performing arts, social practices, rituals, festive events, knowledge and practices concerning nature and the universe or the knowledge and skills to produce traditional crafts, that is "a social ensemble of many different, complex and interdependent manifestations. This is now reflecting the diversity of cultural manifestations." Bouchenaki 2003).

Cultural landscape: This represents combined works of nature and by humans, which express a long and intimate relationship between people and their natural environment (UNESCO, 2007).

Culture: UNESCO defines culture as the set of distinctive spiritual, material, intellectual and emotional features of society or a social group, that encompasses, not only art and literature, but lifestyles, ways of living together, value systems, traditions and beliefs (UNESCO 2001). Whereas it is not always possible to measure such beliefs and values directly, it is possible to measure associated behaviours and practices. As such, the *UNESCO Framework for Cultural Statistics* defines culture through the identification and measurement of the behaviours and practices resulting from the beliefs and values of a society or a social group (UNESCO 2009).This has relevance when assessing climate change vulnerability.

Deforestation: The cutting, clearing, burning or removal of trees and rainforests.

Ecosystem: A distinct system of interacting living organisms, together with their physical environment. The extent of an ecosystem may range from very small spatial scales to, ultimately, the entire earth.

Emissions: The release of pollutants into the atmosphere.

Energy Efficiency: The use of less energy to produce the same level of output or service.

Fossil fuels: Composed of organisms that have been dead for millions of years, these fuels include coal, oil and natural gas. When fossil fuels are burnt to produce energy, they also release carbon dioxide and other pollutants.

Feedback: An interaction mechanism between processes, where the result of an initial process triggers changes in a second process and that in turn influences the initial one.

Positive Feedback Loops: These are situations where the initial change causes ongoing increasing change. An example would be runaway climate change (see below). A positive feedback intensifies the original process, and a negative feedback reduces it.

First Voice: The voice, both literal and metaphorical, of the actual carriers and custodians of cultures and their related heritage resources all over the world. The emerging notion of the First Voice is however most often associated with indigenous peoples at the present time, and it sits well ideologically with the better-known constructs of 'First Nations', 'First Peoples' or 'First Inhabitants'. The long struggle to ensure respect and recognition for the cultural rights of indigenous peoples required such critical positioning". (Galla 2008)

Food security: Refers to the availability of food and people's access to it. A household is food secure when its occupants do not live in hunger or fear of starvation.

Forcing: This is an induced change to a system.

Geothermal Energy: Heat from within the Earth, produced by the decay of radioactive particles, which can be exploited to produce steam or hot water to generate low-emission energy.

Geo-engineering: Technological efforts to reduce global warming by stabilising the climate system through intervention in the energy balance of the earth.

Global Warming: Often used interchangeably with "climate change," global warming refers to the rise of average global surface temperatures as currently caused by the greenhouse effect.

Gradual environmental degradation: Deterioration in environmental quality, such as reductions in arable land, desertification, sea level rise, etc., associated with climate change.

Greenhouse Gases: Greenhouse gases (GHGs) are gases in the atmosphere which keep the earth warm enough to sustain life by trapping heat from the sun. However, too much greenhouse gases will cause the earth to overheat by trapping too much heat, resulting in drastic changes to the Earth's climate. Carbon dioxide is the most abundant greenhouse gas. Others include methane (CH_4) and Nitrous Oxide (N_2O).(Water vapour acts in a secondary way as an important greenhouse gas)

Greenhouse effect: A natural environmental system where gases in the atmosphere (greenhouse gases) trap the sun's energy (The Climate Institute 2010). (The sun radiates the earth with short wavelength energy (i.e. visible light), which greenhouse gases do not impede, and then the earth re-radiates long wave radiation (i.e. far infrared/microwave), which greenhouse gases impede.).

Greenwash: Is green whitewash; it is the common practice of companies, politicians and governments disingenuously portraying their products and policies as environmentally friendly. Greenwash is bogus environmentalism. Public relations' initiatives by a business, organisation, or government that purports to show concern for the environmental impact of its activities.

Greenwashing: The practice of expressing concern about global warming and the environment while taking no actions to actually stabilize climate or preserve the environment. Hansen. http://www.stormsofmygrandchildren.com/climate_catastrophe_solutions.html

International Human Dimensions Programme on Global Environmental Change (IHDP): This programme was founded by the International Council for Science and the International Social Science Council in 1996. IHDP's mission is to generate scientific knowledge on coupled human-environment systems, achieve comprehensive understanding of global environmental change processes and their consequences for sustainable development, and make contributions to explore: the anthropogenic drivers of global environmental change; the impact of such change on human welfare, and societal responses to mitigate and adapt to global environmental change.

Intergovernmental Panel on Climate Change (IPCC): The world's leading body for the assessment of climate change, established in 1988 by the World Meteorological Organisation and the United Nations Environmental Programme. The group includes both scientists and government officials who work together to assess information relevant to the understanding of climate change. They are the international authority on climate change issues. However their last 'Assessment' is universally regarded as being dangerously out of date. The IPCC has been accused of reflecting the conservative views of the government officials at the expense of the evidenced-based science.

Internally Displaced People: While there is no legal definition a widely recognized United Nations report, "Guiding Principles on Internal Displacement", uses the following definition: Persons or groups of persons who have been forced or obliged to flee or to leave their homes or places of habitual residence, in particular as a result of or in order to avoid the effects of armed conflict, situations of generalized violence, violations of human rights or natural or human-made disasters, and who have not crossed an internationally recognized State border (Francis Deng and United Nations 1998).

Kyoto Protocol: An international agreement negotiated in 1997, that set binding targets for industrialised countries to reduce their greenhouse gas emissions before 2012. Australia did not ratify Kyoto until 2007, and the United States still has not ratified the treaty. The Kyoto treaty will expire in 2012.

Maladaptation: This has been defined by the IPCC as "Any changes in natural or human systems that inadvertently increase vulnerability to climatic stimuli" or as "an adaptation that does not succeed in reducing vulnerability but increases it instead."

Methane: A greenhouse gas which is many times more potent at trapping heat in the atmosphere than CO_2, but which remains in the atmosphere for a much shorter period of time, approximately 9-15 years. Methane is often associated with agriculture (livestock and soil bacteria), but is also produced from coal mining, industrial processes, landfills, natural gas systems and other sources.

Migrant: International migrants are those who leave their country to settle in another country, voluntarily or involuntarily and temporarily or permanently. Voluntary migrants normally leave their country in search of a higher standard of living and quality of life elsewhere, typically referred to as economic migration. Involuntary migrants include victims of human trafficking, whose special situation is addressed by a number of international legal instruments (IOM 2008).

Mitigation: Is intervention to reduce the sources or enhance the sinks of greenhouse gases (IPCC). This is distinct from adaptation which involves taking action to minimize the local damaging effects of climate change without influencing climate change's overall progression. See Chapter 3.

Ocean Acidification: An increase in the acidity of the Earth's oceans caused by the absorption of CO_2 in the oceans' waters. Ocean acidification damages and can ultimately destroy reefs and ocean life.

Ppm: Parts per million.

REDD (Reducing Emissions from Deforestation and forest Degradation): United Nations programme designed to reduce deforestation by giving financial value to forests and the carbon stored within them. REDD will provide money for programs in developing countries that voluntarily reduce emissions from deforestation.

Reforestation: Replanting of forests that have previously been cleared.

Refugee: Under the 1951 United Nations Convention Relating to the Status of Refugees and later expanded through a 1967 Protocol relating to the Status of Refugees, a refugee is a person who owing to well-founded fear of persecution for reasons of race, religion, nationality, membership of a particular social group or political opinions, is outside the country of his nationality and is unable or, owing to such fear, is unwilling to avail himself

———. 2011a. Operational Guidelines for the Implementation of the World Heritage Convention. . edited by I. C. f. t. P. o. t. W. C. a. N. Heritage.

———. 2011b. Adoption of retrospective Statements of OUV. WH Committee 35th session, Paris, UNESCO Headquarters, 19 – 29 June 2011 Item 8 of the Provisional Agenda: Establishment of the World Heritage List and of the List of World Heritage in Danger. .

———. 2011c. World Heritage Committee Thirty-fifth session Paris, UNESCO Headquarters 19-29 June 2011 Item 13 of the Provisional Agenda: Revision of the Operational Guidelines. WHC-11/35.COM/13.

———. 2011d. WH Committee 35th session, Paris, UNESCO Headquarters, 19 - 29 June 2011. Item 5 of the Provisional Agenda: Reports of the WC Centre & the Advisory Bodies. 5E. World Heritage Convention and sustainable development.

———. 2011e. From green economies to green societies. UNESCO's committment to sustainable heritage., ed UNESCO. Place Published. http://unesdoc.unesco.org/images/0021/002133/213311e.pdf (accessed.

———. 2011f. World Heritage 35 COM. WHC-11/35.COM/12B. Paris, 6 May 2011. UNESCO. WHC.World Heritage Committee. 35th session. Paris, 19-29 June 2011. 12B. Report of the expert meeting on decision-making procedures of the statutory organs of the WH Convention (Manama, Bahrain, 15-17 December 2010). A set of recommendations is proposed in Part D for adoption by the WH Committee. Draft Decision.

———. *Indigenous Peoples, Marginalized Populations and Climate Change Workshops.* 2011g [cited. Available from http://www.unesco.org/new/en/unesco/events/natural-sciences-events/?tx_browser_pi1[showUid]=3394&cHash=303dea31ef.

———. *Possible outline for inventorying elements of the intangible cultural heritage* 2011i [cited. Available from http://www.unesco.org/culture/ich/index.php?lg=en&pg=00266.

———. *Tsodilo* 2011j [cited. Available from http://whc.unesco.org/en/list/1021.

———. *Global Strategy for a Representative, Balanced and Credible World Heritage List* 2011k [cited. Available from http://whc.unesco.org/en/globalstrategy.

———. 2012a. UNESCO's Input to the Rio+20 Compilation Document.

———. *The Ruins of Kilwa Kisiwani and Ruins of Songo Mnara, United Republic of Tanzania.* 2012h [cited. Available from http://whc.unesco.org/en/list/144.

———. *World Heritage and Sustainable Tourism Programme* 2012j [cited. Available from http://whc.unesco.org/en/tourism/.

UNESCO, ICCROM, ICOMOS, and IUCN. 2010. *Managing Disaster Risks for World Heritage. Manual.(Online).*

UNESCO, SCOPE, and UNEP. *Third Pole Environment* 2011 [cited. Available from http://www.unesco.org/new/fileadmin/MULTIMEDIA/HQ/SC/pdf/sc_env_Third_Pole_EN.pdf.

UNESCO, ICCROM, ICOMOS and IUCN 2010. *Preparing World Heritage Nomintions. Manual. 1st Edition.*

UNESCO, Social and Human Sciences. *Climate Change: Moving Towards a Universal Ethical Framework* 2010ll [cited. Available from http://www.unesco.org/new/en/social-and-human-sciences/themes/science-and-technology/climate-change/.

UNESCOPRESS. *"Migration and Climate Change" A UNESCO publication on one of the greatest challenges facing our time* 2011 [cited. Available from http://www.unesco.org/new/en/media-services/single-view/news/migration_and_climate_change_a_unesco_publication_on_one_of_the_greatest_challenges_facing_our_time/.

UNFCCC. *Ad Hoc Working Group on Long-term Cooperative Action under the Convention Twelfth session. Tianjin, 4–9 October 2010, Item X of the provisional agenda Negotiating text Note by the secretariat* 2010a [cited. Available from http://unfccc.int/resource/docs/2010/awglca12/eng/14.pdf.

———. *Nairobi work programme on impacts, vulnerability and adaptation to climate change (NWP), Understanding vulnerability, fostering adaptation* 2010b [cited. Available from http://unfccc.int/adaptation/nairobi_work_programme/items/3633.php.

———. *Case Studies Summary* 2011a [cited. Available from http://unfccc.int/ttclear/jsp/CaseStudy.jsp.

———. *Technology Mechanism* 2011b [cited. Available from http://unfccc.int/ttclear/jsp/TechnologyMechanism.jsp.

UNHCHR, (Office of the United Nations High Commissioner for Human Rights). *International Covenant on Economic, Social and Cultural Rights* 1966 [cited. Available from http://www2.ohchr.org/english/law/cescr.htm.

UNHCR, (UN High Commissioner for Refugees). *Climate change, natural disaters and human displacement: a UNHCR perspective, 14 August 2009* 2009 [cited. Available from http://www.unhcr.org/refworld/docid/4a8e4f8b2.html

UNSCO. 2008. DECISIONS ADOPTED AT THE 32nd SESSION OF THE WORLD HERITAGE COMMITTEE (QUEBEC CITY, 2008).

US Environmental Protection Authority. *Extreme Events* 2009 [cited. Available from http://www.epa.gov./climatechange/effects/extreme.html.

Wangchuk, Sonam. 2005a. Editorial. *Ladags Melong.*

———. 2005b. Ladakh towards a Vision. *Ladags Melong*, 18.

———. 2007. Interview with Sonam Wangchuk, Director of Students Educational and Cultural Movement of Ladakh (SECMOL).

WBGU. 2009. Solving the climate dilemma: the budget approach. Special Report. Berlin 2009.WBGU. German Advisory Council on Global Change. .

Webster, P J , G J Holland, J A Curry, and H -R Chang. 2005. Changes in Tropical Cyclone Number, Duration, and Intensity in a Warming Environment,. *Science* 309 (5742):1844-1846.

Westerling, A L , H G Hidalgo, D R Cayan, and T W Swetnam. 2006. Warming and Earlier Spring Increase Western U.S. Forest Wildfire Activity, . *Science* 313 (5789):940-943.

WHC, UNESCO -. 2007. Policy Document on the Impacts of Climate Change on World Heritage Properties. . In *Decision WHC-07/ 16.GA/10, adopted by 16th General assembly 2007*. Geneva: UNESCO.

Wikipedia. *Albedo* 2011 [cited. Available from http://en.wikipedia.org/wiki/ Albedo.

Wilkinson, Marion, and Deborah Smith. 2007. Window closing on planet's chances,. *Sydney Morning Herald*, , 7.4.2007.

Wilson, Meredith, Chris Ballard, and Douglas Kalotiti. 2007. Chief Roi Mata's Domain. Challenges facing a World Heritage-nominated property in Vanuatu. . In *Cultural Heritage Management in the Pacific*. Cairns: ICOMOS.

Wold, Chris, and Erica Thorson. 2006. Response to the Position of the United States on Climate Change with Respect to the World Heritage Convention. (March 14, 2006). http://www.elaw.org/system/files/u.s.climate.response_to_US_Position.pdf.

World People's Conference on Climate Change and the Rights of Mother Earth. *Building the People's World Movement for Mother Earth. Bolivia, April 2010.* 2010 [cited. Available from http://pwccc.wordpress.com/.

World Resources Institute. *CAIT.Climate Analysis Indicators Tool.* 2005 [cited. Available from http://cait.wri.org/.

WWF, Global Footprint Network and the Zoological Society of London. *2010 Living Planet Report, Biodiversity, biocapacity, Development.* 2010 [cited. Available from http://www.footprintnetwork.org/en/ index.php/GFN/page/2010_living_planet_report/.

Zeppel, Heather, and Narelle Beaumont. 2011. Green Tourism Futures: Climate Change Responses by Australian Government Tourism Agencies. ACSBD Working Paper No. 2.

CPSIA information can be obtained at www.ICGtesting.com
Printed in the USA
BVOW10s0618020316

438604BV00033B/2/P

9 781612 290713